Risk, Systems and Decisions

AF888911

Series Editors

Igor Linkov, U.S. Army ERDC, Vicksburg, USA

Jeffrey Keisler, College of Management, University of Massachusetts, Boston, USA

James H. Lambert, University of Virginia, Charlottesville, USA

Jose Rui Figueira, CEG-IST Instituto Superior Técnico, University of Lisbon, LISBOA, Portugal

Health, environment, security, energy, technology are problem areas where man-made and natural systems face increasing demands, giving rise to concerns which touch on a range of firms, industries and government agencies. Although a body of powerful background theory about risk, decision, and systems has been generated over the last several decades, the exploitation of this theory in the service of tackling these systemic problems presents a substantial intellectual challenge. This book series includes works dealing with integrated design and solutions for social, technological, environmental, and economic goals. It features research and innovation in cross-disciplinary and transdisciplinary methods of decision analysis, systems analysis, risk assessment, risk management, risk communication, policy analysis, economic analysis, engineering, and the social sciences. The series explores topics at the intersection of the professional and scholarly communities of risk analysis, systems engineering, and decision analysis. Contributions include methodological developments that are well-suited to application for decision makers and managers.

Igor Linkov • Alexander Kott
Editors

Cyber Resilience: Applied Perspectives

Second Edition

Springer

Editors
Igor Linkov
Engineer Research and Development Center
US Army Corps of Engineers
Concord, MA, USA

Alexander Kott
U.S. Army Research Laboratory
Adelphi, MD, USA

ISSN 2626-6717 ISSN 2626-6725 (electronic)
Risk, Systems and Decisions
ISBN 978-3-031-90108-9 ISBN 978-3-031-90109-6 (eBook)
https://doi.org/10.1007/978-3-031-90109-6

© The Editor(s) (if applicable) and The Author(s), under exclusive license to Springer Nature Switzerland AG 2019, 2025

This work is subject to copyright. All rights are solely and exclusively licensed by the Publisher, whether the whole or part of the material is concerned, specifically the rights of translation, reprinting, reuse of illustrations, recitation, broadcasting, reproduction on microfilms or in any other physical way, and transmission or information storage and retrieval, electronic adaptation, computer software, or by similar or dissimilar methodology now known or hereafter developed.
The use of general descriptive names, registered names, trademarks, service marks, etc. in this publication does not imply, even in the absence of a specific statement, that such names are exempt from the relevant protective laws and regulations and therefore free for general use.
The publisher, the authors and the editors are safe to assume that the advice and information in this book are believed to be true and accurate at the date of publication. Neither the publisher nor the authors or the editors give a warranty, expressed or implied, with respect to the material contained herein or for any errors or omissions that may have been made. The publisher remains neutral with regard to jurisdictional claims in published maps and institutional affiliations.

This Springer imprint is published by the registered company Springer Nature Switzerland AG
The registered company address is: Gewerbestrasse 11, 6330 Cham, Switzerland

If disposing of this product, please recycle the paper.

Contents

Contributors

David Perez Abreu CISUC, Department of Informatics Engineering, University of Coimbra, Coimbra, Portugal

Zachary A. Collier Department of Management, Radford University, Radford, VA, USA

Marilia Curado CISUC, Department of Informatics Engineering, University of Coimbra, Coimbra, Portugal

Joel Dawson Oak Ridge National Laboratory, Oak Ridge, TN, USA

José de Arimatéia da Cruz Georgia Southern University, Savannah, GA, USA

U.S. Army War College, Carlisle, PA, USA

Daniel DiMase Aerocyonics, Inc., East Greenwich, RI, USA

Ekzhin Ear Department of Computer Science, University of Colorado Colorado Springs, Colorado Springs, CO, USA

Jason E. Ellis DEVCOM Army Research Laboratory, Adelphi, MD, USA

Nathaniel Evans Oak Ridge National Laboratory, Oak Ridge, TN, USA

Bruno Faria CISUC, Department of Informatics Engineering, University of Coimbra, Coimbra, Portugal

Alexander A. Ganin Department of Systems and Information Engineering, University of Virginia, Charlottesville, VA, USA

Jelle Groenendaal Crisislab, Renswoude, The Netherlands

Madison L. Hassler Department of Systems & Information Engineering, University of Virginia, Charlottesville, VA, USA

Ira Helsloot Crisislab, Renswoude, The Netherlands

Governance of Safety and Security, Radboud University Nijmegen, Nijmegen, The Netherlands

Travis D. Howard United States Navy (Retired), Washington, DC, USA

Andrew Jin Engineer Research and Development Center, US Army Corps of Engineers, Concord, MA, USA

Jesse M. Keenan U.S. Army Corps of Engineers' Engineer Research and Development Center, Vicksburg, MS, USA

School of Architecture and Built Environment, Tulane University, New Orleans, LA, USA

Jeremy Kepner MIT Lincoln Laboratory, Cambridge, MA, USA

Alexander Kott U.S. Army Research Laboratory, Adelphi, MD, USA

Catherine A. Lambert Contractor to Keisler Consulting, Boston, MA, USA

James H. Lambert Department of Systems & Information Engineering, University of Virginia, Charlottesville, VA, USA

Igor Linkov Engineer Research and Development Center, US Army Corps of Engineers, Concord, MA, USA

Alexander Mermelstein Price School of Public Policy, University of Southern California, Los Angeles, CA, USA

Center for Risk and Economic Analysis of Threats and Emergencies, University of Southern California, Los Angeles, CA, USA

Noah Miller Price School of Public Policy, University of Southern California, Los Angeles, CA, USA

Center for Risk and Economic Analysis of Threats and Emergencies, University of Southern California, Los Angeles, CA, USA

Edmundo Monteiro CISUC, Department of Informatics Engineering, University of Coimbra, Coimbra, Portugal

Brian J. Murphy Pennsylvania State University, University Park, PA, USA

Travis W. Parker ICF International, Columbia, MD, USA

Kelsey Poinsatte-Jones U.S. Army Corps of Engineers, Engineering Research and Development Center, Concord, MA, USA

Jose Luis Castanon Remy Department of Computer Science, University of Colorado Colorado Springs, Colorado Springs, CO, USA

Adam Rose Price School of Public Policy, University of Southern California, Los Angeles, CA, USA

Center for Risk and Economic Analysis of Threats and Emergencies, University of Southern California, Los Angeles, CA, USA

Sidney Smith DEVCOM Army Research Laboratory, Adelphi, MD, USA

Logan Sturm Oak Ridge National Laboratory, Oak Ridge, TN, USA

Ryan Styles Oak Ridge National Laboratory, Oak Ridge, TN, USA

Benjamin D. Trump U.S. Army Corps of Engineers, Engineering Research and Development Center, Concord, MA, USA

Joachim Vandekerckhove University of California, Irvine, CA, USA

Karima Velasquez CISUC, Department of Informatics Engineering, University of Coimbra, Coimbra, Portugal

Michael J. Weisman DEVCOM Army Research Laboratory, Adelphi, MD, USA

Aaron Werth Oak Ridge National Laboratory, Oak Ridge, TN, USA

Shouhuai Xu Department of Computer Science, University of Colorado Colorado Springs, Colorado Springs, CO, USA

Quanyan Zhu New York University, Brooklyn, NY, USA

Fundamental Concepts of Cyber Resilience: Introduction and Overview

Andrew Jin, Alexander Kott, and Igor Linkov

1 Cyber Resilience, Foundations and Interconnections

Society has become deeply reliant on cyber systems, transforming nearly every aspect of daily life—from commerce and healthcare to critical infrastructure and national defense. The rapid expansion of digital technologies, accelerated by the adoption of cloud computing, the proliferation of Internet of Things (IoT) devices, and the rise of remote and hybrid work, has created an interconnected ecosystem where almost everything is linked to the internet. This interconnectedness has expanded the attack surface, as each new device, service, or system presents potential vulnerabilities for bad actors to exploit or for bugs to propagate. Critical sectors like energy, healthcare, and finance are highly dependent on digital operations, making disruptions more damaging. Even when disruptions do not result from malicious intent or targeted attacks, they can have broad-scale consequences. The 2024 Cloudstrike disruption has shown that even small incompatibilities between software can cause disruptions that shut down everything from emergency services to healthcare to airlines (U.S. Government Accountability Office 2024).

The proliferation of digital technologies changes the types of attacks and failures that systems are vulnerable to. This enables not only a higher magnitude of exploits, such as direct denial of the surface with ever-increasing requests per second, but also increased sophistication, such as BIOS-level rootkit attacks (Cybersecurity & Infrastructure Security Agency 2023a) As the capabilities of generative AI have grown

A. Jin · I. Linkov (✉)
Engineer Research and Development Center, US Army Corps of Engineers, Concord, MA, USA
e-mail: igor.linkov@usace.army.mil

A. Kott
U.S. Army Research Laboratory, Adelphi, MD, USA

© The Author(s), under exclusive license to Springer Nature Switzerland AG 2025
I. Linkov, A. Kott (eds.), *Cyber Resilience: Applied Perspectives*, Risk, Systems and Decisions, https://doi.org/10.1007/978-3-031-90109-6_1

rapidly, so have the vulnerabilities in these systems, both in inducing hallucination and extracting data.

Attack targets have become equally diverse, ranging from individuals to multinational corporations and national government agencies. At the individual level, data aggregation has scaled, with recent records showing over 2.9 billion records breached. As Operations Technologies (OT) and Information Technologies (IT), even well-protected OT systems can be impacted by vulnerabilities in IT systems, such as in the Colonial Pipeline hack (Cybersecurity & Infrastructure Security Agency 2023b). Moreover, advanced persistent threats by nation-state actors continue to grow, especially in the context of warfare (Kott et al. 2015). Cyberwarfare has escalated in recent conflicts, such as in the Russia-Ukraine war, where nation-state actors have increasingly deployed cyberattacks to target critical infrastructure and disrupt civilian life alongside conventional warfare (Willett 2023).

The increased attack surface, growing complexity of systems, and increased sophistication of threats have made it challenging to identify and quantify all potential threats, vulnerabilities, and consequences. Thus, traditional risk assessment, which comprises a calculation of the product of threats, vulnerabilities, and consequences for hazards and their subsequent exposures, has become increasingly difficult in cybersecurity (Kaplan and Garrick 1981). For highly complex and interconnected systems, it becomes prohibitively difficult to conduct a risk assessment that adequately accounts for the potential cascading effects that could occur through an outage or loss spilling over into other systems (Collier et al. 2014a; DiMase et al. 2015; Ganin et al. 2016). Given the rapid evolution of threats to cyber systems, new management approaches are needed that address risk across all interdependent domains (i.e., physical, information, cognitive, and social) of cyber systems (Linkov et al. 2013a, b).

Further, the unpredictability, extreme uncertainty, and rapid evolution of potential cyber threats can leave risk assessment efforts unable to address cybersecurity concerns for critical infrastructural systems adequately. For this reason, the traditional approach of hardening cyber systems against identified threats has proven to be very difficult. While cybersecurity professionals have taken steps to harden systems, such as air gapping technologies, even these can be imperfect or impractical in protecting systems. Therefore, similar to how biological systems develop immunity to respond to infections, cyber systems must adapt to ever-changing threats to vital system functions and bounce back from the effects of the attacks (Linkov et al. 2014).

For these reasons, cyber resilience refers to the ability of the system to prepare, absorb, recover, and adapt to adverse effects, especially those associated with cyberattacks. Here, depending on the context, we use the term cyber resilience to refer mainly to the resilience property of a system or network; sometimes, we also use the term to refer to the features or components of the system that enable cyber resilience.

2 Resilience and Systems

Cyber resilience should be considered in the context of complex systems that comprise not only physical and information but also cognitive and social domains (Smith 2005). Cyber Resilience ensures that a system can recover by designing and implementing interconnected hardware, software, and sensing components of cyberinfrastructure (Fig. 1). Cyber resilience, thus, is a system property that enables the system to sustain operations while ensuring mission execution. Resilient systems are able to sustain operations not only during normal, anticipatable interruptions but also during unforeseen circumstances.

Resilience has been a growing area of focus for infrastructure. The National Cybersecurity Strategy has called for "a more secure, resilient, privacy-preserving, and equitable digital ecosystem through strategic investments and coordinated, collaborative action." The North Atlantic Treaty Organization (NATO) has called for "resilient civil communications systems—ensuring that telecommunications and cyber networks can function even under crisis conditions, with sufficient back-up capacity" as a baseline for national resilience preparedness (NATO 2023). Yet, no universal definition of resilience has emerged.

Resilience has roots in many disciplines and integrates ecological, social, psychological, organizational, and engineering perspectives and definitions (Florin 2016). Resilience engineering, for example, has been defined as "the ability of systems to anticipate and adapt to the potential for surprise and failure" and has been associated with a shift in the safety paradigm, acknowledging that system coping is important when prevention is impossible (Hollnagel et al. 2006). Ecological resilience, by contrast, refers to the ability of the system to absorb and withstand shocks, emphasizing persistence (Holling 1996). Resilience is used as a metaphor to describe how systems react to stressors, and to bridge the gap in understanding

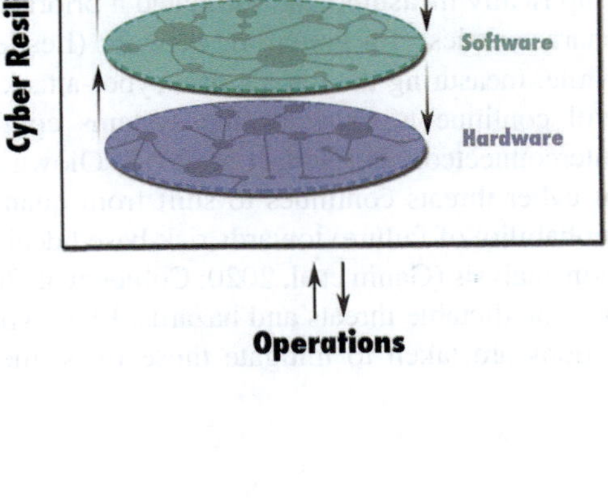

Fig. 1 The cyber resilience domains comprise sensing, hardware, and software nonparallel components that collectively contribute to sustaining system operations

between fields, resilience needs to be discussed less abstractly, separating the metaphor from the science. Finding ways to accurately quantify and operationalize the resilience of cyber systems is the key subject of this book.

3 Conflation of Resilience, Risk, and Robustness

Resilience is defined by the National Academies of Science (NAS) as "the ability to prepare and plan for, absorb, recover from, and more successfully adapt to adverse events," is emerging as one of the most widely used by various organizations and governance agencies (Larkin et al. 2015). Even so, what is often measured is the probability of the failure of system components in response to different threat scenarios with the goal of hardening the system so the compromise does not happen. This is not exactly resilience. As systems become increasingly complex, cyber incidents will occur, whether the result of malicious action or not. Cyber resilience characterizes what happens after an adverse event and requires preparedness for both known and unknown threats.

While cyber resilience and cyber risk are often conflated, they are both critically important to ensuring the operations of cyber systems. The literature on cyber risk most commonly defines cyber risk in terms that combine the likelihood of an undesirable event and a measure of the impact of the event. Although several approaches to risk assessment exist, the methods adopted by US regulatory agencies are based primarily on the risk = threat × vulnerability × consequence model. For example, NIST's description from NIST Publication SP 800-30 (NIST) states: "Risk is a function of the likelihood of a given threat source's exercising a particular potential vulnerability, and the resulting impact of that adverse event on the organization (NIST 2002). To determine the likelihood of a future adverse event, threats to an IT system must be analyzed in conjunction with the potential vulnerabilities and the controls in place for the IT system." ISO's definition of IT risk is similar: the potential that a given threat will exploit vulnerabilities of an asset or group of assets and thereby cause harm to the organization. It is measured in terms of a combination of the probability of occurrence of an event and its consequence (ISO/IEC 27005 2018).

Cyber risk is generally well understood. The likelihood of a cyber-attack can be empirically measured and estimated a priori with a degree of accuracy from known characteristics of a system or network (Leslie et al. 2018; Gil et al. 2014). Meanwhile, measuring the impact of a cyber-attack is an area of active development and will continue to grow as the systems continue to increase in complexity and interconnectedness (Kott et al. 2017; Olowu et al. 2021). While the measurement of cyber threats continues to shift from quantifying risk in specific units (like the probability of failure) towards risk-based decision-making using multi-criteria decision analysis (Ganin et al. 2020; Collier et al. 2014b), these measurements still reflect only predictable threats and hazards. Even when individual risks are identified, and actions are taken to mitigate those risks, the residual risk still remains. In other words, risk mitigation focuses on known attacks, while resilience focuses also on recovering from unknown attacks.

Robustness is another concept often confused with resilience. Robustness is closely related to risk. Robustness denotes the degree to which a system can withstand an unexpected internal or external attack or change without degradation in the system's performance. To put it differently, assuming two systems—A and B—of equal performance, the robustness of system A is greater than that of system B if the same unexpected impact on both systems leaves system A with greater performance than system B. We stress the word unexpected because the concept of robustness focuses specifically on performance, not only under ordinary, anticipated conditions (which a well-designed system should be prepared to withstand) but also under unusual conditions that stress its designers' assumptions. For example, in IEEE Standard 610.12.1990, "Robustness is defined as the degree to which a system operates correctly in the presence of exceptional inputs or stressful environmental conditions" (IEEE 2003). Similarly, "robust control refers to the control of unknown plants with unknown dynamics subject to unknown disturbances" (Chandrasekharan 1996). The length of recovery typically depends on the extent of damage and potentially a point beyond which recovery is impossible. Resilience and robustness are closely tied together—robustness determines how much damage is incurred in response to an adverse event, and resilience determines how quickly the system can recover from such damage.

4 Making Business Case for Cyber Resilience

Traditional risk assessment is appealing for cyber risk governance due to the quantitative nature and the single risk value that is output. These characteristics make risk thresholds easy to formalize in policy documents and to regulate consistently at the federal level. However, quantitative risk assessment typically involves quantifying the likelihood of an event's occurrence and the vulnerability to the event. Emerging cyber realities and technologies present new threats with uncertain intensity and frequency, and the vulnerabilities and consequences in terms of the extent of casualties, economic losses, time delays, or other damages are not yet fully understood or modeled. As a result, risk calculations become more uncertain and generate costly solutions since multiple, often hypothetical, threat scenarios could point to many vulnerabilities and catastrophic system failures that are unaffordable to mitigate, absorb, or recover. Furthermore, users and other stakeholders may prefer accepting some loss in the performance of one part of the system over any degradation in another (Bostick et al. 2017). One outcome can be significant funding spent in ways that do not align with stakeholder values, resulting in dissatisfaction with performance despite the expense or even litigation that interferes with the risk reduction efforts.

A key risk management strategy is to identify critical components of a system that are vulnerable to failure and, subsequently, to harden them (Roege et al. 2014). This

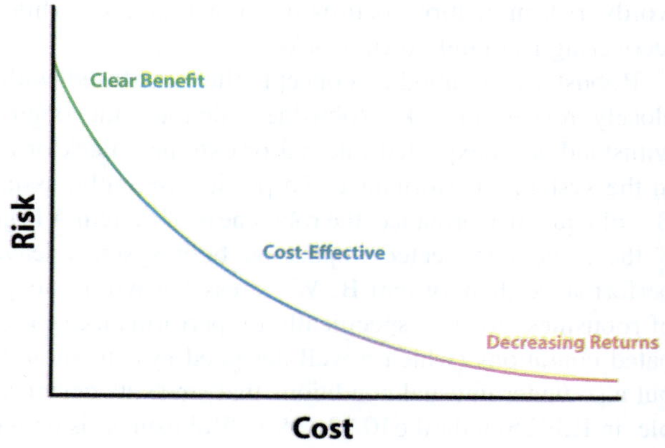

Fig. 2 Conceptual diagram of the cost of buying-down risk in cyber systems

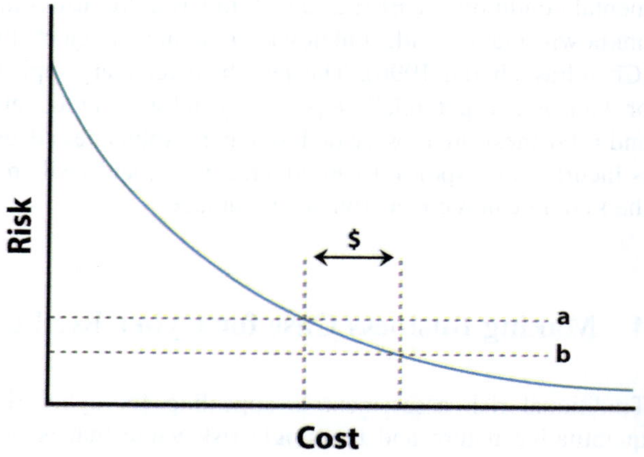

Fig. 3 Existential threats and mega-costs are avoided when resources are reallocated to resilience—e.g., accepting risk level (**a**) instead of risk level (**b**)

approach can be appropriate for many isolated cyber systems, but when the nature of the threat is unknown, as discussed above, it is difficult to identify all, or even most, of the critical components; it becomes increasingly expensive to act conservatively to harden or protect all parts of the system against all types of threats (Fig. 2). The result has been a stagnation in investment. As risk mitigation plans become more expensive and are delayed while funding is sought, infrastructure and societal systems are left largely unprepared for emerging and uncertain threats. Furthermore, there are fewer and fewer isolated systems in our world, and the degree of interdependency and interconnectedness can be challenging to characterize and quantify.

In exchange for accepting the current levels of risk rather than demanding greater preventative and protective measures, funding can be reallocated toward resilience enhancement efforts. For systems that have already completed cost-effective risk reduction measures, Fig. 3 shows the funding that can be reallocated toward resilience by accepting risk level (a) over risk level (b). Parallel with these public changes, the academic community should be called on to develop decision models that identify the optimal investment in risk reduction versus resilience and recovery improvement.

5 Assessment of Resilience

The diversity of approaches to measuring resilience continues to grow. Estay et al. (2020) conducted a systematic review of Cyber Resilience Frameworks (CRFs), analyzing 208 articles across 36 industries and 25 research areas. The study categorizes CRFs by their general methodology finding those algorithmic assessments, architecture, game theoretic, machine learning, simulation, risk assessment, or statistical, strategy, or survey-based assessments. Developers of the tools span a wide range of entities, including academic, private, government, and boundary organizations that bridge across research, policy, and practitioner realms, as well as potential users themselves. While some standardization has been conceptualized, the resilience of systems fundamentally depends on the ability of those systems to execute their missions. Potential users include state and city managers, industry process administrators, and utility operators, many lacking the expertise to choose among the rapidly accumulating products in this emerging field. Thus, resilience often becomes a tradeoff between simpler assessments that may give a broader understanding of resilience and more complex, more detailed assessments.

Figure 4 shows two categories to conceptualize the approaches currently described in the literature to measuring resilience, including metric-based and model-based approaches. Metric-based approaches use measures of individual properties of system components or functions to assess overall system performance. In contrast, model-based approaches use system configuration modeling and scenario analysis to predict system evolution. Often, metrics provide a simplified model for resilience, provide vital information about a given system, and are generally acquired by analyzing relevant attributes of that system.

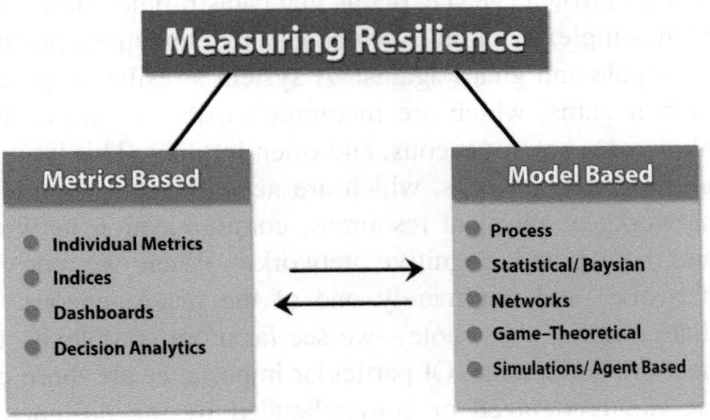

Fig. 4 Metric-based and model-based approaches for Resilience Assessment. Multiple tools have been developed to address resilience in systems in both methodological groups

Model-based approaches focus on representing the real world and defining resilience using mathematical or physical concepts. Modeling requires knowledge of key aspects of a system, including critical functions, mission objectives, temporal patterns, thresholds, system memory, and system adaptation. Process models require a detailed understanding of the physical approaches within a system to simulate event impacts and system recovery; they are challenging to construct and information-consuming. Recently, efforts to implement digital twins of systems have given rise to opportunities to iterate more quickly on recovery scenarios to assess resilience. Statistical approaches alternatively require a lot of data on system performance. Bayesian models combine features of process and statistical models. Network models require a presentation of the system as interconnected networks whose structure depends on the system's function. Alternatively, the game theoretical/agent-based approach focuses on the model performance of the system based on a limited set of rules defined by modelers.

6 Foundations for Improving Resilience

The resilience of a system, a network, or an organization is influenced by a number of factors in a complex and often contradictory manner. We describe four overarching considerations relevant to improving cyber resilience: managing complexity, controlling propagation, providing active agents, and conducting analysis and wargaming. Then, we discuss several broad paradigms of implementing cyber resilience and illustrate each with examples taken from the 2022 Russo-Ukrainian War. This war exhibited the most ferocious cyber hostilities known during this book's writing. Finally, we mention that such paradigms may be implemented and operated in two fashions—Resilience by Design and Resilience by Intervention—and explain the differences between the two.

1. **Manage Complexity**

The resilience of a system or network depends greatly on the complexity of links within the system (Kott and Abdelzaher 2014; Ganin et al. 2016). In his pioneering work, Perrow (1984) explains that catastrophic failures of systems emerge from the high complexity of links, which lead to interactions that the system's designer cannot anticipate and guard against. A system's resilience precautions can be defeated by hidden paths, which are incomprehensible to the designer because the links are numerous, heterogeneous, and often implicit. This issue is particularly important in multi-genre networks, which are networks that combine several distinct genres—networks of physical resources, communication networks, information networks, and social and cognitive networks. When we consider an entire multi-genre network—and not merely one of the heterogeneous, single-genre sub-networks that comprise the whole—we see far more complexity of the paths connecting the network's elements. Of particular importance are those paths within the system that are not recognized or comprehended by the designer. Indeed, the designer can usually devise a mechanism to prevent the propagation of failure through obvious links. Many, however, are not obvious, either because there are simply too many paths to consider—and the numbers rapidly increase once we realize that the paths between elements of a communication system, for example, may also pass through a social or an information network. For example, subtle feedback links may reduce the resilience of organizational decision-making (Kott 2006). Therefore, greater complexity should be avoided in most cases unless it directly supports resilience functions.

2. **Control Propagation**

To enhance the system's ability to absorb the impact of a cyber-compromise, the designer should guard against cascaded failures. Such failures are non-independent in that one triggers another. A network prone to significant "domino effects" will likely sustain severe damage in response to even modest disturbances, limiting the scope within which efficient absorption and recovery (and hence resilient operation) remains possible. Therefore, the dependencies or links between nodes should be, whenever possible, designed to minimize the likelihood that a failure propagates easily from one node to another. Ideally, links should both passively and actively filter out failure propagation.

One possible form of such filtering is buffering. In data and commodity flow networks, the function of the network is to offer its clients access to a set of delivered items. In such networks, buffers (e.g., caching, local storage) constitute a resilience mechanism that obviates the need for continued access to the original source. Should the original source become unreachable, one can switch to a local supplier. Hence, operations can continue despite interruptions of the global supply network as long as access to a local cache (buffer) is available. Local access is an especially valuable solution in the case of a data flow network, where the commodity (namely, the data content) is not consumed by user access, in the sense that a local distributor can continue to serve a content item to new users irrespective of its use by others. Much work on network buffering has been done to increase the resilience of data access to fluctuations in resource availability. For example, buffering (or caching) has been used to restore connectivity and performance upon topology changes in ad hoc networks, as well as to reduce access latency in disruption-tolerant networks.

Other ways to control propagation and guard against cascading failures include adding multiple functional capacities to each node (usually implying the need for additional resources), processing more input sources (requiring more resources for the acquisition of inputs and processing), by a combination of multiple parallel processing mechanisms, and by providing redundant resources can help both to absorb the effect of a cyber attack and to restore the system. Redundancy, however, should be used with caution. If the designer adds identical redundant software or hardware, the same malware would be able to compromise multiple redundant resources. If diversity is introduced, and the redundant resources are significantly different, the complexity grows with its potential negative impact on resilience.

3. **Provide Active Agents**

Active agents—human or artificial—should be available to take active measures to absorb, recover, and adapt. In order to be effective, the agents must have plans, processes, and preparation. Where appropriate and necessary, human operators or users of the system should play the role of active resilience agents. Wherever possible, however, the system designer should consider introducing a set of artificial partly autonomous intelligent agents that can conduct the absorption and recovery actions in an autonomous manner (Kott et al. 2018). To perform the absorb and recover actions in the presence of the adversary malware deployed on the friendly system, the artificial agent often has to take destructive actions, such as deleting or quarantining certain software. Such destructive actions are carefully controlled by the appropriate rules of engagement. Developers of the agent should attempt to design its actions and planning capability to minimize the associated risk. This risk has to be balanced against the destruction caused by the adversary if the agent's action is not taken. Because the adversary and its malware know that the agent exists and is likely to be present on the system, the malware seeks to find and destroy the agent. Therefore, the agent possesses techniques and mechanisms for maintaining a degree of stealth, camouflage, and concealment. More generally, the agent takes measures that reduce the probability that the malware will detect the agent. The agent is mindful of the need to exercise self-preservation and self-defense. When humans are active resilience agents, these human agents must be appropriately trained, prepared, and motivated to be effective. They should have skills, resources, and processes available to them to absorb, recover, and adapt. The human organization should be properly structured, clearly define roles and responsibilities, develop collective skills, and conduct team training and rehearsals.

Ideally, agents should be able to perform one of the multiple functions depending on context, and the same function could be performed by one of several agents. For example, network storage agents (buffers, caches) can use their space to store any of a set of possible items. Also, the same item can be stored by any of multiple agents. For example, individuals in an organization can allocate their time to any of a set of possible projects. Similarly, the same project can be performed by any of multiple individuals. The combination of agents' versatility and redundancy significantly improves network functions' resilience by facilitating reconfiguration to adapt to perturbations. Intuitively, the higher the versatility of the individual agents and the higher the degree to which they are interchangeable, the more resilient the system is to perturbation because it can reallocate functions to agents more flexibly to restore its performance upon resource loss. Another helpful dimension is the agent's capacity, the number of different functions an agent can simultaneously perform. Capacity quantifies, for example, the number of items a storage node can simultaneously hold, or the number of projects a given individual can simultaneously work on.

4. **Conduct Analysis and Wargaming**

All resilience-enhancing measures and actions can also cause unanticipated effects, leading to an overall reduction in resilience. Therefore, rigorous, high-fidelity analysis is a must. A resilience-enhancing measure should not be designed or introduced without an appropriate analysis capable of revealing potential negative impacts and systemic effects (Kott et al. 2017). Comparative analytical studies should be conducted with and without the proposed measure. High-fidelity, simulation-based analysis is particularly appropriate. The fidelity of the simulation should be sufficient to replicate multiple modes of propagation, modes of interactions, feedback channels, and second and third-order effects.

Because adversary actions and counteractions play a great role in cyber resilience, the analysis must also include the adversary. If the adversary specifically tailors his or her techniques and procedures—and possesses the necessary capabilities—to defeat the absorption and recovery efforts specifically, the system's resilience will suffer accordingly. The designer of the system should consider the likely adversary's capabilities, intent, tactics, techniques, and procedures and design the mechanisms and processes of absorption and recovery in a manner that is more likely to withstand the adversary's actions. Game-theoretic analysis and war gaming—manual or computerized—should be conducted to optimize the proposed measures (Colbert et al. 2017).

7 Paradigms of Cyber Resilience: Lessons from the Russo-Ukrainian War

Cyber resilience supports system recovery through several paradigms or mechanisms. These fall mainly into several broad categories, following (Ross et al. 2019) rearranged by the authors, which we discuss below. We bring an example from cyber activities observed in the Russo-Ukrainian War for each of these broad paradigms. At the time of writing this book, the largest and most intense manifestation of cyber hostilities has been this conflict, and, as such, it is an apt case study of resilience.

7.1 *Discarding and Replacement*

Discarding and replacement remove system elements if indications and the degree of damage warrant doing so. After the removal, those elements are replaced with undamaged ones, or the system can operate without them, although potentially with degraded capabilities. Discarding and replacement starts by identifying the compromised components or subsystems of the overall system and, if impossible or too difficult to fix, discarding them. Then, they are replaced with different components that have not yet been compromised and, hopefully, will be more difficult to compromise. This was the paradigm that Ukraine followed—not necessarily by design or with a prior intent—when the country rapidly replaced the hopelessly compromised Viasat system with Starlink.

On February 24, 2022, about an hour before Russian ground forces started to enter the territory of Ukraine, Russian hackers working for the intelligence organization GRU initiated a denial-of-service attack that forced many Viasat models offline. This was followed by an attack on a ground-based network that allowed malware to enter a management segment of the communications satellite K-SAT network. The malware called AcidRain was involved in the process. From the management segment, the cyber actors issued commands to thousands of modems; the commands overwrote data in the flash memory of the modem and made them incapable of performing key functions.

Although some of Viasat's equipment was quickly restored, thousands of replacement modems had to be brought into Ukraine and re-installed. The Ukraine decision-makers elected to follow essentially the "discard and replace" approach– discard Viasat and replace it with Starlink. Only 4 days after the Viasat hack, Starlink terminals began arriving in Ukraine, eventually reaching tens of thousands.

While the attack on Viasat did cause a significant disruption of Ukraine government communications in the early days and weeks of the Russian invasion, there is no confirmation that specifically military channels were affected. More broadly, there is no evidence that the attack yielded any tangible military benefits to the Russian invasion. Furthermore, by replacing the Viasat systems with Starlink, Ukraine's military and government not only restored the pre-invasion capacity but actually improved the military-specific functions of their satellite communications. In particular, Starlink has reportedly enabled the control of Ukrainian drones, both aerial and maritime, in ways that the Viasat system might be ill-suited to support (Bateman 2022; O'Neill 2022).

7.2 *Deflection and Absorption*

Deflection and absorption refer to transferring threat events or their effects to different system elements or to systems other than those that were targeted or initially affected, accepting a level of damage to some of the system elements while also taking actions to reduce the impacts (or repair the damage) to other system elements.

In the 2022 Russo-Ukraine war, the deflection and absorption paradigm of resilience was exemplified by Ukraine's advantage of having highly decentralized, even fragmented telecommunication networks. A successful attack on one of the many operators could readily absorbed by shifting the load to other, unaffected operators. Meanwhile, an agile, collaborative, and highly motivated workforce of IT professionals across multiple enterprises would repair the affected operators.

The diverse and decentralized patchwork of Ukraine's telecommunication providers—hundreds of separate and geographically distant mobile and Internet providers—turned out highly advantageous from the perspective of its resilience. The telecommunication sector did not have any obvious choke points, dominant hubs, or exceptionally important key players such that no Russian cyber attack on any one of the providers could cripple the entire national telecommunications capacity (Bateman 2022). This decentralized, highly entrepreneurial infrastructure also demanded—and produced—a broadly competent workforce with diverse skills to rapidly repair various telecommunication systems. This workforce of IT and network professionals also turned out to be highly collaborative, mobile, and enthusiastic. Whenever one of the providers would fall victim to a Russian cyber attack, engineers from other providers would assist in re-establishing the victim's operations (Herbert 2022). To provide Ukrainian users with reliable connectivity, telecom operators went so far in their collaboration as to establish universal national roaming such that a subscriber of an operator disrupted by a cyber-attack would see no impact in service as they could use the networks of other providers (Giles 2023).

7.3 *Reconfiguration, Resource Re-allocation, and Relocation*

Numerous actions can fall under the umbrella of reconfiguration, re-allocation, and relocation. Such courses of action may include changing individual subsystems or components to modify their functionality or behavior, changing the allocation of resources to tasks or functions, or changing the location of resources that provide functionality or information, either by moving the assets or by transferring functional responsibility.

One manifestation of this broad paradigm was Ukraine's successful relocation of its data to cloud servers outside Ukraine, particularly in Europe, where Russia had few opportunities to disrupt its operation. Furthermore, whenever Russia succeeded in destroying software and data maintained on servers within Ukraine proper, defenders could rapidly restore and reconfigure their systems using the backups kept on European servers.

Just a few days before the Russian invasion, Ukraine passed a law allowing the government to store data in clouds, including those outside Ukraine. Clouds are decentralized, networked systems of highly secure servers where data are distributed geographically worldwide, duplicated, and strongly encrypted. They are accessible to their owners whenever they need them. In today's Ukraine, the cloud servers are generally located outside of Ukraine's borders and cannot be destroyed by Russia by either physical or cyber-attacks.

Given the unrelenting attacks of Russia that included both cyber and physical destruction of IT facilities, Ukraine started to migrate rapidly, i.e., relocate their data, applications, and software backups to the cloud. Only in the first few months after the start of the full-scale war did hundreds of government, academic, and commercial organizations complete major migrations (Burt 2022; White 2022; Amazon Staff n.d.; Stupp 2022). These migrations helped to improve the resilience of IT facilities. A telecommunications company, for example, may use clouds to recover from cyber attacks that destroy or corrupt its software and data. The company would periodically back up its data and software in the cloud. When a cyber-attack destroys or corrupts its data or software, the company could retrieve its most recent backup from the cloud and reconstitute its IT infrastructure.

While we cannot know for sure, this approach likely contributed to the high degree of resilience demonstrated by Ukrainian telecommunications companies. For example, in March 2022, Russian cyber actors successfully targeted a major Ukrainian provider, Ukrtelecom. The company experienced the loss of an estimated 87% of Internet connectivity, a significant, deep drop in its functionality. Nevertheless, merely 15 h later, Ukrtelecom restored its full service (Brewster 2022; Beals 2022).

Russian attempts to deliver a major cyber blow to Ukraine have continued relentlessly. In another example, in December of 2023, nearly 2 years after the Ukrtelecom event, another major telecommunications company, Kyivstar, found itself a victim of a sophisticated cyber attack. Likely delivered by Sandstorm, a unit of the same GRU mentioned earlier, the attack wiped out thousands of virtual servers, drastically degrading the Kyivstar services.

Despite Kyivstar's well-funded cyber security operations, Russian cyber actors were able to place their malware within the Kyivstar infrastructure several months in advance, undetected. The resulting attack was considered the first so powerful that it "completely destroyed the core of a telecoms operator." The security clearly failed. Yet, just a week later, Kyivstar fully restored its operations. Unlike security, resilience prevailed (Balmforth 2024; Coker 2024).

7.4 Dynamic Generation of Information, Services, and Connections

Here, the autonomous cyber-resilience system may refresh information and services periodically or generate information on demand and delete it when no longer needed. It may also establish connections on demand and terminate connections when no longer needed.

If they were to apply this paradigm, Ukrainian cyber defenders or their international supporters would continually create new services and connection links to minimize the chance that the adversary would compromise the existing ones. Certainly, cloud backups, which we discussed above, are examples of such continually created and updated resources.

One example of the dynamic generation of communication channels is known. Reportedly, immediately before the Russian invasion of 2022, Ukraine and the United States established a direct, highly secure communication line between an entity within the Ukrainian military and the U.S. European Command.

Presumably, unlike other communication channels that existed and were utilized earlier between Ukraine and its Western allies, this new line would be unknown to

Ukraine's adversary and unlikely to be quickly compromised. As the full-scale war unfolded, this line was used to keep Ukraine's military in reliable contact with the U.S. military (Harris et al. 2022).

7.5 *Resilience by Design vs. Resilience by Intervention*

Resilience can be implemented and operated in two ways: Resilience by Design (RBD) and Resilience by Intervention (RBI). The differences between RBI and RBD of a given system are as follows. RBD is likely to be found where strong links exist between the system's resilience elements and those components responsible for producing the system's primary functions. On the other hand, if the links are established occasionally and selectively based on circumstances, we are probably looking at RBI. For example, in the case of a built-in autonomous intelligent agent responsible for cyber-resilience, the agent is tightly integrated within the system in an RBD fashion.

Another key difference between RBD and RBI relates to authority. When actions associated with resilience are taken, such as resisting, minimizing, and mitigating a degradation caused by a successful cyber-attack, who has the authority to direct these actions? Who decides whether, when, where, and how these actions are performed? When these decision rights are internal to the system, it does not require an external actor's permission (or direction) to change system properties or reorganize itself. This authority is important to seize opportunities quickly to minimize or recover from disruption. A system with an internal decision-making authority can interpret new information within itself or its environment and adjust accordingly. The system benefits from avoiding potential inefficiencies or delays from waiting for an external actor to yield permission or direction for change. If the decision authority resides primarily outside of the system, we suspect RBI. If, on the other hand, the decisions are made primarily within the system (whether by humans or by a technical decision-making element), RBD is more likely. Although different, RBD and RBI are not incompatible—both approaches have advantages and disadvantages and may co-exist.

8 Preview of the Book

Resilience, as a field, continues to develop as researchers, stakeholders, and governments adapt to a changing threat landscape. This book focuses on building from foundational concepts of cyber-resilience to deepen our understanding of resilience concepts, capture key interdependencies of cyber-systems that impact resilience, and strengthen our knowledge of how to implement policies, both technical and institutional, to improve resilience. Part "Fundamentals of Cyber Resilience" contains chapters describing the fundamentals of cyber resilience, introducing theoretical and practical approaches for framing and assessing cyber resilience, including when a system of interest relies on dependent cyber systems. Part "Cyber-Social and Cyber-Physical Interactions" addresses dependencies that extend outside of the cyber domain, with the first half focusing on cyber-social interactions and the second half on cyber-physical interactions. Part "Governance and Applications of Cyber Resilience" takes a step back to consider governance for cyber resilience and case studies that show cyber resilience implemented in practice.

8.1 First Part: Fundamentals of Cyber Resilience

The first part of this book focuses on framing what is considered cyber-resilience, defining the bounds of cyber-resilient systems, and how to map existing decision-making theories to key concepts of resilience. Specifically, it leans towards developing quantifiable metrics for resilience.

Part "Fundamentals of Cyber Resilience" begins with a chapter by Zhu maps existing system theory concepts to cyber resilience assessment and improvement. This chapter presents three major systems-view theories—game theory, control theory, and learning theory—as three major concepts to assess and improve cyber system behavior towards resilience. Zhu maps key methods in game and control theories to the specific task of modeling resilience and game and learning frameworks to tasks oriented to wargame effectively by proactively anticipating adversarial behaviors and resilient strategies. These theoretical approaches can then be applied in practice in organizations using defined resilience assessment approaches.

Where the first chapter takes a top-down approach for framing cyber resilience, the following chapter from Moukafih et al. describes a bottom-up approach for assessing cyber risk and resilience. Quantifying risks to cyber systems is difficult given their complexity and interdependencies across systems, making it challenging to assess risk. This chapter first reviews nine existing approaches for cyber risk assessment, performing a gap analysis to reveal flaws in most established approaches. It then extends traditional threat and vulnerability risk assessment (TVRA) approaches to more complex cyber systems, proposing a cyber risk assessment approach centered on quantifying cyber systems' threats, vulnerabilities, and risks. This methodology allows users to track cybersecurity requirements across components and systems and establishes a cyber risk assessment approach that fuses theoretical ideas with real-world applications.

While the first two chapters generally acknowledge the complexity of cyber dependencies, the third chapter from Evans et al. aims to quantify specifically cyber dependencies and their risks. This chapter formally defines different types of cyber dependencies, their risks, and methods for assessing and managing such risks across multiple connected cyber assets. While this chapter focuses specifically on how cyber-dependencies—where an asset's operation depends on information transmitted via electronic or informational links—it introduces methods of identifying and assessing dependencies throughout cyber-systems that will become the focus of

Part "Cyber-Social and Cyber-Physical Interactions". Much like the previous chapter, Evans et al.'s approach for assessing risk first defines potential risks associated with different threat scenarios before quantifying the consequence of loss of each hazard. This chapter details several risk assessment methods, including dependency curves, pathway analysis, model checking, and systems dynamics approaches. Such tools are particularly applicable in complex systems with many dependencies between individual cyber systems.

The last chapter in this part from Remy et al. looks at the other side of the resilience coin, seeking to quantify and reduce system non-resilience. This is particularly useful given many of the issues identified in other chapters, namely that cyber-attacks can be very complex, impacting a wide attack surface and evolving over time. The methodology presented here first models the system of interest in detail, including all components and their interactions, before introducing a set of attacks using different threat models and clearly defining the attack flow of each threat. With the system and threat model constructed, the authors propose a set of definitions and requirements that can serve as metrics of non-resilience within system models, applying the approach to the case study of a communications satellite system.

8.2 *Second Part: Cyber-Social and Cyber-Physical Interactions*

While nearly all the chapters in the first part acknowledge the importance and relevance of cyber dependencies in understanding cyber risk and resilience, the analyses previously presented tend to limit themselves to the cyber domain, even if considering multiple cyber assets or systems. However, in the context of a larger organization, cyber systems depend heavily on staff, personnel, and the social dimension of a system to adequately respond to and recover from threats.

Part "Cyber-Social and Cyber-Physical Interactions" begins with Howard and da Cruz describing the importance of this social dimension of an organization from the first moment when an organization is considering whether and how to invest in cyber resilience. Unlike many engineering-oriented analyses of resilience, the authors focus on not just the timeline of when an adverse event occurs but also the decisions that lead up to organizations prioritizing resilience in their system design and implementation. While cyber systems can be engineered to be redundant, adaptation and leadership (among other social traits) are necessary factors that help achieve cyber resilience across an organization through the willingness to adopt various strategies to improve cyber resilience. Chapter 6 on the roles of decision-makers when planning to invest in or adopt improvements for cyber resilience, identifying nine socio-technical influences, including workforce skill and organizational policies that impact their decisions to innovate. This is combined with an innovation-decision process model (IDPM) describing how an organization decides to invest in innovation and applied to the Colonial Pipeline cyberattack as a case study.

Whereas Chap. 7 addresses the importance of the social dimension of cyber resilience at the point of technology adoption, people play an important role in cyber resilience across the system lifecycle. Thus, the second chapter in this part by Groenendaal and Helsloot examines the role of resilient decision-making in cyber incident response. For organizations to improve their cyber resilience performance, the organizational, structural, and social components of cyber incident response must be considered, as well as the resilience of their decisions. This study defines three levels of cyber resilience: the resilience performance of an organization, the resilience of cyber incident response decision-making, and actual decisions made for cyber resilience. All three concepts are necessary to understand resilient decision-making in cyber incident response; they are all related, yet they are separate, and improving one area of resilience may not improve another. It is necessary to consider all three to fully capture the social role in cyber resilience.

Whereas Chap. 8 describes how people within an organization can improve resilience, Chap. 8 focuses on the vulnerabilities human agents introduce to the system, especially considering how humans are typically the most significant driver of cyber threats. Because much of cyber risk mitigation is rooted in allowing trusted entities to access sensitive cyber assets and blocking untrusted sources, threat mitigation becomes much more challenging when the threat is an insider with access to cyber resources. Therefore, this chapter uses a rule-based paradigm to minimize cyber threats, particularly from insiders in an organization, using a behavioral science lens instead of a computer science lens. Because rules are the primary way of coordinating individual activity with a variety of goals to a coordinated purpose, both over- and under-regulation of an organization via rules can be exploited by insider threats. This chapter uses both a toy model and a percolation model of system interactions to illustrate this point, helping quantify the relationship between rules and risk for a better understanding of insider threats and potential mitigation approaches.

In Chap. 9, we shift the focus of the part from cyber-social to cyber-physical interactions. The remainder of this part examines the key role of physical networks in assessing and improving cyber-resilience. Abreu et al. apply a cyber-physical lens to cyber resilience by considering the spectrum of internet-connected smart devices that extends from a central off-site cloud server through edge computing to bring computing closer to physical devices. Given the added cyber and physical vulnerabilities across the Cloud-to-Things continuum, this chapter discusses metrics and methods for assessing cyber-physical resilience mainly in terms of their quality of resilience (QoR), proposing metrics including mean up time (MUT), mean downtime (MDT), and their synthesis in the idea of availability. It further describes how several tools, such as machine learning (ML), increase system intelligence when applied in paradigms such as self-organizing networks (SONs) or intent-based networks (IBNs). This added intelligence and ability to respond to threats and detect anomalies can strengthen both the fault tolerance and security of cyber-physical systems, in tur increasing their resilience.

The following chapter in this part, from Weisman et al., builds on this need to understand cyber-physical interactions to fully quantify cyber resilience. With the added dependencies of cyber-physical systems introducing new vulnerabilities, such systems can be infiltrated or inhibited through both cyber and physical pathways, meaning that examining either domain in isolation lacks the necessary level of detail. This study proposes a simulation testbed integrating both cyber and physical elements to quantify the performance of cyber attackers and defense mechanisms in modern vehicle test cases. This testbed is composed of a module simulating cyber-attacks, paired with Toyota's Portable Automotive Security Testbed with Adaptability (PASTA) and the Army's Active Defense Framework (ADF) to simulate competing cyber-attack and cyber-defense forces and model their impact on the performance of a simulated vehicle test case. The simulation testbed and functionality-based resilience quantification approach are extendable to assess the resilience of any cyber-physical system, as this method can simulate the physical effects of cyber degradation by modeling both systems through separate programs that respond to one another. This chapter applies a generalizable quantitative measurement of cyber resilience (QMOCR) approach to quantify system functionality and measure resilience as the area under the curve (AUC) when functionality is degraded.

8.3 *Third Part: Governance and Applications of Cyber Resilience*

The first two parts of this book examine the criticality of cyber, physical, and social dependencies in understanding and improving cyber resilience. Cyber resilience extends beyond organizational boundaries and requires additional methods to ensure cyber resilience at a societal scale. This final part examines the role of governance in cyber resilience and presents case studies as applications for resilience.

The first chapter in this part from Keenan et al. discusses cyber resilience governance in the context of cyber-physical infrastructure systems. Such infrastructure networks meet many of the basic needs of both civilian and military objectives, making them prime targets for attack from hostile actors, especially in the current conditions of world conflict (e.g., the 2022 Russo-Ukraine War). To meet recent NATO objectives for securing military-civilian infrastructure from disruption from hostile actors, this chapter describes current US government policies for cyber-physical infrastructure resilience and proposes guidance for developing a National Resilience Plan (NRP). This chapter centers on governance and planning for the resilience of critical cyber-physical infrastructure both within and between NATO member states.

The second chapter of this part, from Evans and Sturm, is similarly motivated by the need to strengthen the regional resilience of critical infrastructure. Because these systems are heavily dependent on one another and often cross multiple jurisdictional boundaries, there is a need to assess cyber resilience at such a scale. This chapter identifies several existing methods and tools for conducting regional cyber resiliency assessments (CRAs), such as the Cyber Security Evaluation Tool (CSET), and proposes its own methodology to assess cyber resilience across infrastructure systems within a region. In the context of governance, CRAs are intended to improve and facilitate communication between critical infrastructure owners/operators, local state and federal governments, and other stakeholders to improve a region's resilience through discussion and collaboration. The authors demonstrate the applicability of CRAs by presenting a case study of the Department of Homeland Security's (DHS) Regional Resilience Assessment Program (RRAP) focused on Ashburn, VA, a region containing an incredibly high volume of data servers in the area.

The third chapter in this part, from Collier et al., focuses on the resilience of supply chains for cyber-physical systems. As opposed to many of the previous chapters that focus on the resilience of a cyber-physical system itself, this perspective considers the life cycle of developing such products through their supply chains. Supply chains are a very useful and interesting case study in resilience governance, given that cyber-physical systems are composed of elements sourced from diverse industry stakeholders crossing national boundaries. This chapter first describes the complexity and risks of such complexity in electronics supply chains. Then, it moves towards identifying metrics for supply chain resilience, advancing a resilience matrix (RM) approach for generating and categorizing resilience metrics across all temporal phases of resilience, as well as different organizational domains (physical, information, cognitive, and social). These methods are applied to the test case of medical device supply chains to illustrate their broad applicability.

The next chapter, from Evans and Werth, centers on the idea that organizations cannot rely only on passive defensive measures for security and resilience but rather that effective resilience requires active defense strategies. These strategies may be automated and built into the cyber domain of a system, or they may require organizational and social elements to properly respond and recover. This chapter examines a broad range of potential strategies, such as honeypots and distributed ledger technology, categorizing approaches in terms of how reactive or proactive they are for resilience. Organizations will need to assess their infrastructure, personnel, and potential threats to fully understand how different active defense approaches apply to their case.

The last chapter from Rose et al. aims to incorporate an economic analysis component when considering the cyber resilience of an organization. Whereas many of the previous chapters seek to quantify resilience in terms of cyber functionality and degradation, this study extends to quantify such losses in economic terms using the concept of business interruption (BI). This framing aims to bridge the gap between an engineering concept of resilience and a social science perspective. The chapter identifies 11 resilience tactics, such as import substitution and maintaining inventories, that can strengthen organizational resilience. It then advances an operational metric of organizational resilience, defining static resilience in terms of the amount of BI prevented by a given resilience tactic, expressed as a percentage of maximum BI with no resilience strategies. Lastly, the authors survey resilience decision-makers to quantify the effectiveness of different strategies in terms of minimizing BI through resilience.

References

Amazon Staff (n.d.) Safeguarding Ukraine's data to preserve its present and build its future. https://www.aboutamazon.com/news/aws/safeguarding-ukraines-data-to-preserve-its-present-and-build-its-future. Accessed 2 Apr 2024

Balmforth T (2024) Exclusive: Russian hackers were inside Ukraine telecoms giant for months. Reuters. https://www.reuters.com/world/europe/russian-hackers-were-inside-ukraine-telecoms-giant-months-cyber-spy-chief-2024-01-04/

Bateman J (2022) Russia's wartime cyber operations in Ukraine: military impacts, influences, and implications. https://policycommons.net/artifacts/3348853/russias-wartime-cyber-operations-in-ukraine/4147737/

Beals M (2022) Ukrainian military internet provider suffers cyberattack. The Hill. https://thehill.com/policy/international/600154-ukrainian-military-internet-provider-suffers-cyberattack/

Bostick TP, Holzer TH, Sarkani S (2017) Enabling stakeholder involvement in coastal disaster resilience planning. Risk Anal 37(6):1181–1200. https://doi.org/10.1111/risa.12737

Brewster T (2022) 'Most severe' cyberattack since Russian invasion crashes Ukraine internet provider. https://www.forbes.com/sites/thomasbrewster/2022/03/28/huge-cyberattack-on-ukrtelecom-biggest-since-russian-invasion-crashes-ukraine-telecom/?sh=2b38791a7dc2

Burt J (2022) Microsoft gives away $400m in cloud support to Ukraine. https://www.theregister.com/2022/11/04/microsoft_ukraine_tech_support/

Chandrasekharan PC (1996) Robust control of linear dynamical systems. https://cir.nii.ac.jp/crid/1130282268974880384

Coker J (2024) Ukraine blames Russian sandworm hackers for Kyivstar attack. Infosecurity Magazine. https://www.infosecurity-magazine.com/news/ukraine-russian-sandworm-kyivstar/

Colbert EJM, Kott A, Knachel L, Sullivan DT (2017) Modeling cyber-physical war-gaming. US Army Research Laboratory. https://apps.dtic.mil/sti/citations/AD1038105

Collier ZA, Linkov I, DiMase D, Walters S, Tehranipoor M, Lambert J (2014a) Risk-based cybersecurity standards: policy challenges and opportunities. Computer 47:70–76

Collier ZA, Walters S, DiMase D, Keisler JM, Linkov I (2014b) A semi-quantitative risk assessment standard for counterfeit electronics detection. SAE Int J Aerosp 7(2014-01-9002):171–181

Cybersecurity & Infrastructure Security Agency (2023a) A call to action: bolster UEFI cybersecurity now. August 3. https://www.cisa.gov/news-events/news/call-action-bolster-uefi-cybersecurity-now

Cybersecurity & Infrastructure Security Agency (2023b) The attack on colonial pipeline: what we've learned & what we've done over the past two years. May 7. https://www.cisa.gov/news-events/news/attack-colonial-pipeline-what-weve-learned-what-weve-done-over-past-two-years

DiMase D, Collier ZA, Heffner K, Linkov I (2015) Systems engineering framework for cyber physical security and resilience. Environ Syst Decis 35(2):291–300. https://doi.org/10.1007/s10669-015-9540-y

Estay S, Daniel A, Sahay R, Barfod MB, Jensen CD (2020) A systematic review of cyber-resilience assessment frameworks. Comput Secur 97(October):101996. https://doi.org/10.1016/j.cose.2020.101996

Florin M-V (2016). IRGC resource guide on resilience. https://infoscience.epfl.ch/record/228206/files/IRGC.%20(2016).%20Resource%20guide%20on%20resilience.%20Book.pdf

Ganin AA, Massaro E, Gutfraind A, Steen N, Keisler JM, Kott A, Mangoubi R, Linkov I (2016) Operational resilience: concepts, design and analysis. Sci Rep 6(1):1–12

Ganin AA, Quach P, Panwar M, Collier ZA, Keisler JM, Marchese D, Linkov I (2020) Multicriteria decision framework for cybersecurity risk assessment and management. Risk Anal 40(1): 183–199. https://doi.org/10.1111/risa.12891

Gil S, Kott A, Barabási A-L (2014) A genetic epidemiology approach to cyber-security. Sci Rep 4(1):1–7

Giles K (2023) Russian cyber and information warfare in practice. Chatham House, London. https://www.researchgate.net/profile/Keir-Giles/publication/376521850_Russian_cyber_and_information_warfare_in_practice_Lessons_observed_from_the_war_on_Ukraine/links/657c10afea5f7f02057364ba/Russian-cyber-and-information-warfare-in-practice-Lessons-observed-from-the-war-on-Ukraine.pdf

Harris S, DeYoung K, Khurshudyan I, Parker A, Sly L (2022) Road to war: U.S. struggled to convince allies, and Zelensky, of risk of invasion. Washington Post. https://www.washingtonpost.com/national-security/interactive/2022/ukraine-road-to-war/

Herbert J (2022) Ukraine IT sector: resilient, agile, and hopefully here to stay. Blog – Everest Group. https://www.everestgrp.com/blog/ukraine-it-sector-resilient-agile-and-hopefully-here-to-stay-blog.html

Holling CS (1996) Engineering resilience versus ecological resilience. Engineering within Ecological Constraints 31(1996):32

Hollnagel E, Woods DD, Leveson N (2006) Resilience engineering: concepts and precepts. Ashgate Publishing, Ltd. https://books.google.com/books?hl=en&lr=&id=rygf6axAH7UC&oi=fnd&pg=PP1&dq=).+Resilience+engineering:+Concepts+and+precepts.&ots=ir3ERW8-9d&sig=I5uVJCDn9U37OOlz4fAqUbMRM-w

IEEE (2003) 610.12–1990. https://ieeexplore.ieee.org/document/159342

ISO/IEC 27005 (2018) Information technology–security techniques–information security risk management

Kaplan S, Garrick BJ (1981) On the quantitative definition of risk. Risk Anal 1(1):11–27. https://doi.org/10.1111/j.1539-6924.1981.tb01350.x

Kott A (2006) Information warfare and organizational decision-making. Artech House. https://cds.cern.ch/record/1251596

Kott A, Abdelzaher TF (2014) Resiliency and robustness of complex systems and networks. Adapt Dyna Resilient Syst 67:67–86

Kott A, Alberts DS, Wang C (2015) Will cybersecurity dictate the outcome of future wars? Computer 48(12):98–101

Kott A, Ludwig J, Lange M (2017) Assessing mission impact of cyberattacks: toward a model-driven paradigm. IEEE Secur Privacy 15(5):65–74

Kott A, Theron P, Drašar M, Rzadca K, LeBlanc B, Pihelgas M, Mancini L, De Gaspari F (2018) Reference architecture of an autonomous agent for cyber defense of complex military systems. In: Jajodia S, George Cybenko VS, Subrahmanian VS, Wang C, Wellman M (eds) Adaptive autonomous secure cyber systems. Springer, Cham, pp 1–21. https://doi.org/10.1007/978-3-030-33432-1_1

Larkin S, Fox-Lent C, Eisenberg DA, Trump BD, Wallace S, Chadderton C, Linkov I (2015) Benchmarking agency and organizational practices in resilience decision making. Environ Syst Decis 35(2):185–195. https://doi.org/10.1007/s10669-015-9554-5

Leslie NO, Harang RE, Knachel LP, Kott A (2018) Statistical models for the number of successful cyber intrusions. J Def Model Simul Appl Methodol Technol 15(1):49–63. https://doi.org/10.1177/1548512917715342

Linkov I, Eisenberg DA, Bates ME, Chang D, Convertino M, Allen JH, Flynn SE, Seager TP (2013a) Measurable resilience for actionable policy. Environ Sci Technol:130903081548008. https://doi.org/10.1021/es403443n

Linkov I, Eisenberg DA, Plourde K, Seager TP, Allen J, Kott A (2013b) Resilience metrics for cyber systems. Environ Syst Decis 33(4):471–476. https://doi.org/10.1007/s10669-013-9485-y

Linkov I, Bridges T, Creutzig F, Decker J, Fox-Lent C, Kröger W, Lambert JH, Levermann A, Montreuil B, Nathwani J (2014) Changing the resilience paradigm. Nat Clim Chang 4(6): 407–409

NATO (2023) Resilience, civil preparedness and article 3. NATO August 2. https://www.nato.int/cps/en/natohq/topics_132722.htm

NIST, SP (2002) 800–30 risk management guide for information technology systems. National Institute for Standards and Technology, Gaithersburg

O'Neill PH (2022) Russia hacked an American satellite company one hour before the Ukraine invasion. MIT Technol Rev, May

Olowu TO, Dharmasena S, Hernandez A, Sarwat A (2021) Impact analysis of cyber attacks on smart grid: a review and case study. In: Tyagi H, Chakraborty PR, Powar S, Agarwal AK (eds) New research directions in solar energy technologies. Springer, Singapore, pp 31–51. https://doi.org/10.1007/978-981-16-0594-9_3

Perrow C (1984) Normal accidents: living with high risk technologies. Princeton University Press, Princeton, NJ

Roege PE, Collier ZA, Mancillas J, McDonagh JA, Linkov I (2014) Metrics for energy resilience. Energy Policy 72:249–256

Ross R, Pillitteri V, Graubart R, Bodeau D, McQuaid R (2019) Developing cyber resilient systems: a systems security engineering approach. National Institute of Standards and Technology. https://csrc.nist.gov/CSRC/media/Publications/sp/800-160/vol-2/draft/documents/sp800-160-vol2-draft-fpd.pdf

Smith EA (2005) Effects based operations: applying network centric warfare in peace, crisis, and war. Command and Control Research Program (CCRP), Office of the Assistant Secretary of Defense, Washington, DC

Stupp C (2022) Ukraine has begun moving sensitive data outside its borders. WSJ. https://www.wsj.com/articles/ukraine-has-begun-moving-sensitive-data-outside-its-borders-11655199002

U.S. Government Accountability Office (2024) CrowdStrike chaos highlights key cyber vulnerabilities with software updates. U.S. GAO. September 12. https://www.gao.gov/blog/crowdstrike-chaos-highlights-key-cyber-vulnerabilities-software-updates

White R (2022) How the cloud saved Ukraine's data from Russian attacks. https://www.c4isrnet.com/2022/06/22/how-the-cloud-saved-ukraines-data-from-russian-attacks/

Willett M (2023) The cyber dimension of the Russia–Ukraine war. In: Survival: October–November 2022. Routledge, pp 7–26. https://www.taylorfrancis.com/chapters/edit/10.4324/9781003422211-1/cyber-dimension-russia%E2%80%93ukraine-war-marcus-willett

Fundamentals of Cyber Resilience

Foundations of Cyber Resilience: The Confluence of Game, Control, and Learning Theories

Quanyan Zhu

1 Introduction to Cyber Resilience

Cyber resilience of a network or organization refers to its ability to prepare for, respond to, and recover from cyber threats or incidents (Ganin et al. 2016; Zhao et al. 2022; Rieger et al. 2009, 2012). It represents a departure from classical cybersecurity measures, including cryptography, firewalls, and intrusion detection systems, which aim to prevent attackers from infiltrating or causing harm to the network. Instead, cyber resilience focuses on the capacity to respond to and recover from threats. Cyber resilience and cybersecurity are complementary to each other. Despite efforts and investments in cybersecurity for protection, perfect security cannot be guaranteed. There will always be threats that cannot be fully mitigated. These threats may include zero-day attacks, which are unknown to the network, or black swan events that are costly to secure against. Therefore, with limited resources and budgets, networks must decide how to protect themselves from threats to ensure the continuity of their applications and missions.

As illustrated in Fig. 1, resilience provides a complementary approach (Zhu and Başar 2015; Zhu and Xu 2020; Linkov et al. 2023). When threats are difficult to prevent successfully, there is a need to find ways to respond to them after they have gained initial access to the network, deter further infiltration by attackers, and recover from any damage incurred. Additionally, it is important to recognize that the success of security mechanisms is probabilistic. A higher probability of success will require a greater allocation of resources. Despite the careful deployment of cybersecurity measures, there will still be a probability, albeit small, for attackers to succeed in bypassing cryptographic protocols and firewalls to gain access to the network.

Q. Zhu (✉)
New York University, Brooklyn, NY, USA
e-mail: qz494@nyu.edu

© The Author(s), under exclusive license to Springer Nature Switzerland AG 2025

I. Linkov, A. Kott (eds.), *Cyber Resilience: Applied Perspectives*, Risk, Systems and Decisions, https://doi.org/10.1007/978-3-031-90109-6_2

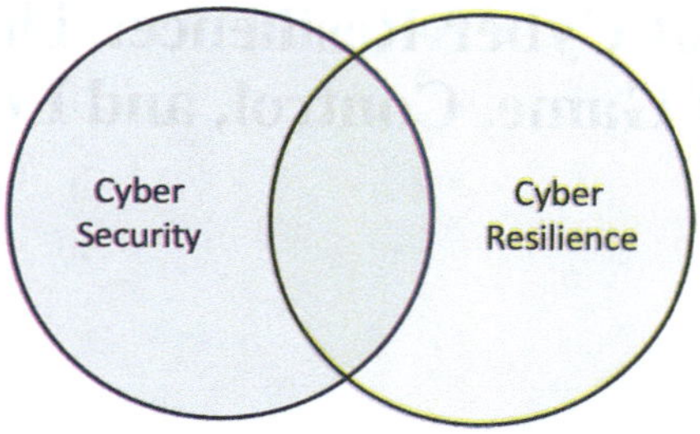

Fig. 1 Cybersecurity and cyber resilience focus on different sets of issues. Cybersecurity primarily concentrates on safeguarding assets from unauthorized access, attacks, and damages. On the other hand, cyber resilience aims to both deter attackers and recover the network from the initial success of attacks. Both strive to enhance the confidentiality, integrity, and availability (CIA) of information and resources. Cyber resilience specifically aims to mitigate the impact on the CIA triad despite the initial compromise. Despite their differences, there are overlaps between the two in terms of the techniques used to provide security and resilience. For instance, detection plays a critical role in both security and resilience efforts

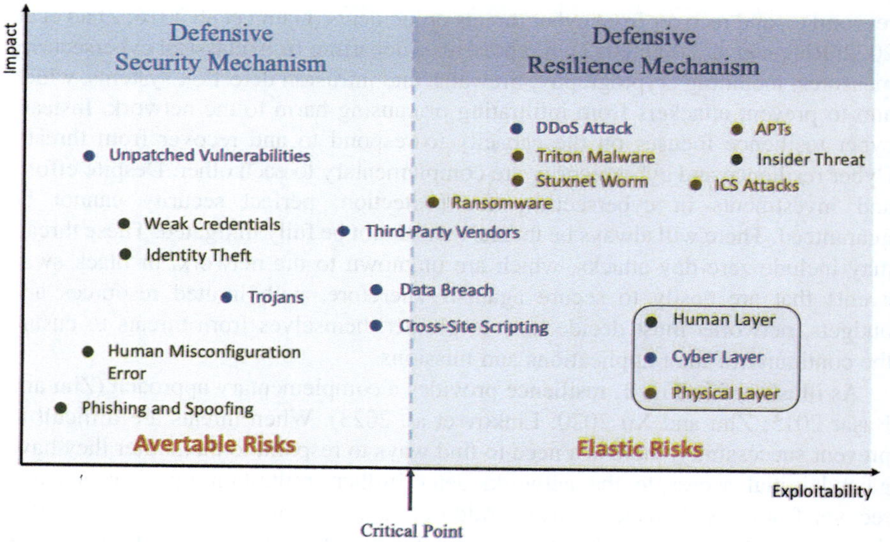

Fig. 2 The vulnerabilities and their associated threats on the attack surface are represented in two dimensions: their exploitability and their impact. They span across cyber, physical, and human layers of networks or organizations. These vulnerabilities are categorized into two types of risks. One type is avertable risks, which are managed by defensive security mechanisms, while the other type is elastic risks, which are addressed through resilience mechanisms

The tradeoffs between cyber resilience and cybersecurity stem from economic and technological considerations. As illustrated in Fig. 2, common threats that occur frequently or are easily exploited by attackers often require defensive security mechanisms to protect the network with a high probability of success. Conversely, sophisticated threats that demand advanced skill sets and occur less frequently can be mitigated through defensive resilience. The decision on resource allocation across the entire threat landscape depends on factors such as budget, application domains, and available technology. For instance, mission-critical applications like nuclear power plants or oil and gas manufacturing plants require investment not only in designing security mechanisms but also in crafting resilience mechanisms to counter highly sophisticated attackers and mitigate the exploitation of unknown vulnerabilities.

The decision regarding which threats to defend against is not solely limited by budgetary constraints; it also involves human resources, such as the time and effort required for defense, as well as the availability of technologies that can facilitate or enable such defense. For instance, defending against Advanced Persistent Threat (APT) attacks requires a coordinated effort from various aspects of missions, including human, cyber, and physical assets, to ensure assurance. This defense may require AI technologies, which may still be in their infancy, particularly those that can be employed to mitigate the exploitation of human cognitive vulnerabilities leading to successful social engineering attacks.

Within the set of exploitable vulnerabilities, we categorize them into two types of risks: avertable risk and elastic risk. Elastic risks are those triggered by events that can only be mitigated elastically through resilience strategies. Conversely, avertable risks are triggered by events that can be prevented with high probability using security mechanisms.

2 Cyber Risks and Resilience Mechanisms

The first step toward understanding cyber resilience is to comprehend and quantify cyber risks. Just like all other risks, cyber risks are characterized by three primary components. First and foremost, cyber risk stems from the existence of vulnerabilities within the network. These vulnerabilities can manifest as software bugs, human biases, interface errors, system integration failures, inadequate preparedness, or hardware fragility. They serve as sources of susceptibility or exposure for both the network and the organization to identified threats.

These vulnerabilities create risks because they represent opportunities for threats to exploit them. Hence, the threat constitutes the second component of the risk. A threat refers to an event or circumstance created by an adversary to exploit vulnerabilities within the system, triggering further events that lead to harm, loss, or damage. For instance, an attacker may craft a phishing email to exploit the cognitive vulnerability of a user, leading to the disclosure of passwords or access credentials to the attacker, or the unwitting download of malware to spread within the network. Alternatively, an attacker may exploit a buffer overflow bug in software to execute malicious code.

The intersection of vulnerabilities and threats forms what we term the attack surface. This represents the collection of vulnerabilities and entry points within the

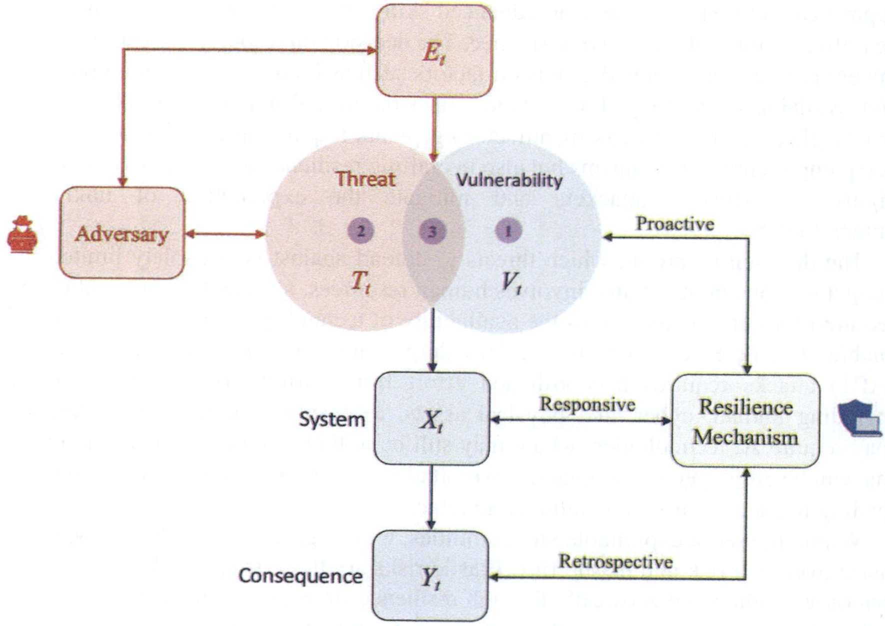

Fig. 3 Cyber risk at time t arises from vulnerabilities within the system itself, denoted by the set V_t, and potential threats posed by adversaries T_t. The intersection of these threats and vulnerabilities forms the attack surface of the network. An adversary can launch an event, E_t, to exploit a vulnerability on the attack surface, resulting in changes to the system states. The consequences are expressed as outcomes of these changes in system state. The goal of resilience mechanisms is to mitigate the impact of such attacks. There are three types of resilience mechanisms: proactive, responsive, and retrospective. Proactive mechanisms aim to reduce the attack surface and create a more resilient network that is harder for attackers to exploit. Responsive mechanisms involve equipping the network with adaptive and automated responses to attack behaviors in real-time. Retrospective mechanisms focus on reducing the impact after the consequences of the attack, Y_t, are realized and observed. They can involve restoring the system to its previous state or recovering financially from losses incurred

network or organization that attackers can potentially leverage to gain unauthorized access, disrupt operations, steal data, or carry out other malicious activities.

When an adversarial event exploits a vulnerability within the attack surface, consequences ensue. Consequences represent the outcomes of the interaction between threats and vulnerabilities. They can encompass performance degradation, disruptions to operations, financial losses, reputational damage, legal liabilities, and other adverse effects.

The vulnerabilities, threats, and consequences constitute the landscape of cyber risks, all of which can evolve over time. As illustrated in Fig. 3, the set of vulnerabilities at time t is denoted by V_t, while the set of threats is denoted by T_t. One immediate consequence of the exploitation by an adversary event, E_t, is the change of the network system state, X_t, which subsequently influences the consequences for the network, denoted by Y_t.

A vulnerability (3) is one that resides on the attack surface. For each vulnerability on the surface, there exists an event that leads to exploitation. However, a vulnerability, as exemplified in (2), is a threat that does not result in an event capable of exploiting a vulnerability within the network. This could be due to the network being well protected against that threat or the threat being irrelevant to the system under focus. A vulnerability (2) is one that cannot be exploited by a threat, either because the adversary is unaware of it or the vulnerability is difficult to exploit.

A cyber resilience mechanism is one that supports the network system in dynamically reducing its risk over time through three distinct types of cyber resilience mechanisms. One such mechanism is the proactive approach, which focuses on directly changing the attack surface to reduce the risk. Proactive measures may involve creating secure network policies, such as network segmentation, patch management, and zero trust access control to minimize potential threat vectors. For example, network segmentation involves dividing a network into smaller, isolated segments, organizations can limit the scope of potential attacks. Network segmentation allows for stricter access controls and segmentation of sensitive resources, preventing lateral movement by attackers within the network. Regularly updating and patching software and systems helps address known vulnerabilities and weaknesses. Patch management programs ensure that critical security patches are applied promptly to reduce the risk of exploitation by attackers. Zero-trust access control verifies the identity of users and devices attempting to access resources, regardless of their location or network connection. This involves implementing strong authentication methods such as multi-factor authentication (MFA) and biometric authentication to ensure that only authorized individuals gain access. By proactively implementing these policies, the set of vulnerabilities V_t or threats T_t can be reduced, subsequently shrinking the attack surface.

Another one is responsive resilience mechanisms that involve continuous monitoring and assessment of the threat landscape to identify emerging risks and vulnerabilities and dynamic defense strategies that evolve in real-time based on contextual factors and situational awareness. By leveraging predictive analytics and threat intelligence, organizations can anticipate potential attack vectors and vulnerabilities, allowing them to strengthen their defenses before adversaries can exploit them. The dynamic adaptability enables organizations to quickly adjust their security posture in response to evolving threats and changing environmental conditions. Leveraging machine learning and control algorithms, these adaptive defenses can autonomously analyze incoming threats, assess their impact, and dynamically allocate resources to mitigate them effectively. One notable example is adaptive cyber deception (Jajodia et al. 2016; Pawlick et al. 2018; Zhang et al. 2020; Pawlick and Zhu 2017a), aimed at thwarting attackers from achieving their objectives by deploying decoys, such as honeypots or other traps, to divert their attention from critical assets. The effectiveness of cyber deception schemes relies on their adaptability, achieved through continuous learning of attackers' objectives and tactics. By dynamically configuring and deploying deception techniques based on this acquired knowledge, organizations can enhance their ability to mislead and deter adversaries effectively. Moving target defense is another example that can also be tailored to counteract attackers' evolving tactics and behaviors. Moving target defense such as network address randomization, where IP addresses and network configurations are frequently shuffled to prevent attackers from mapping out the network topology and identifying vulnerable entry points. System administrators can employ dynamic service migration, automatically moving critical services and applications between different hosts or cloud instances to evade detection and mitigate the impact of targeted attacks.

The third mechanism is retrospective, which focuses on restoring the system after an attacker has successfully compromised it. One immediate retrospective mechanism is cleaning up the network following a shutdown. Consider a manufacturing plant that falls victim to a sophisticated ransomware attack. As a result of the attack, critical systems controlling production lines and machinery are rendered inoperable. To contain the spread of the ransomware and prevent further damage, the plant is forced to initiate a complete shutdown of its operations. This entails halting production, isolating infected systems, and disconnecting the compromised network from external connections. The restoration process involves restoring encrypted data from backups, rebuilding compromised systems, and implementing enhanced security measures to prevent future incidents. The recovery process incurs a significant investment of time, resources, and financial costs.

Another retrospective mechanism is post-incident analysis and remediation efforts aimed at understanding the root cause of the breach and implementing measures to prevent similar incidents in the future. This involves conducting thorough forensic investigations to identify the entry point of the attack, the techniques used by the adversary, and any vulnerabilities exploited. By analyzing the attack vector and identifying weaknesses in the security infrastructure, organizations can implement targeted security enhancements and patches to strengthen their defenses against future threats. Cyber insurance is another example of a retrospective mechanism by organizations to mitigate the financial fallout of a cybersecurity incident (Liu and Zhu 2023; Zhu 2018). In the aftermath of a data breach or network intrusion, businesses may face a multitude of costs, including forensic investigations, legal fees, regulatory fines, and loss of revenue. Cyber insurance policies provide a financial safety net by offering coverage for these expenses and compensating organizations for the losses incurred as a result of the incident. By having a cyber insurance policy in place, organizations can alleviate the financial burden of cybersecurity incidents and safeguard their long-term viability.

While retrospective mechanisms are often a last resort, they become necessary when facing a formidable adversary. Proper preparation and design are essential to minimize recovery time and mitigate financial and technical risks to the network.

3 Network and Dynamic Effects on Cyber Risks

A large network can consist of multiple interconnected subnetworks. The risk within one subnetwork can be influenced by others, as they are interconnected. A threat to one subnetwork can lead to consequences within that subnetwork and may also impact other connected subnetworks. As illustrated in Fig. 4, the states of other subnetworks are aggregated as Σ_t, and the well-being of these other subnetworks can affect the state of the subnetwork X_t. The interdependencies among subnetworks pose a significant challenge in managing cyber risks. Risk in one subnetwork can propagate to other subnetworks and vice versa. These interdependencies can arise from three sources. Firstly, threats can exploit the connectivity between subnetworks, making it more convenient for attackers to deploy threats. For instance, attackers may compromise neighboring subnetworks before targeting the intended one. In this way, the threat space T_t is expanded for the targeted network as threats to connected networks also affect its security. Secondly, vulnerabilities in the connections with neighboring subnetworks bring in new vulnerabilities of the subnetwork itself. Additionally, vulnerabilities in the nodes connected to it become potential threats that the subnetwork must address. In this way, the vulnerability space V_t is expanded for the targeted network. Thirdly, interdependencies can arise through consequences. Risks can propagate through state influence, where the state of one subsystem influences others through physical dynamics. In this way, the consequence X_t or Y_t is directly affected.

Such interdependencies have been vastly investigated in critical infrastructure networks (Chen and Zhu 2019b; Huang et al. 2017, 2018; Zimmerman et al. 2016, 2017) and communication networks (Farooq and Zhu 2018a, b, 2019; Kieras et al. 2022). Defending against threats originating from within the network itself is challenging, and preparing for risks from connected networks is even more complex. Subnetworks may face threats from other networks for which they are not directly

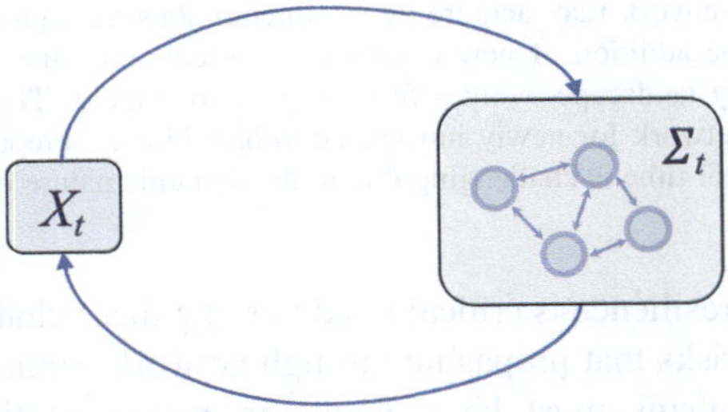

Fig. 4 In a networked environment, the system states are influenced not only by its own risk factors but also by the system states of other nodes connected to it. Σ_t denotes the aggregate state of other nodes, while X_t represents the state of the system itself. The risk of a network system arises from not only local internal risks but also interdependent risks due to these connections. Studies on interdependent risks have been explored in the literature. One way to understand such risks is through game-theoretic methods, which have been studied for systemic risks (Acharya et al. 2017) and holonic risks (Liu et al. 2022; Van Brussel et al. 1998)

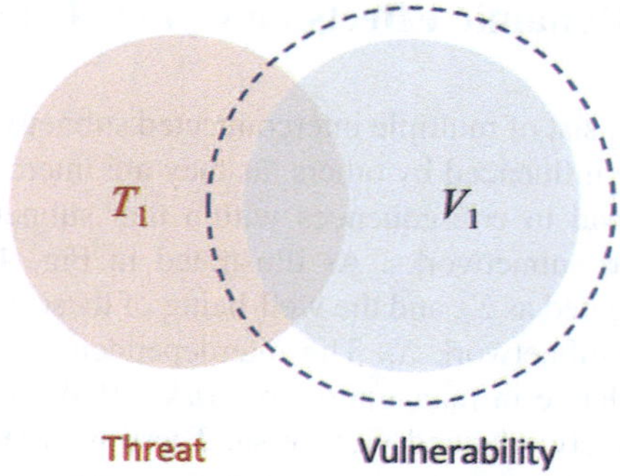

Fig. 5 The vulnerability set V_1 can be significantly larger, unbeknownst to the network defender, due to incomplete knowledge of vulnerabilities. This implies that the attack surface can be much broader than what is currently known. Attackers may exploit vulnerabilities on the attack surface that are unknown to defenders. These types of vulnerabilities must be addressed through cyber resilience strategies

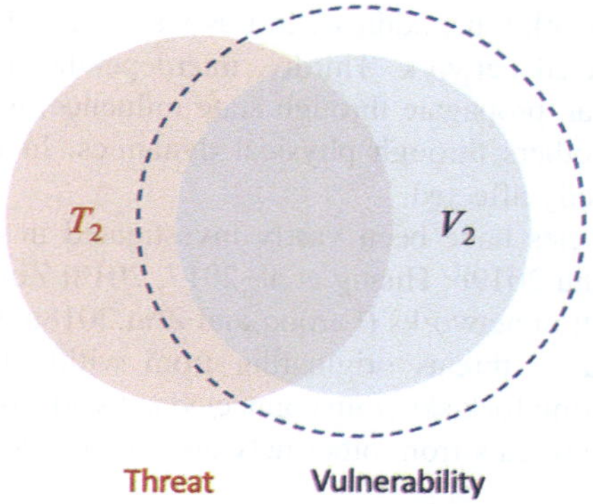

Fig. 6 At stage 2, the network may acquire new vulnerabilities through the installation of new software, patching, or the addition of new interfaces. Consequently, the attack surface expands accordingly. The security landscape evolves from stage 1 to stage 2. There is a need for cyber resilience to ready the network for newly introduced vulnerabilities. Preparing to protect against these vulnerabilities in real-time is challenging due to the dynamic nature of network changes

prepared. Therefore, resilience is critical in addressing these challenges, particularly due to the unknown risks that propagate through network connections.

Resilience mechanisms must be dynamic in nature as they operate within dynamic environments. Firstly, threats evolve over time. Unlike natural threats, adversaries can discover new vulnerabilities and acquire new knowledge to expand the attack surface. Figure 5 depicts the security landscape at time 1, denoted by threat set T1 and vulnerability set V_1. Figure 6 also illustrates the security landscape at time 2, with the corresponding threat set T2 and vulnerability set V_2. It is evident that the threat set expands as adversaries learn and enhance their skill sets over time.

Conversely, the defender's vulnerability set increases as the network introduces new vulnerabilities due to software and patch updates, as well as changes in connectivity. Consequently, the attack surface dynamically grows from time 1 to time 2. Such growth can be strategically directed by the actions taken by both parties. The attacker may choose actions to expand its set of attack vectors by probing the network, while defenders can take actions to increase or decrease their vulnerabilities through the implementation of defense mechanisms over time. Therefore, resilience mechanisms must adapt over time to address the evolving landscape of threats and the dynamics of the system itself. This adaptation encompasses not only responsive mechanisms but also proactive and retrospective ones. One distinction is that these different forms of adaptation may evolve at different time scales. Proactive mechanisms often rely on long-term strategic planning, trend analysis, and predictive modeling, which may unfold over weeks or months. Responsive mechanisms, by contrast, operate on rapid time scales, ranging from milliseconds to minutes, as they detect and mitigate threats in real time (e.g., through automated containment or anomaly detection). Retrospective mechanisms, such as forensic analysis or post-incident audits, function on intermediate time scales, typically activated after significant events have occurred. Importantly, these temporal differences require careful coordination: long-term decisions about architecture and policy must be informed by short-term operational signals and retrospective lessons. Recognizing and aligning these multi-temporal processes is critical for building a coherent and effective resilience strategy that can keep pace with the fluid and adversarial nature of networked systems.

Resilience mechanisms must also consider incomplete knowledge and uncertainties associated with threats and vulnerabilities. The complete set of vulnerabilities, as depicted in Fig. 2, is not always known to the agents. Agents possess varying degrees of knowledge regarding vulnerabilities. One example is zero-day vulnerabilities, where the network defends itself without awareness of their existence in the software or hardware. If an attacker exploits zero-day vulnerabilities, they can create unexpected events that the defender did not anticipate, leaving the network defenseless. In such cases, resilience mechanisms become the last resort to mitigate such exploitation. The asymmetric information or knowledge between the attacker and defender adds another layer of complexity that resilience mechanisms must address. In addition to knowledge asymmetry, uncertainties exist in the network environment such as user behaviors, traffic patterns, and topology changes. For example, in a battlefield communication network (Farooq and Zhu 2017, 2018c, 2021), uncertainties in terrain features, such as hills, forests, and urban structures, can lead to signal interference, propagation losses, and unpredictable coverage areas. In addition, uncertainties in military unit movements, deployments, and formations can disrupt communication links. These uncertainties can affect the probability of threat event success and consequently impact the state of the network. Resilience mechanisms require tools capable of handling such uncertainties. While investing in defense can reduce the probability of certain events succeeding or the attacker advancing to the next level of attacks, the risk may still persist, and resilience is the key approach to mitigate it.

To design dynamic mechanisms, control and learning tools are indispensable. This entails monitoring systems and gathering information about adversaries. Based on this data, reasoning frameworks are developed to respond to the system, as illustrated in Fig. 3. The sequential events initiated by the attacker and the resilience policies implemented by the defender both influence the state of the network. The objective of resilience mechanism design is to regulate the state of the network to achieve acceptable performance in its evolution.

4 Cyber Resilience and Cyber Risk Metrics

Before designing, it is essential to establish metrics for quantifying performance and design criteria. Figure 4 illustrates the evolution of network performance over time, driven by sequential interactions between the attacker and the defender, as depicted in Fig. 3. Performance over time is represented by the measurable performance Y_t across various stages. The evolutionary process is divided into three phases: proactive, responsive, and retrospective. At t_1, the proactive mechanism is implemented. Subsequently, at t_2, an adversary exploits vulnerabilities in the attack surface, leading to performance degradation at t_3. Responsive defensive measures at t_4 restore the network's performance from level M to level D. The retrospective mechanism engages at t_5 to address post-incident damage and restore performance.

Numerous metrics can be proposed to quantify resilience, which can be classified into three categories. The first category is performance-based metrics. These metrics require the monitoring of system states and provide accurate measures of resilience when credible information about system states is available. For example, the total losses incurred between t_2 and t_5, denoted by J_E, reflect the impact of the event initiated by the attacker. Other metrics in this category include M and D, which assess the severity of damage and the effectiveness of recovery during the responsive stage. The second category comprises latency-based metrics, which are dependent on timing. These metrics measure the speed or frequency of reaction against risks.

For instance, the response time $t_4 - t_2$ indicates how quickly recovery measures take effect, while the retrospective time $t_5 - t_2$ reflects the time taken for the system to rectify the problem retrospectively. The third category consists of knowledge-based metrics, which are associated with the known area of the attack surface (AS). Resilience in this context depends on the comprehensiveness of the knowledge about the system. For example, the percentage of the system model that can be expressed deterministically.

Resilience metrics offer a means to gauge risk within our network under a specified resilience mechanism. The risk associated with vulnerability v_E comprises two key components: impact and the probability of successful exploitation. With the integration of resilience mechanisms into the network, impact can be quantified by the total loss J_E resulting from the successful exploitation of v_E. The probability of success, denoted by p_E^t, is influenced by security mechanisms and proactive

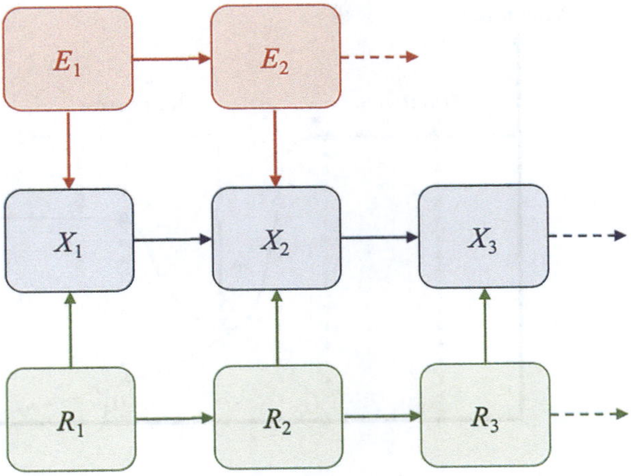

Fig. 7 Sequential interactions between the adversarial exploitation events E_1, E_2 and the actions taken by the resilience mechanisms R_1, R_2, R_3. They affect how the system states evolve over time X_1, X_2, X_3

resilience measures. Risk is expressed as the product $r_E^t = p_E^t \times J_E^t$, and overall risk is evaluated by aggregating r_E^t across all events in the attack surface (AS), which is denoted by

$$r^t = \sum_{E \in AS} r_E^t \tag{1}$$

While this formula does not consider the cumulative effect of events, it serves as a fundamental metric for evaluating cyber risk. It is noteworthy that p_E^t evolves over time and is shaped by the tactics employed by both attackers and defenders.

Figure 6 illustrates that attack surface can dynamically expand when it evolves from stage 1 to stage 2, and hence the sequential interactions between the adversarial behaviors and the resilience mechanisms will lead to evolving risks over time illustrated in Fig. 7. The security and resilience strategies employed by the parties not only impact the probability but also affect the magnitude of impact over time. Thus, the objective of cyber resilience, considering the implemented security strategies, is to mitigate both overall and long-term risks, as captured by (1) A baseline design problem can be finding resilience policies $\{R_t\}$ that minimize $\sum_{t=1}^{H} r^t$ over the horizon from stage 1 to stage H. Conceptually, the sum can be visualized through Fig. 8, which captures 3 stages of the risks. Cyber resilience is measured by the total sum of the loss of performance over time.

It is evident that the problem presents an optimal control challenge, wherein the dynamics are governed by the evolution of cyber states, manipulated by both parties as depicted in Fig. 3. However, solving this problem is inherently difficult. Firstly, it is necessary to incorporate the attacker's behavior into the control problem to more accurately represent the dynamics. To achieve this, adopting a game-theoretic perspective towards this fundamental problem is essential. The evolving interactions over time can be captured through a dynamic game framework involving two players. Secondly, the defender does not possess complete knowledge of the

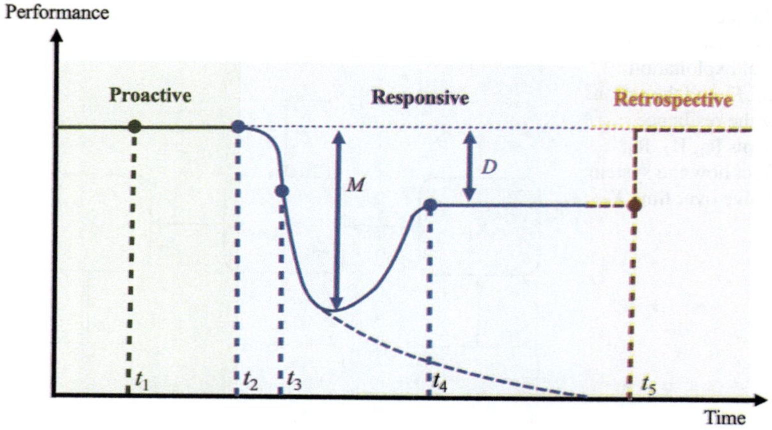

Fig. 8 The metrics of cyber resilience in the time domain divide the performance of the network into three stages: proactive, responsive, and retrospective. In the proactive stage, resilience mechanisms prepare strategies to deter potential attacks. The responsive stage involves online, adaptive responses to adversarial behaviors. The retrospective stage focuses on resolving problems or minimizing further impact after an incident has occurred and caused damage. t_1 represents the time when the preparation mechanism is deployed and implemented. t_2 marks the beginning of an event initiated by the adversary to exploit vulnerabilities on the attack surface. t_3 indicates the point when performance degradation begins. t_4 is the moment when responsive defense measures facilitate the network's recovery to performance level D after a significant performance drop to level M. Without the responsive resilience mechanism, performance degradation would have continued. The retrospective mechanism becomes active at t_5, aiming to restore performance after the incident has caused damage

dynamics and the attacker's behaviors. Despite employing game-theoretic models, there persists a challenge in accurately determining the parameters of the underlying game. Thus, learning becomes indispensable in bridging the gap between theoretical game frameworks and practical implementation. Consequently, the framework of cyber resilience and its design is built upon the intertwined principles of control, game theory, and learning, which are essential for assessing, analyzing, and designing for cyber resilience.

5 Theoretical Foundations

Following the discussion above, control theory, game theory, and learning provide theoretical foundations for understanding cyber resilience. Control theory addresses the dynamic nature of threat landscapes and network systems by offering tools to shape the metrics associated with cyber resilience. Game theory naturally captures the adversarial and strategic interactions between defenders and attackers, providing models and tools to analyze and predict the outcomes of these interactions and to design resilience mechanisms and associated strategies to achieve performance objectives despite strategic exploitations and disruptions by attackers. Learning is indispensable for enabling the system to assimilate data, information, and knowledge, allowing the integration of control and game-theoretic algorithms with them. Learning leads to implementable algorithms despite unknowns and uncertainties in the model. In this section, we discuss these foundations separately and argue that their confluence provides theoretical foundations for quantifying, analyzing, predicting, and designing cyber resilience for networks.

5.1 Control-Theoretic Foundations

The formulated control problem provides a baseline for understanding and designing cyber resilience mechanisms. In Fig. 9, we illustrate the evolution of the attack surface over time. The attacker exploits vulnerabilities on this surface, progressing stage by stage towards the targeted asset. A resilience mechanism can eliminate the vulnerability targeted by the attacker in stage 2, thereby deterring their progress. An example is the moving target defense, which alters the attack surface by changing network attributes such as IP addresses, ports, or software configurations. Importantly, this process is dynamic. A successful attack at stage 1 results in a new attack surface at stage 2. To shift the attack surface or eliminate targeted vulnerabilities, defenders must understand the current situation, including the vulnerability set, successful attacker exploits, and the network's new state. Additionally, defenders can analyze the attacker's objectives and intended assets to determine how best to alter the attack surface. However, reconfiguring and changing networks to shift the

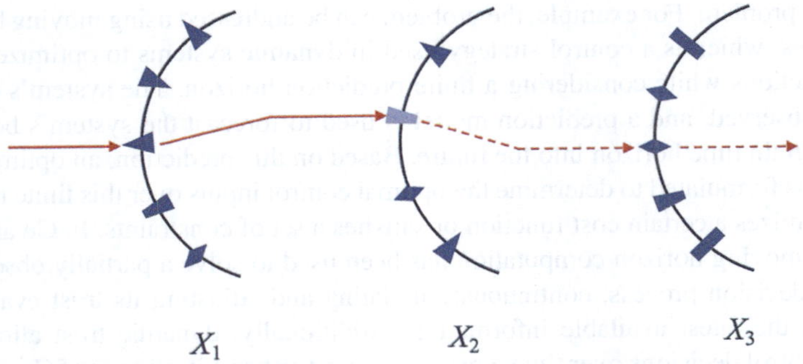

Fig. 9 The evolution of the security landscape results in different attack surfaces at various stages. The attacker can exploit the initial vulnerability at stage 1, causing the system state to transition from X_1 to X_2. This attack can be thwarted by resilience mechanisms if proactive measures eliminate a vulnerability (targeted by the attacker) from the attack surface at stage 2, thereby disrupting the entire attack progression. Reactive mechanisms also enable the system to reconfigure and present a different attack surface (e.g., through moving target defense or cyber deception), deterring the attacker from further progression. Resilience mechanisms, by nature, must consider multi-stage dynamic behavior, with control theory being an appropriate framework for effectively addressing this aspect

attack surface incurs costs. Thus, there is a need to find implementable approaches that maintain system performance while minimizing defense costs.

This approach coincides with the concept of feedback in control theory, where sensing, control, and actuation form the core components. This provides an essential building block for cyber resilience mechanisms. There is a need for situational awareness to comprehend the network's state, reasoning to configure responses subject to system constraints, and triggers to initiate changes. The concept of feedback loop can be further extended to an OODA loop consisting of the stages of "Observe", "Orient", "Decide", and "Act", (Zager and Zager 2017; Bodström and Hämäläinen 2018). The "Observe" stage involves gathering information about the network, including identifying threats, vulnerabilities, and changes in the landscape. Observations are made through various means, such as monitoring network traffic, analyzing system logs, and gathering threat intelligence. The "Orient" stage, the collected information is analyzed and interpreted to understand the current situation and assess its implications. This includes evaluating the capabilities of adversaries, identifying potential risks, and understanding the context in which decisions need to be made. In the "Decide" stage, based on the observations and orientation, decisions are made regarding the appropriate course of action. This stage involves evaluating different options, weighing their potential outcomes, and selecting the most effective response strategy. Finally, in the "Act" stage, the decided course of action is implemented. This may involve deploying security measures, executing response plans, or initiating countermeasures to mitigate threats and vulnerabilities. The control-inspired OODA loop allows for continuous improvement and adaptation, enabling organizations to stay ahead of evolving threats and maintain a resilient posture in the face of cyber challenges.

Besides the concept of feedback, many control tools can become available to solve the problem. For example, the problem can be addressed using moving horizon techniques, which is a control strategy used in dynamic systems to optimize future control actions while considering a finite prediction horizon. The system's current state is observed, and a prediction model is used to forecast the system's behavior over a certain time horizon into the future. Based on this prediction, an optimization problem is formulated to determine the optimal control inputs over this finite horizon that minimizes a certain cost function or satisfies a set of constraints. In Ge and Zhu (2023a), moving horizon computation has been used to solve a partially observable Markov decision process, continuously updating and adjusting its trust evaluation based on the latest available information. Additionally, dynamic trust allows for access control decisions over time, creating a zero-trust architecture for 5G networks in the face of evolving threats.

Moreover, Qian et al. (2020) studies the problem of assessing the effectiveness of a proactive defense-by-detection policy with a network-based moving target defense. A moving horizon planning algorithm proposed in Qian et al. (2020) is employed to compute the attacker's strategy for security evaluation. This approach involves continually updating the attacker's plan over a finite time horizon based on real-time information and changing conditions. As the planning horizon advances, the algorithm recalculates the optimal strategy, taking into account the latest insights and observations.

5.2 Game-Theoretic Foundations

As cyber risk emerges from strategic interactions between players, resilience mechanisms aiming to mitigate such risks must integrate a game-theoretic model to effectively capture these interactions (Manshaei et al. 2013; Zhu and Rass 2018a; Rass et al. 2020a). Illustrated in Fig. 3, the dynamics between the defender and the attacker need a dynamic game, as depicted in Fig. 10, denoted by a sequence of games $G_1, G_2, \cdots, G_6$. Each game G_t encapsulates the local interaction at stage t. A fundamental game comprises three key components: the set of players, their action sets, and their payoff functions. The sequences of games can capture different stages of resilience (Zhu and Rass 2018b). For example, proactive resilience mechanisms can be captured by one game, while responsive resilience mechanisms can be captured by another. Retrospective resilience can also be captured by a separate game. Each mechanism involves different types of players and different actions. For example, responsive resilience involves the sequential tactics and procedures of the attacker aimed at advancing toward the target. Proactive resilience involves the attacker's strategy to exploit vulnerabilities on the attack surface, while retrospective resilience involves analyzing the entirety of the attacker's operation. Hence, the defender's tactics and strategies will vary according to the stage involved.

Each described game represents the outcome of the interplay between threats posed by adversaries and defense mechanisms employed by defenders. The equilibrium of these games, often described using Nash equilibrium, which is a point of strategies where no players have incentives to deviate from unilaterally. It serves as a primary metric to gauge the consequence of these interactions, as depicted in Fig. 3. This equilibrium offers insights into the predicted outcome Y_t. Here, the risk r can be thoroughly assessed, not just on an average basis across all threat events, but also

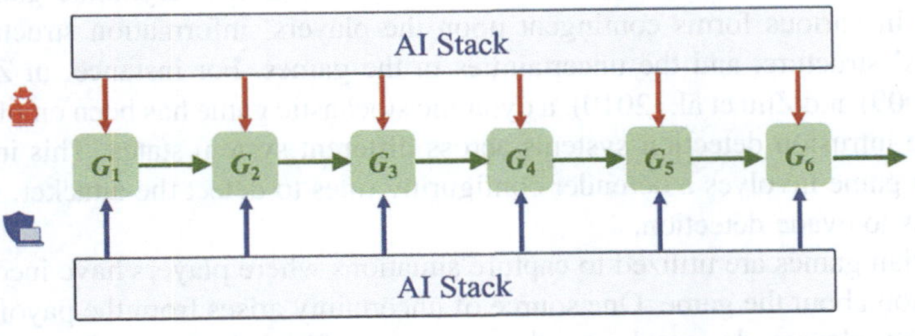

Fig. 10 The interactions between the attacker and the defender utilizing resilience mechanisms can be depicted through a sequence of games at various stages. The actions taken by both parties shape the progression of the games in subsequent steps. Resilience mechanisms can be augmented by an AI stack that facilitates the analysis of adversarial behaviors and strategically adjusts its own policies throughout the stages of the games

considering the strategic maneuvers of the attacker, who intelligently selects the most impactful vulnerability and associated event E_t at time t to achieve their goal $G_A(E_t, R_t)$. Here, G_A is the attacker's reward function, which takes as input the threat event chosen at the attacker at time t and the defender's resilience action R_t at time t. It yields a reward that indicates the proximity toward the goal.

Similarly, the defender's objective is to select a resilience mechanism R_t at time t to achieve their own goal $G_R(E_t, R_t)$. Here G_R is the defender's resilience reward function, which is influenced not only by the defense action R_t but also by the attacker's action E_t. Both players aim to maximize their reward by choosing respective policies. The equilibrium outcome, denoted by E_t^* and R_t^*, is a point where no players can benefit from deviating from their actions. As suggested in Fig. 2 this outcome will further result in the equilibrium state X_t^* and the associated consequence Y_t^*. This consequence Y_t^* serves as a key metric for risk evaluation within this paradigm. We call this risk as strategic cyber risk. Its evaluation considers the strategic behaviors of both attackers and defenders, offering meaningful predictions when agents behave rationally and engage in sufficient reasoning over time. Strategic risk is particularly valuable when event probabilities are unavailable, a common occurrence in cybersecurity scenarios. Obtaining data to assess how frequently attackers exploit the attack surface can be challenging, and often the precise attack surface is not fully understood. Consequently, strategic risk offers an alternative approach. Its evaluation needs an attacker model that quantifies attacker actions and motivations. Risk assessment involves considering potential outcomes given both the rational behaviors of attackers and the network's best effort defense. While assuming rationality and understanding exact incentives may be difficult, these assumptions provide a foundational framework. Nevertheless, this framework can be enhanced through the incorporation of bounded rationality models (Chen and Zhu 2019c; Hu and Zhu 2023a, b; Thakoor et al. 2020; Rass et al. 2020b) and learning methodologies (Zhu et al. 2011a, 2012; Zhu and Ishii 2022), allowing for continual refinement and improvement.

The games depicted in Fig. 10 are interleaved, with the outcome of a game at stage t determining the subsequent game. The dynamic game is characterized by a transition kernel that outlines the transition between games. Dynamic games can manifest in various forms contingent upon the players' information structure, the dynamics' structure, and the uncertainties in the games. For instance, in Zhu and Başar (2009) and Zhu et al. (2010), a dynamic stochastic game has been employed to configure intrusion detection systems across different system states. This intrusion detection game involves a defender configuring rules to detect the attacker, who, in turn, aims to evade detection.

Bayesian games are utilized to capture situations where players have incomplete information about the game. One source of uncertainty arises from the payoff of the game. The players do not know the exact payoff of the game but rather the probabilities of the outcomes. Sometimes, the players may not know the exact game they are playing, including the actions they can take and the other players' actions. Harsanyi has provided methods to understand such games with uncertainties by adopting a Bayesian approach (Harsanyi 1967). The unknowns of a player are characterized by the notion of type, representing the private information or characteristics that affect one player's payoffs or strategies in the game. Each player's type is drawn from a probability distribution known to the player but not necessarily to others. It is often assumed that the players share beliefs about the distribution of types or the uncertainty in the game. This assumption is known as the common prior, which serves as a basis for reasoning and decision-making in the game.

Another approach is based on Aumann's knowledge models (Aumann and Brandenburger 1995; Aumann 1974, 1976; Aumann and Heifetz 2002). It makes use of epistemic logic to deal with reasoning about knowledge and belief. Epistemic logic provides formal tools for representing and reasoning about agents' knowledge and beliefs, as well as the interactions between them. One key notion of Aumann's approach is the notion of an information partition. This partition divides the set of possible states of the world into distinct subsets, each corresponding to a particular piece of information available to players. The partition captures the structure of players' information and determines how it influences their beliefs and strategies. In Aumann's approach, common knowledge is defined as the information that is not only known by an individual agent but is also known to be known by all agents, and so on iteratively. It connects with the common prior assumption in Harsanyi's model. It has been shown that Harsanyi's Bayesian games and Aumann's epistemic approach are equivalent representations of games of incomplete information.

In cases where one agent holds more information than the other, asymmetric information games arise, often used to model cyber deception contexts. For example, they have been used in honeypot deployment and design (Horak et al. 2017), where the attacker is assumed to have incomplete information about the honeypot's location, but the defender has complete information about it. On the other hand, in the application of designing insider threat mitigation programs, the attacker has complete knowledge of whether they are an insider or not, but the defender has incomplete information about who the insiders are (Huang and Zhu 2020, 2021). Asymmetric information games provide a modeling framework to understand the information advantage of the players and find ways to either reduce this advantage gap or leverage it to mitigate cyber risks.

One special class of asymmetric information games is signaling games (Casey et al. 2016; Zhang and Zhu 2018; Mohebbi Moghaddam et al. 2015; Pawlick and Zhu 2017b; Hu and Zhu 2022), wherein a sender observes the true state of the world, crafts a message, and sends it to a receiver to elicit a response aligned with the sender's intentions. The receiver, upon observing the message, must estimate the true state of the world based on the sender's strategies. Lack of consistency arises if the receiver's belief about the true state of the world is not supported by the messages sent by the sender as per the sender's strategy. Hence, the equilibrium concept of this game involves not only strategies that do not allow unilateral deviation but also the sender's associated belief with these strategies. This equilibrium concept is called Bayesian Nash equilibrium. This class of games has contributed to understanding information manipulation, man-in-the-middle attacks, and misinformation over networks.

For example, in Xu and Zhu (2015a), Pawlick et al. (2015) and Zhu and Xu (2020), a man-in-the-middle attack between an industrial control system and a cloud is described. An industrial control system sends a computation task to the cloud for heavy computations, including image processing, control computation, or state estimation. The cloud then sends back the results to the control systems, which will utilize the results to actuate the system. The cloud-enabled control system leverages the computational power at the edge to reduce the computation burden on the system itself, which has limited real-time computation power due to hardware constraints. However, the attacker can manipulate the result and send a false result to mislead the control system into a failure mode of operation. One approach to creating resilience in case cryptography fails is to implement a checking mechanism to verify whether the sent message is coherent. Utilizing a belief system can aid in this verification process. The signaling game framework naturally captures the fact that the attacker has complete information and sends a message to the control system, which then must decide whether to utilize the result or discard it. Developing a strategy to screen the inputs provides a layer of resilience to the system against such attacks.

Another example is the zero-trust access control problem (Ge and Zhu 2022, 2023b; Li et al. 2023b). In this scenario, the attacker knows their type, but the network does not. The network must decide whether to grant or revoke access based on the behavior or footprint of the user. This situation corresponds to a dynamic signaling game where the attacker has complete information about their type, but the network does not. A game-theoretic approach offers a strategy for authentication based on the dynamic formation of a belief process, which relies on continuous monitoring. This plays a critical role in providing zero-trust security to modern networks, where access is determined through continuous monitoring.

Game theory not only captures scenarios of interactions and provides a way to determine the optimal strategy to defend against threats but also facilitates the design and deployment of overarching resilience mechanisms. In this regard, mechanism design theory (Myerson 1989; Borgers 2015) becomes essential as it enables us to shape the equilibrium or outcome of the game to a desirable one through the design of the game's rules. One such framework is Stackelberg games (Liu and Zhu 2022; Maharjan et al. 2013; Pawlick and Zhu 2016). Stackelberg games typically consist of a leader and several followers. The leader, acting as a designer, anticipates the outcomes of the game played by the followers. The followers interact with each other based on the input from the leader. The leader then designs an appropriate input to the game to shift the outcome to achieve the goal of the design. Such designs have been utilized in insurance schemes for computer networks (Zhang et al. 2017; Zhang and Zhu 2019, 2021). The insurer must design a scheme that enhances the network's security ecosystem and resilience. To achieve this, the insurer needs to assess risks by considering the interactions between attackers and defenders. The equilibrium of the game framework indicates the risks associated with the network under the designed insurance mechanism. The objective of the design is to minimize the insurer's costs while aligning with the cyber risks of the insuree. This framework has been extended to dynamic and network settings.

Bayesian mechanism design theory plays another pivotal role in shaping game outcomes to achieve cyber resilience. It has been employed to understand and influence player behaviors within a novel game-theoretic framework known as the duplicity game (Huang and Zhu 2021), aimed at designing effective deception mechanisms. This framework, comprising a generator, an incentive modulator, and a trust manipulator, collectively referred to as the GMM mechanism, is formulated as a mathematical programming problem. Its objectives include computing the optimal GMM mechanism, quantifying the upper limit of enforceable security policies, and characterizing conditions for user identifiability and manageability in terms of cyber attribution and user management.

Dynamic game modeling provides a foundational framework for understanding the cyber environment, as illustrated in Fig. 3. Various games capture distinct types of interactions within this environment. The dynamic game framework needs an investigation with a forward-looking perspective, meaning actions taken now must consider not only the current stage but also subsequent stages of the game.

Equilibrium analysis of dynamic games within this forward-looking context often employs dynamic programming principles or moving-horizon schemes, which are fundamental control-theoretic techniques utilized to tackle long-term problems. The control framework introduced in 1 is extended to a dynamic game framework with a more intricate modeling of adversarial behaviors, thereby creating a multi-player control problem. Such analysis yields a risk trajectory of the network environment in the face of adversaries. This analysis facilitates the measurement of network risks based on the prescribed attack model. The sequence of stage-by-stage risks corresponds to the cyber resilience metric depicted in Fig. 8. In essence, at each stage, a game can assess risks by predicting the consequences of interactions between threats and defenses.

The strategies associated with dynamic games offer a means to manage the evaluated risks under given attack models. If the assessed risks meet the performance criterion, the game strategies can inform the design of cyber resilience mechanisms at the operational and tactical levels, informing network actions how to act to maintain an acceptable level of risk. However, if the assessed risks do not meet the performance criterion, mechanism design theory enables a further step towards planning and designing resilience mechanisms to ensure the game arrives at a desirable set of outcomes with acceptable risks. These designed games correspond to the design of networks, rules, and policies, thereby informing the strategic-level planning and deployment of resilience mechanisms.

5.3 *Learning Foundations*

Despite advances in game-theoretic modeling and related computational and mechanism tools, challenges persist in utilizing game theory for cyber resilience. One key challenge is the imprecise modeling or capture of threat models. Attack behaviors can surpass prescribed models, leading to inaccurate risk assessments and potentially

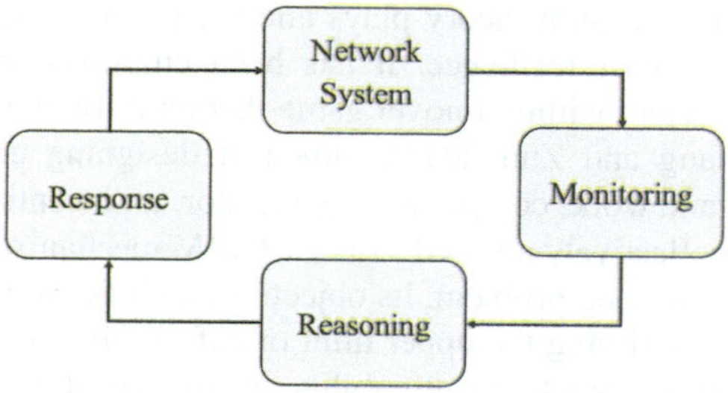

Fig. 11 Reinforcement learning offers a paradigm for achieving adaptive resilience mechanisms. The network system undergoes continuous monitoring, with a reasoning framework employed to update network policies in response to events within the network. This adaptation aims to mitigate risks to the network system

misleading designs due to misalignment between actual threats and prescribed ones. This issue is inevitable as networks face zero-day threats. Even if models initially capture exact attack behavior, the rapidly evolving threat landscape poses challenges for modeling to keep pace.

In addition to the epistemic uncertainties associated with threat models, aleatoric uncertainties related to game parameters such as payoffs, action sets, common priors, and dynamics remain challenging to precisely determine. Therefore, learning emerges as a crucial approach to enhance the game and control framework, enabling data-driven, autonomous, and agile defense strategies for resilience. There is a pressing need to develop an AI stack, as depicted in Fig. 10, which utilizes dynamic game models to facilitate advanced reasoning about attackers. This stack should also incorporate computational and control techniques to compute lookahead policies, allowing for online observations to adapt resilience policies dynamically and defend against attackers.

Recent advances in learning and AI have introduced numerous models and approaches to cybersecurity. For instance, leveraging large language models can expedite information processing and learning, facilitating rapid parsing of network log and traffic data for improved decision-making. Deep neural networks have also notably enhanced intrusion detection. Among learning paradigms, reinforcement learning emerges as a vital element for enabling autonomous and data-driven resilience mechanisms. Figure 11 illustrates the feedback structure of reinforcement learning. In this paradigm, the network system is subject to continuous monitoring, ensuring a continuous watch over its operations and interactions. This monitoring is followed by a sophisticated reasoning framework, designed to analyze incoming data and promptly identify any deviations or anomalies within the network's behavior. Leveraging this analytical capability, the framework dynamically updates network policies in real-time, tailoring responses to specific network events as they unfold. This adaptive approach not only enables fast detection of potential threats but also facilitates proactive risk mitigation strategies. This paradigm aligns well with the control framework detailed in Section 5.1, where a feedback structure is integral for enabling dynamic responses and automated mechanisms.

In Ge et al. (2023), meta-learning reinforcement has been used to create network defense adaptable to various attack scenarios. Meta-learning is a machine learning technique that focuses on learning how to learn. It involves training models on a variety of tasks or datasets, with the goal of enabling the model to quickly adapt to new tasks or scenarios with minimal additional training. In the context of defense strategies, meta-learning is used to enhance the adaptability and generalization capabilities of the system in response to evolving and diverse attack scenarios. A meta policy is first learned through meta-learning. This meta policy encapsulates a high-level strategy for making trust-related decisions based on partial observations of the system and potential attack scenarios. Along with the meta policy, an adaptation mapping is learned using meta-learning techniques. This mapping allows the system to quickly adjust the meta policy based on new information or unseen attack scenarios, enabling rapid adaptation to changing threat landscapes. By leveraging meta-learning, the defense system can efficiently adapt to new and unknown attack scenarios using insights gained from previous training data. This adaptive capability reduces the need for extensive retraining or manual intervention when faced with novel threats.

In Li et al. (2024), Conjectural Online Learning (COL) has been proposed as a learning scheme designed for generic Asymmetric Information Stochastic Games (AISGs) where decision-making entities have limited or asymmetric information. COL utilizes a forecaster-actor-critic (FAC) architecture to update strategies in an online manner by incorporating first-order beliefs and subjective forecasts of opponent strategies. In the context of security, COL can be applied to enhance cyber resilience in IT infrastructures. For example, in defending against Advanced Persistent Threats (APTs), COL can help in formulating strategies for the defender to protect the infrastructure and respond to intrusions effectively. By utilizing Bayesian learning and online adaptability, COL can enable defenders to adjust their strategies based on evolving threats and nonstationary attack patterns, leading to improved resilience against cyber threats.

In Li and Zhu (2024), the interconnections among foundation models (FMs), game models, and learning are explored. FMs serve as essential components for constructing customized machine learning models tailored to diverse cybersecurity applications. Game-theoretic models provide a foundational framework for comprehensively modeling adversarial interactions, facilitating the encapsulation of adversarial knowledge and domain-specific insights. The synergy between GMs and FMs can yield proactive and automated cyber defense mechanisms. These mechanisms not only aim to secure networks against attacks but also endeavor to enhance resilience against well-planned operations. It has been highlighted that one promising direction is the multi-agent neurosymbolic conjectural learning (MANSCOL) approach, which empowers defenders to predict adversarial behaviors, design adaptive defensive deception tactics, and synthesize knowledge for operational-level synthesis and adaptation. FMs assume a pivotal role in various functions within MANSCOL, encompassing reinforcement learning, knowledge assimilation, formation of conjectures, and contextual representation.

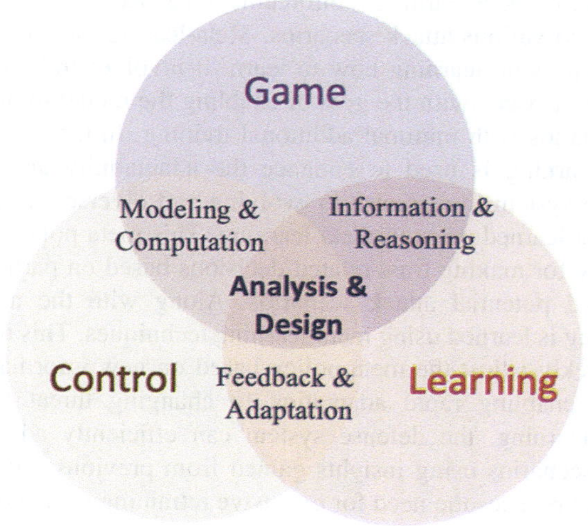

Fig. 12 The convergence of game, control, and learning methods is essential. Game and control methods establish the groundwork for modeling and computations in cyber risk and its mitigation. Control and learning methods are pivotal in the development of feedback-driven and adaptive resilience mechanisms. Game and learning methods offer data-driven strategic reasoning mechanisms utilizing both symbolic and data information. Their convergence enables the analysis and design capabilities crucial for cyber resilience

5.4 Confluences of Control, Games, and Learning Methods

Control, game, and learning methods each play pivotal roles in bolstering cyber resilience. Control-theoretic methods offer a dynamic systems perspective, framing cyber resilience as a continual process where defense mechanisms must operate over time across multiple stages to mitigate risks stemming from the attack surface. Conversely, game-theoretic methods distinguish cyber resilience from resilience to natural disasters by modeling adversarial behaviors of attackers. These methods facilitate the assessment of strategic risks faced by defenders under the strategic behaviors of attackers. Additionally, leveraging mechanism design and computation tools from game theory and control theory enables the dynamic design of cyber resilience mechanisms. Their confluences are central to the theoretic foundations of cyber resilience as illustrated in Fig. 12.

Furthermore, learning methods facilitate the integration of data and symbolic information from models, enabling the creation of an implementable AI stack. This stack enables analysis of adversarial behaviors, real-time response of defenses, and adaptive design and planning of mechanisms. By integrating these three methods, machine intelligence can be harnessed to address threats in an automated, adaptive, and agile manner, fostering new innovations in the realm of AI and cyber resilience solutions.

Recent advances in predictive learning with adversarial thinking exemplify the synthesis of control, game, and learning methods. By effectively learning the adversarial model, valuable insights into adversaries' tactics are gained, enabling the development of robust response strategies. Predictive control methods, such as rollout and forward-looking strategies, complement predictive learning by facilitating rapid adaptation and decision-making in dynamic environments.

This advance is closely tied to the concept of conjectural learning (Li et al. 2023a, 2024), wherein agents hypothesize about adversaries' behavior based on observed actions and outcomes. Incorporating conjectural learning enhances the ability to anticipate and respond to adversarial maneuvers, thereby improving overall resilience and adaptive capacity.

Additionally, meta-learning, which involves learning to learn, plays a significant role in this synthesis (Pan et al. 2023a; Ge et al. 2023). By leveraging meta-learning techniques, adaptability and optimization of response strategies based on evolving threats and changing environmental conditions are enhanced. Meta-learning facilitates the extraction of meaningful patterns and insights from past experiences, enabling more effective decision-making and strategy formulation.

An essential aspect of this synthesis involves coordinating and managing tradeoffs among different stages of the resilience process. Each stage, from proactive preparation to reactive response and retrospective analysis, presents its own set of challenges and considerations. Effective coordination between these stages is crucial for ensuring a cohesive and comprehensive approach to resilience. By carefully balancing tradeoffs and leveraging insights from each stage, resource allocation can be optimized, actions prioritized, and overall resilience effectiveness maximized.

5.5 *Cyber Deception as a Case Study*

Modern advances in cyber deception for cyber resilience is an example of the integration of control, games, and learning within the cybersecurity domain. By deploying deceptive techniques, organizations can manipulate the behavior of adversaries, influence the dynamics of the cyber environment, and enhance their resilience against threats.

Control mechanisms play a crucial role in cyber deception by governing the deployment and management of deceptive techniques and assets. Through control mechanisms, organizations can strategically design and implement deceptive environments to manipulate adversary behavior and mitigate threats (Fang and Zhu 2021, 2022; Peng and Zhu 2019). Control aspects ensure that deceptive measures are deployed effectively and efficiently, maximizing their impact on adversaries while minimizing the risk to legitimate systems and data. In the context of resilience, control enables defenders to adapt and optimize their deception strategies in response to evolving threats and changing environmental conditions, thereby enhancing overall cyber resilience.

Game theory provides a strategic framework for understanding and influencing the interactions between defenders and adversaries in the cyber domain (Pawlick et al. 2019, 2021). In the context of cyber deception, game theory principles guide the design and implementation of deceptive strategies to outmaneuver adversaries and increase the defender's advantage. By presenting adversaries with strategic dilemmas and uncertainties, defenders can exploit decision-making vulnerabilities and manipulate adversary behavior to their advantage. Game theory also facilitates the analysis of adversary strategies and the identification of optimal deception strategies for enhancing resilience against cyber threats.

Learning mechanisms enable defenders to adapt and improve their deception strategies based on insights gained from past experiences and interactions with adversaries (Li et al. 2021, 2022; Huang et al. 2022). Through machine learning algorithms and data analytics, defenders can analyze adversary behavior, identify patterns, and refine deception tactics over time. Learning enables defenders to anticipate evolving threats, proactively adjust deception strategies, and continuously enhance resilience against cyber attacks. By incorporating learning mechanisms into cyber deception and resilience strategies, organizations can improve their ability to detect, deter, and mitigate threats in dynamic and evolving cyber environments.

The integration of control, game theory, and learning principles enhances the effectiveness of cyber deception and resilience strategies. Control mechanisms govern the deployment and management of deceptive assets, ensuring strategic placement and utilization. Game theory principles guide the design of deceptive environments to exploit adversary decision-making vulnerabilities. Learning mechanisms enable defenders to adapt and optimize deception strategies based on insights gained from past interactions with adversaries. Together, these integrated approaches enable organizations to enhance their cyber deception capabilities and resilience against evolving cyber threats.

6 Future Directions

Cyber resilience encompasses a diverse array of challenges. This domain transcends mere technicalities, extending to the intersections of cyber, physical, and human components within systems. Many of these challenges are interdisciplinary, necessitating a systems-scientific approach for resolution. Furthermore, within the realm of tools, there exists a wealth of opportunities at the intersection of game theory, control theory, and machine learning. Numerous methods can draw inspiration from emerging applications and be developed to tackle a wide range of issues.

On the methodological front, non-equilibrium learning, large language models, and foundation models are emerging as promising tools to enhance the intelligence associated with cyber resilience mechanisms. Nonequilibrium learning is a novel approach in machine learning and artificial intelligence that diverges from traditional equilibrium-based models (Pan et al. 2023b, c). Unlike equilibrium-based approaches that assume stable states and predictable outcomes, nonequilibrium learning focuses on dynamic, non-stationary environments where conditions are constantly evolving. This approach is particularly relevant in cybersecurity, where adversaries continually adapt their tactics to circumvent defenses. Nonequilibrium learning algorithms are designed to adapt to changing conditions in real-time, enabling defenders to detect and respond to emerging threats more effectively. By leveraging nonequilibrium learning, organizations can enhance their ability to anticipate and mitigate cyber attacks in dynamic and unpredictable environments.

Large language models (LLMs), such as Generative Pre-trained Transformer (GPT) models, have revolutionized natural language processing. These models, trained on vast amounts of text data, possess advanced capabilities in generating human-like text and understanding contextual nuances. In cybersecurity, LLMs are increasingly being applied to analyze and understand adversarial behavior in strategic games. By training LLMs on extensive datasets of cyber threat intelligence and historical attack patterns, researchers can develop models capable of predicting adversary strategies and identifying potential vulnerabilities. These LLM-based approaches provide valuable insights into the strategic dynamics of cyber conflicts, enabling defenders to anticipate and counteract adversarial actions more effectively. Foundation models with strategic learning represent a paradigm shift in the development of AI and machine learning algorithms for cybersecurity (Li and Zhu 2024). Unlike traditional models that focus solely on optimizing specific tasks or objectives, foundation models with strategic learning incorporate strategic decision-making capabilities inspired by game theory and multi-agent systems. These models are designed to analyze and predict the behavior of multiple actors in complex cyber environments, taking into account their strategic interactions and incentives. By integrating strategic learning into foundation models, researchers can develop more robust and adaptive cybersecurity solutions capable of addressing the strategic challenges posed by sophisticated adversaries. These models enable defenders to anticipate adversary tactics, identify strategic vulnerabilities, and devise effective countermeasures to enhance cyber resilience.

Besides the advances in methodologies, there are also increasingly challenging cyber risk issues that we need to address. One key challenge associated with cyber resilience is the interdependencies among the risks within the subnetworks. Developing resilience mechanisms for the entire network itself is not a scalable solution. Instead, creating decentralized mechanisms for individuals is more plausible. However, in this case, the resilience mechanisms need to coordinate with each other due to the interdependencies, while simultaneously controlling and managing the cyber risks within their own network. The cyber risk of the entire network under this cyber resilience architecture is called holonic risk. Holons are entities that constitute the whole system, but each is capable of autonomous functions (Van Brussel et al. 1998). Holonic risks refer to the risk of the whole network, in which each constituent subnetwork autonomously manages its risk. This type of risk is becoming more common as the network grows larger and more heterogeneous, and each subnetwork is owned by different entities who have the right to control it. One way to deal with such issues relies on game-theoretic, control-theoretic, and learning tools. The interdependencies among the subnetworks can be viewed as a game in which individuals respond to each other. In this case, a subnetwork not only needs to dynamically respond to threats but also to the behaviors of other subnetworks, bringing a new layer of complexity into the cyber resilient framework. Similarly, the associated learning involves not only learning the behaviors of adversaries but also the behaviors of other networks through communications and cooperation.

The consequences of cyber risks in critical infrastructures and industrial networks often extend to risks within physical systems. Consequently, the risk is not solely confined to the cyber domain but incorporates both cyber and physical domains, making it a cyber-physical challenge. Thus, associated resilience mechanisms must not only address cyber threats but also consider the implications for physical systems. Therefore, the design of resilience mechanisms needs to be physics-informed, i.e., integrating insights from physics to ensure compliance with operational level constraints. Additionally, it is essential to develop physical-level resilient mechanisms, such as controllers, to mitigate impacts on physical systems. Consequently, cyber resilience must be jointly designed with physical resilience, emphasizing a co-design paradigm for networks (Zhu and Xu 2020; Huang et al. 2021; Xu and Zhu 2018; Yuan et al. 2013; Huang and Zhu 2020). This approach is crucial and has been applied in understanding cyber-physical resilience across various domains, including nuclear systems (Zhao et al. 2019, 2020; Smidts et al. 2022), power systems (Zhu and Başar 2011, 2015; Chen and Zhu 2016; Zhu 2019; Zhu et al. 2011b), industrial control systems (Collier et al. 2016; Chen et al. 2021; Zhao et al. 2024; Xu and Zhu 2015b, 2020; Farooq and Zhu 2018a), multi-agent systems (Zhao et al. 2022; Nugraha et al. 2020, 2021; Rieger et al. 2012; Chen and Zhu 2019a), and interdependent infrastructures (Chen and Zhu 2019b; Chen et al. 2019; Huang et al. 2018; Huang and Zhu 2020; Rinaldi et al. 2001).

Incorporating emerging methods into cybersecurity research and practice holds great promise for advancing our understanding of cyber threats and improving our ability to defend against them. By leveraging nonequilibrium learning, large language models, and foundation models with strategic learning, organizations can stay ahead of evolving threats and improve their cyber resilience in an increasingly complex and dynamic threat landscape.

7 Conclusions

Dealing with cyber risk in today's world is increasingly challenging. Mitigating cyber risk should not only involve cyber security defense mechanisms but also cyber resilience mechanisms, which complement each other to strengthen the cyber risk posture of the network. Understanding cyber risk and the associated resilience mechanisms requires the development of fundamental principles, metrics, and tools to analyze and design networks to achieve resilience goals. This chapter introduces game theory, control theory, and learning theory as the foundation for cyber resilience. Game theory has been shown to be suitable for modeling strategic interactions. Control theory has been pivotal in dealing with dynamics and designing dynamic defenses. Learning theory plays an important role in bridging theory and practice through autonomous adaptation and dealing with uncertainties. The confluence of game theory, control theory, and learning theory emerges as a cornerstone for cyber resilience, enabling organizations to improve their cyber resilience posture. As cyber threats continue to escalate in complexity and frequency, the interdisciplinary framework presented in this chapter serves as a roadmap for organizations seeking to enhance their cyber resilience capabilities and navigate the intricate dynamics of modern cybersecurity challenges.

References

Acharya VV, Pedersen LH, Philippon T, Richardson M (2017) Measuring systemic risk. Rev Financ Stud 30(1):2–47

Aumann RJ (1974) Subjectivity and correlation in randomized strategies. J Math Econ 1(1):67–96

Aumann RJ (1976) Agreeing to disagree. Ann Stat 4(6):1236–1239

Aumann R, Brandenburger A (1995) Epistemic conditions for Nash equilibrium. Econometrica 63: 1161–1180

Aumann RJ, Heifetz A (2002) Incomplete information. In: Handbook of game theory with economic applications, vol 3. North Holland, Amsterdam, pp 1665–1686

Bodström T, Hämäläinen T (2018) A novel method for detecting apt attacks by using ooda loop and black swan theory. In: Computational data and social networks: 7th international conference, CSoNet 2018, Shanghai, China, December 18–20, 2018, proceedings 7. Springer, Cham, pp 498–509

Borgers T (2015) An introduction to the theory of mechanism design. Oxford University Press, Oxford

Casey W, Morales JA, Wright E, Zhu Q, Mishra B (2016) Compliance signaling games: toward modeling the deterrence of insider threats. Comput Math Organ Theory 22:318–349

Chen J, Zhu Q (2016) A game-theoretic framework for resilient and distributed generation control of renewable energies in microgrids. IEEE Trans Smart Grid 8(1):285–295

Chen J, Zhu Q (2019a) Control of multilayer mobile autonomous systems in adversarial environments: a games-in-games approach. IEEE Trans Control Netw Syst 7(3):1056–1068

Chen J, Zhu Q (2019b) A game-and decision-theoretic approach to resilient interdependent network analysis and design. Springer, Cham

Chen J, Zhu Q (2019c) Interdependent strategic security risk management with bounded rationality in the internet of things. IEEE Trans Inf Forensics Secur 14(11):2958–2971

Chen J, Touati C, Zhu Q (2019) A dynamic game approach to strategic design of secure and resilient infrastructure network. IEEE Trans Inf Forensics Secur 15:462–474

Chen J, Zhu Q, Başar T (2021) Dynamic contract design for systemic cyber risk management of interdependent enterprise networks. Dyn Games Appl 11:294–325

Collier ZA, Panwar M, Ganin AA, Kott A, Linkov I (2016) Security metrics in industrial control systems. In: Cyber-security of SCADA and other industrial control systems. Spinger, Cham, pp 167–185

Fang S, Zhu Q (2021) Fundamental limits of obfuscation for linear Gaussian dynamical systems: an information-theoretic approach. In: 2021 American control conference (ACC). IEEE, Piscataway, pp 4574–4579

Fang S, Zhu Q (2022) Fundamental stealthiness–distortion trade-offs in cyber-physical systems. In: Security and resilience in cyber-physical systems: detection, estimation and control. Springer, Cham, pp 37–60

Farooq MJ, Zhu Q (2017) Secure and reconfigurable network design for critical information dissemination in the internet of battlefield things (IoBT). In: 2017 15th international symposium on modeling and optimization in mobile, ad hoc, and wireless networks (WiOpt). IEEE, Piscataway, pp 1–8

Farooq MJ, Zhu Q (2018a) Adaptive and resilient revenue maximizing dynamic resource allocation and pricing for cloud-enabled IoT systems. In: 2018 Annual American control conference (ACC). Piscataway, IEEE, pp 5292–5297

Farooq MJ, Zhu Q (2018b) A multi-layer feedback system approach to resilient connectivity of remotely deployed mobile internet of things. IEEE Trans Cognit Commun Netw 4(2):422–432

Farooq MJ, Zhu Q (2018c) On the secure and reconfigurable multi-layer network design for critical information dissemination in the internet of battlefield things (IoBT). IEEE Trans Wirel Commun 17(4):2618–2632

Farooq MJ, Zhu Q (2019) Modeling, analysis, and mitigation of dynamic botnet formation in wireless IoT networks. IEEE Trans Inf Forensics Secur 14(9):2412–2426

Farooq J, Zhu Q (2021) Resource management for on-demand mission-critical internet of things applications. Wiley, Hoboken

Ganin AA, Massaro E, Gutfraind A, Steen N, Keisler JM, Kott A, Mangoubi R, Linkov I (2016) Operational resilience: concepts, design and analysis. Sci Rep 6(1):1–12

Ge Y, Zhu Q (2022) MUFAZA: multi-source fast and autonomous zero-trust authentication for 5G networks. In: MILCOM 2022 – 2022 IEEE military communications conference (MILCOM). IEEE, Piscataway, pp 571–576

Ge Y, Zhu Q (2023a) GAZETA: game-theoretic zero-trust authentication for defense against lateral movement in 5g IoT networks. IEEE Trans Inf Forensics Secur. https://doi.org/10.1109/TIFS.2023.3326975

Ge Y, Zhu Q (2023b) Zero trust for cyber resilience. arXiv preprint arXiv:231202882

Ge Y, Li T, Zhu Q (2023) Scenario-agnostic zero-trust Defense with explainable threshold policy: a meta-learning approach. In: IEEE INFOCOM 2023 – IEEE conference on computer communications workshops (INFOCOM WKSHPS). IEEE, Piscataway, pp 1–6. https://doi.org/10.1109/infocomwkshps57453.2023.10225816. 2303.03349

Harsanyi JC (1967) Games with incomplete information played by "Bayesian" players, I–III Part I. The basic model. Manag Sci 14(3):159–182

Horak K, Zhu Q, Bosansky B (2017) Manipulating adversary's belief: a dynamic game approach to deception by design for proactive network security. In: Decision and game theory for security: 8th international conference, GameSec 2017, Vienna, Austria, October 23–25, 2017, Proceedings. Springer, Cham, pp 273–294

Hu Y, Zhu Q (2022) Evasion-aware Neyman-Pearson detectors: a game-theoretic approach. In: 2022 IEEE 61st conference on decision and control (CDC). IEEE, Piscataway, pp 6111–6117

Hu Y, Zhu Q (2023a) Detection in human-sensor systems under quantum prospect theory using Bayesian persuasion frameworks. In: 2023 IEEE statistical signal processing workshop (SSP). IEEE, Piscataway, pp 36–40

Hu Y, Zhu Q (2023b) Game of travesty: decoy-based psychological cyber deception for proactive human agents. arXiv preprint arXiv:230913403

Huang L, Zhu Q (2020) A dynamic games approach to proactive defense strategies against advanced persistent threats in cyber-physical systems. Comput Secur 89:101660

Huang L, Zhu Q (2021) Duplicity games for deception design with an application to insider threat mitigation. IEEE Trans Inf Forensics Secur 16:4843–4856

Huang L, Chen J, Zhu Q (2017) A large-scale Markov game approach to dynamic protection of interdependent infrastructure networks. In: International conference on decision and game theory for security. Springer, Cham, pp 357–376

Huang L, Chen J, Zhu Q (2018) Distributed and optimal resilient planning of large-scale interdependent critical infrastructures. In: 2018 winter simulation conference (WSC). IEEE, Piscataway, pp 1096–1107

Huang Y, Xiong Z, Zhu Q (2021) Cross-layer coordinated attacks on cyber-physical systems: a LQG game framework with controlled observations. In: 2021 European control conference (ECC). IEEE, Piscataway, pp 521–528

Huang Y, Huang L, Zhu Q (2022) Reinforcement learning for feedback-enabled cyber resilience. Annu Rev Control 53:273–295

Jajodia S, Subrahmanian V, Swarup V, Wang C (2016) Cyber deception. Springer, Cham

Kieras T, Farooq J, Zhu Q (2022) IoT supply chain security risk analysis and mitigation: modeling, computations, and software tools. Springer Nature, Cham

Li T, Zhu Q (2024) Symbiotic game and foundation models for cyber deception operations in strategic cyber warfare. arXiv preprint arXiv:240310570

Li T, Peng G, Zhu Q (2021) Blackwell online learning for Markov decision processes. In: 2021 55th annual conference on information sciences and systems (CISS), pp 1–6. https://doi.org/10.1109/ciss50987.2021.9400319

Li T, Peng G, Zhu Q, Baar T (2022) The confluence of networks, games, and learning a game-theoretic framework for multiagent decision making over networks. IEEE Control Syst 42(4): 35–67. https://doi.org/10.1109/mcs.2022.3171478

Li T, Lei H, Zhu Q (2023a) Self-adaptive driving in nonstationary environments through conjectural online lookahead adaptation. In: 2023 IEEE international conference on robotics and automation (ICRA), pp 7205–7211. https://doi.org/10.1109/icra48891.2023.10161368, 2210.03209

Li T, Pan Y, Zhu Q (2023b) Decision-dominant strategic defense against lateral movement for 5g zero-trust multi-domain networks. arXiv preprint arXiv:231001675

Li T, Hammar K, Stadler R, Zhu Q (2024) Conjectural online learning with first-order beliefs in asymmetric information stochastic games. arXiv preprint arXiv:240218781

Linkov I, Ligo A, Stoddard K, Perez B, Strelzoffx A, Bellini E, Kott A (2023) Cyber efficiency and cyber resilience. Commun ACM 66(4):33–37. https://doi.org/10.1145/3549073

Liu S, Zhu Q (2022) Stackelberg risk preference design. arXiv preprint arXiv:220612938

Liu S, Zhu Q (2023) Cyber insurance for cyber resilience. arXiv preprint arXiv:231202921

Liu S, Zhao Y, Zhu Q (2022) Herd behaviors in epidemics: a dynamics-coupled evolutionary games approach. Dyn Games Appl 12(1):183–213

Maharjan S, Zhu Q, Zhang Y, Gjessing S, Başar T (2013) Dependable demand response management in the smart grid: a Stackelberg game approach. IEEE Trans Smart Grid 4(1):120–132

Manshaei MH, Zhu Q, Alpcan T, Başar T, Hubaux JP (2013) Game theory meets network security and privacy. ACM Comput Surv (CSUR) 45(3):1–39

Mohebbi Moghaddam M, Manshaei MH, Zhu Q (2015) To trust or not: a security signaling game between service provider and client. In: Decision and game theory for security: 6th international conference, GameSec 2015, London, UK, November 4–5, 2015, proceedings 6. Springer, Cham, pp 322–333

Myerson RB (1989) Mechanism design. Springer, Cham

Nugraha Y, Cetinkaya A, Hayakawa T, Ishii H, Zhu Q (2020) Dynamic resilient network games with applications to multiagent consensus. IEEE Trans Control Netw Syst 8(1):246–259

Nugraha Y, Cetinkaya A, Hayakawa T, Ishii H, Zhu Q (2021) Rolling horizon games for cluster formation of resilient multiagent systems. In: 2021 60th IEEE conference on decision and control (CDC). IEEE, Piscataway, pp 4829–4834

Pan Y, Li T, Li H, Xu T, Zheng Z, Zhu Q (2023a) A first order meta Stackelberg method for robust federated learning. In: Adversarial machine learning frontiers workshop at 40th international conference on machine learning. https://doi.org/10.48550/arxiv.2306.13800

Pan Y, Li T, Zhu Q (2023b) Is stochastic mirror descent vulnerable to adversarial delay attacks? A traffic assignment resilience study. In: 2023 62nd IEEE conference on decision and control (CDC), pp 8328–8333. https://doi.org/10.1109/cdc49753.2023.10384003, 2304.01161

Pan Y, Li T, Zhu Q (2023c) On the resilience of traffic networks under non-equilibrium learning. In: 2023 American control conference (ACC), pp 3484–3489. https://doi.org/10.23919/acc55779.2023.10156139

Pawlick J, Zhu Q (2016) A Stackelberg game perspective on the conflict between machine learning and data obfuscation. In: 2016 IEEE international workshop on information forensics and security (WIFS). IEEE, Piscataway, pp 1–6

Pawlick J, Zhu Q (2017a) Proactive defense against physical denial of service attacks using Poisson signaling games. In: International conference on decision and game theory for security. Springer, Cham, pp 336–356

Pawlick J, Zhu Q (2017b) Strategic trust in cloud-enabled cyber-physical systems with an application to glucose control. IEEE Trans Inf Forensics Secur 12(12):2906–2919

Pawlick J, Farhang S, Zhu Q (2015) Flip the cloud: cyber-physical signaling games in the presence of advanced persistent threats. In: Decision and game theory for security: 6th international conference, GameSec 2015, London, UK, November 4–5, 2015, proceedings 6. Springer, Cham, pp 289–308

Pawlick J, Colbert E, Zhu Q (2018) Modeling and analysis of leaky deception using signaling games with evidence. IEEE Trans Inf Forensics Secur 14(7):1871–1886

Pawlick J, Colbert E, Zhu Q (2019) A game-theoretic taxonomy and survey of defensive deception for cybersecurity and privacy. ACM Comput Surv (CSUR) 52(4):1–28

Pawlick J, Zhu Q et al (2021) Game theory for cyber deception. Springer, Cham

Peng G, Zhu Q (2019) Game-theoretic analysis of optimal control and sampling for linear stochastic systems. In: 2019 57th annual allerton conference on communication, control, and computing (Allerton). IEEE, Piscataway, pp 647–654

Qian Z, Fu J, Zhu Q (2020) A receding-horizon MDP approach for performance evaluation of moving target defense in networks. In: 2020 IEEE conference on control technology and applications (CCTA). IEEE, Piscataway, pp 1–7

Rass S, Schauer S, König S, Zhu Q (2020a) Cyber-security in critical infrastructures, vol 297. Springer, Cham

Rass S, Schauer S, König S, Zhu Q, Rass S, Schauer S, König S, Zhu Q (2020b) Bounded rationality. In: Cyber-security in critical infrastructures: a game-theoretic approach. Springer, Cham, pp 99–114

Rieger CG, Gertman DI, McQueen MA (2009) Resilient control systems: Next generation design research. In: 2009 2nd conference on human system interactions. IEEE, Piscataway, pp 632–636

Rieger C, Zhu Q, Başar T (2012) Agent-based cyber control strategy design for resilient control systems: concepts, architecture and methodologies. In: 2012 5th international symposium on resilient control systems. IEEE, Piscataway, pp 40–47

Rinaldi SM, Peerenboom JP, Kelly TK (2001) Identifying, understanding, and analyzing critical infrastructure interdependencies. IEEE Control Syst Mag 21(6):11–25

Smidts C, Ray I, Zhu Q, Vaddi PK, Zhao Y, Huang L, Diao X, Talukdar R, Pietrykowski MC (2022) Cyber-security threats and response models in nuclear power plants. Springer, Cham

Thakoor O, Jabbari S, Aggarwal P, Gonzalez C, Tambe M, Vayanos P (2020) Exploiting bounded rationality in risk-based cyber camouflage games. In: Decision and game theory for security: 11th international conference, GameSec 2020, College Park, MD, USA, October 28–30, 2020, proceedings 11. Springer, Cham, pp 103–124

Van Brussel H, Wyns J, Valckenaers P, Bongaerts L, Peeters P (1998) Reference architecture for holonic manufacturing systems: PROSA. Comput Ind 37(3):255–274

Xu Z, Zhu Q (2015a) A cyber-physical game framework for secure and resilient multi-agent autonomous systems. In: 2015 54th IEEE conference on decision and control (CDC). IEEE, Piscataway, pp 5156–5161

Xu Z, Zhu Q (2015b) Secure and resilient control design for cloud enabled networked control systems. In: Proceedings of the first ACM workshop on cyber-physical systems-security and/or privacy. ACM, New York, pp 31–42

Xu Z, Zhu Q (2018) Cross-layer secure and resilient control of delay-sensitive networked robot operating systems. In: 2018 IEEE conference on control technology and applications (CCTA). IEEE, Piscataway, pp 1712–1717

Xu Z, Zhu Q (2020) Secure and resilient control of IoT-based 3D printers. In: Modeling and design of secure Internet of Things. Wiley, Hoboken, pp 383–405

Yuan Y, Zhu Q, Sun F, Wang Q, Başar T (2013) Resilient control of cyber-physical systems against denial-of-service attacks. In: 2013 6th international symposium on resilient control systems (ISRCS). IEEE, Piscataway, pp 54–59

Zager R, Zager J (2017) OODA loops in cyberspace: a new cyber-defense model. Small Wars J 21(12). https://archive.smallwarsjournal.com/jrnl/art/ooda-loops-cyberspace-new-cyber-defense-model

Zhang T, Zhu Q (2018) Hypothesis testing game for cyber deception. In: Decision and game theory for security: 9th international conference, GameSec 2018, Seattle, WA, USA, October 29–31, 2018, proceedings 9. Springer, Cham, pp 540–555

Zhang R, Zhu Q (2019) FlipIn: a game-theoretic cyber insurance framework for incentive-compatible cyber risk management of internet of things. IEEE Trans Inf Forensics Secur 15: 2026–2041

Zhang R, Zhu Q (2021) Optimal cyber-insurance contract design for dynamic risk management and mitigation. IEEE Trans Comput Soc Syst 9(4):1087–1100

Zhang R, Zhu Q, Hayel Y (2017) A bi-level game approach to attack-aware cyber insurance of computer networks. IEEE J Sel Areas Commun 35(3):779–794

Zhang T, Huang L, Pawlick J, Zhu Q (2020) Game-theoretic analysis of cyber deception: evidence-based strategies and dynamic risk mitigation. In: Modeling and design of secure Internet of Things. Wiley, Hoboken, pp 27–58

Zhao Y, Huang L, Smidts C, Zhu Q (2019) A game theoretic approach for responding to cyber-attacks on nuclear power plants. In: 11th nuclear plant instrumentation, control, and human-machine Interface technologies, NPIC and HMIT 2019. American Nuclear Society, Westmont, pp 399–410

Zhao Y, Huang L, Smidts C, Zhu Q (2020) Finite-horizon semi-Markov game for time-sensitive attack response and probabilistic risk assessment in nuclear power plants. Reliab Eng Syst Saf 201:106878

Zhao Y, Rieger C, Zhu Q (2022) Multi-agent learning for resilient distributed control systems. arXiv preprint arXiv:220805060

Zhao Y, Chen J, Zhu Q (2024) Integrated cyber-physical resiliency for power grids under IoT-enabled dynamic botnet attacks. arXiv preprint arXiv:240101963

Zhu Q (2018) Cyber insurance. arXiv preprint arXiv:181000290

Zhu Q (2019) Multilayer cyber-physical security and resilience for smart grid. In: Smart grid control: overview and research opportunities. Springer, Cham, pp 225–239

Zhu Q, Başar T (2009) Dynamic policy-based IDS configuration. In: Proceedings of the 48h IEEE conference on decision and control (CDC) held jointly with 2009 28th Chinese control conference. IEEE, Piscataway, pp 8600–8605

Zhu Q, Başar T (2011) Robust and resilient control design for cyber-physical systems with an application to power systems. In: 2011 50th IEEE conference on decision and control and European control conference. IEEE, Piscataway, pp 4066–4071

Zhu Q, Başar T (2015) Game-theoretic methods for robustness, security, and resilience of cyberphysical control systems: games-in-games principle for optimal cross-layer resilient control systems. IEEE Control Syst Mag 35(1):46–65

Zhu Q, Ishii H (2022) Introduction to the special section on learning and security for multi-agent systems. Annu Rev Control 53:249–251

Zhu Q, Rass S (2018a) Game theory meets network security: a tutorial. In: Proceedings of the 2018 ACM SIGSAC conference on computer and communications security. ACM, New York, pp 2163–2165

Zhu Q, Rass S (2018b) On multi-phase and multi-stage game-theoretic modeling of advanced persistent threats. IEEE Access 6:13958–13971
Zhu Q, Xu Z (2020) Cross-layer design for secure and resilient cyber-physical systems. Springer, Cham
Zhu Q, Tembine H, Başar T (2010) Network security configurations: a nonzero-sum stochastic game approach. In: Proceedings of the 2010 American control conference. IEEE, Piscataway, pp 1059–1064
Zhu Q, Tembine H, Başar T (2011a) Distributed strategic learning with application to network security. In: Proceedings of the 2011 American control conference. IEEE, Piscataway, pp 4057–4062
Zhu Q, Wei D, Başar T (2011b) Secure routing in smart grids. In: Workshop on foundations of dependable and secure cyber-physical systems (FDSCPS), pp 55–59
Zhu Q, Tembine H, Başar T (2012) Hybrid learning in stochastic games and its application in network security. In: Reinforcement learning and approximate dynamic programming for feedback control. Wiley, Hoboken, pp 303–329
Zimmerman R, Zhu Q, Dimitri C (2016) Promoting resilience for food, energy, and water interdependencies. J Environ Stud Sci 6:50–61
Zimmerman R, Zhu Q, de Leon F, Guo Z (2017) Conceptual modeling framework to integrate resilient and interdependent infrastructure in extreme weather. J Infrastruct Syst 23(4):04017034

Analysis of Cyber Dependencies for Assessment of Cyber Resilience

Nathaniel Evans, Joel Dawson, and Ryan Styles

1 Introduction

A cyber dependency is a connection between two assets, in which the state of one relies on that of the other. For example, in a networked business system, there may be many desktop computers linked to a central server and data storage unit. These components depend on one another, and loss of any one of them degrades the performance of the system as a whole. Similarly, an industrial control process may involve sensors, data-acquisition computers, control-system computers, and servers; those components operate together and must be linked to one another to provide effective monitoring and process control. Whenever separate components must operate together to complete a business function, there is a cyber dependency. Such dependencies must be identified and understood as part of a cyber resilience assessment.

This chapter addresses:

- Definitions and types of cyber dependencies. Formal definitions and examples are provided.
- Risks and benefits associated with cyber dependencies. Cyber dependencies can have both risks and benefits for resilience—a system with redundancies and distributed functions can be more resilient but at the expense of additional dependencies.
- Methods for assessing cyber dependency risk. Methods for characterizing risks and assessing their significance are described.
- Managing the risk of cyber dependencies. Measures to manage dependency risk through assessment, monitoring, and contingency planning are described.

N. Evans · J. Dawson (✉) · R. Styles
Oak Ridge National Laboratory, Oak Ridge, TN, USA
e-mail: natevans@me.com; dawsonja@ornl.gov; stylesra@ornl.gov

© The Author(s), under exclusive license to Springer Nature Switzerland AG 2025

I. Linkov, A. Kott (eds.), *Cyber Resilience: Applied Perspectives*, Risk, Systems and Decisions, https://doi.org/10.1007/978-3-031-90109-6_3

2 What Is a Cyber Dependency?

2.1 Formal Definitions of Cyber Dependency

Risk analysts have proposed several, similar definitions to help establish methods for assessing the risk of cyber dependencies:

- (Rinaldi et al. 2001): *An infrastructure has a cyber dependency if its state depends on information transmitted through the information infrastructure.*
- (Pederson et al. 2006): *An informational or control requirement between components. For example: a supervisory control and data acquisition (SCADA) system that monitors and controls elements on the electrical power grid. A loss of the SCADA system will not by itself shut down the grid, but the ability to remotely monitor and operate the breakers will be lost.*
- (Clemente 2013): *An informational or control requirement between components.*

2.2 Identifying Cyber Dependencies

Cyber dependencies exist when computer systems depend upon other computer systems. Dependencies exist between multiple types of computer systems and include both information and services. Dependencies upon services range from information processing and storage to system configuration and security controls. Computers also depend upon networks to transmit data and upon specialized storage devices to store data; storage devices might require their own specialized communication networks. Various other types of equipment and computer services are required by computer systems to meet their business objectives.

Cyber dependencies appear when a computer system starts using information or services supplied by another computer or computing device, such as a network. For example, a computer that controls an industrial process might depend upon measurements provided by another computer. Thus, the process-control computer depends upon the measurement computer and also depends upon the networks that allow the two computers to communicate. Other examples include computers that depend upon servers to process or store information. Often, computers will depend upon services provided by other computers for configuration settings or security controls.

These examples illustrate internal dependencies that reside within the organization on equipment that they own and control. External dependencies also appear, typically when organizations subscribe to specialized services provided by software or security vendors.

External dependencies form when organizations need to communicate with other organizations, either to exchange information or to receive essential updates. For example, organizations often need to receive security updates to maintain their security posture. These updates include operating system and application patches as well as anti-malware signatures. This security requirement creates an external dependency on an external organization as well as on the network infrastructure that enables the external communication.

Cyber dependencies are typically formed explicitly, such as the intentional dependency between an industrial control computer and a measurement computer. Other dependencies might form more implicitly, such as a measurement computer that receives its network settings at start-up from a network configuration server. Although both dependencies form intentionally, not by accident, the more explicit dependency has a higher profile. It might be better understood and managed than the implicit dependency.

An implicit dependency may be hidden in the sense that the organization is not aware of it. Organizations typically recognize the critical cyber services that support or provide critical business functions. However, these systems might depend upon other, less obvious cyber services. Because of this dependency, organizations should treat these supporting cyber services as essential systems and include them in their risk management plan.

2.3 *Characterizing Cyber Dependencies*

As noted, a cyber dependency can be defined as an informational or data linkage or connection between two assets, by which the state of one asset influences or is reliant upon the state of the other. An asset has a cyber dependency if its operation depends on information that is transmitted via electronic or informational links (Fig. 1).

Analysis of cyber dependency should consider the assets required—the end point systems—and the characteristics of the data (for information or control) dependency. There are three possible states of data in a dependency relationship: data in motion, data at rest, and data in process. Understanding these data dependencies requires an analysis of the computer and network elements that are required for the transmission, storage, and processing of data (Table 1).

Fig. 1 Example of the normal flow of information. (Petit et al. 2015)

Table 1 Cyber dependencies

End point systems	End point systems characterize computer hardware devices (e.g., desktops, laptops, tablets) and their components
Data in motion	Data in motion characterizes data as they are in transit, including the processes and equipment (e.g., switches, networks, and firewalls) that are used for transfer of data
Data at rest	Data at rest characterizes data being physically stored, including storage capabilities and other storage device requirements
Data in process	Data in process (or data in use) characterizes data being edited or analyzed within a computer process, including data within a database or a computation

Petit et al. (2015)

2.4 Understanding the Risks and Benefits

Cyber dependencies can degrade as well as improve cyber resiliency. Dependencies can degrade resiliency by increasing the number of elements required to provide essential cyber services. When a computer controlling an essential service depends upon other cyber systems, the loss of those dependent systems can stop or degrade that essential service. However, dependencies can also improve resiliency by adding more reliable, fault-tolerant components or enabling the use of redundant, failover systems. In this context, fault tolerance refers to a particular component's ability to operate normally in the presence of failures (Moore and Cho 2019). To understand how dependencies can increase resilience despite increasing the number of components, consider a basic industrial control system composed of a computer that controls and records the status of a process. A single computer could perform all the required services: actuate the controllers and store the status on its internal disks. By distributing these services, the resilience of this basic control system can be increased.

Consider the risk of coupling the two main processes, controlling the actuators and storing the system status. If a fault occurs on the control computer, the organization risks losing not only control of the process but also the historical record of that control system. The control computer is a single point of failure.

Because coupling these services poses risks, control systems often add separate, independent data storage servers known as historians. Adding a separate data storage server does require an additional communications network dependency but enables the possibility of using multiple, redundant historians deployed in an automated, failover configuration. If one historian fails, stored data is preserved in another. This preserves the system against a failure on the part of a historian machine (e.g., a disk-drive fault).

Another way to reduce risk from failure of a historian is to use an external, highly reliable network-connected storage device. These external storage devices use resilient, redundant technologies that provide higher reliability than the internal disks typically used within computer systems.

Although adding dependencies in the form of another computer to store data or a separate data storage technology increases the complexity and the number of systems requiring management, these designs can increase the resiliency of the overall cyber system.

3 Assessing the Risk of Cyber Dependencies

3.1 *Defining the Risk*

Risk assessment considers four components: (1) threat/hazard, (2) vulnerability of a system, (3) the system's resilience, and (4) the consequence or resulting impact of a disruption. Risk is a function of the likelihood of the threat or hazard, the vulnerability (likelihood the threat or hazard will negatively affect the system), the resilience, and the magnitude of the consequences.

Cyber dependencies increase the complexity of assessing cyber risk. Because the threats/hazards that affect any system or network element can propagate downstream to connected, dependent systems, the risk analysis must consider not only the system's internal risk but also the risk of the upstream assets and of the transmission systems upon which it depends.

When assessing cyber dependencies, the analyst must consider the assets and their communication requirements, their data dependencies, the type of their data dependencies, and the type of security threats to which they might be vulnerable.

The assessment should evaluate whether the organization internally provides the required services or whether they are provided by an external source. Note that a given service might depend upon both internal and external resources. For example, an off-site (internal) asset owned by the organization might communicate over an externally provided communication channel.

The analysis should characterize the service provider (internal or external), operating environment, coupling and response behavior, type of failure, infrastructure characteristics, and state of operations, as presented in Petit et al. (2015).

3.2 *Threat Scenarios*

As described in the X.800 recommendation (OECD 2012), four types of scenarios threaten the flow of data from the provider to the dependent asset. The data flow can be interrupted, intercepted, modified, or fabricated, as illustrated in Fig. 2.

During normal operations, a normal flow of data exists between the provider asset (or transmitter) and the user asset (or receiver). Several threats may occur:

1. Flow interruption or denial, when the data do not go to the receiver. The interruption thus affects data availability.

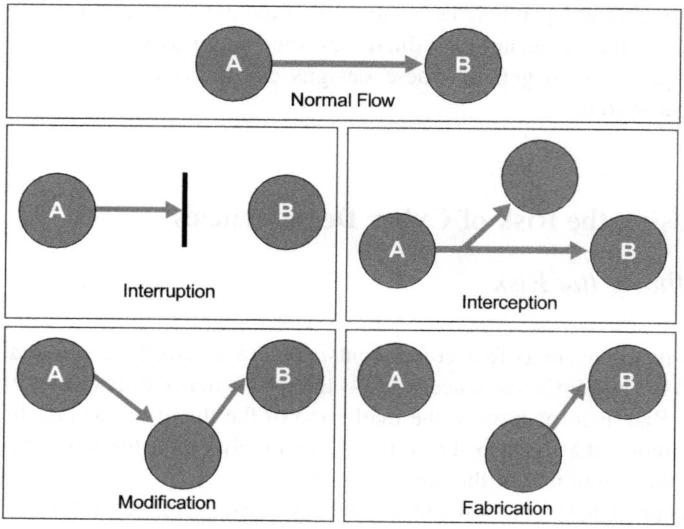

Fig. 2 Normal flow faces four types of threats. (International Communication Union 2004; Petit et al. 2015)

2. Flow interception or sniffing, when the data are read by an unauthorized party in transit between the transmitter and the receiver. The interception thus affects data confidentiality.
3. Flow modification, when the data are changed, corrupted, or deleted before reaching the receiver. The modification thus affects data integrity.
4. Flow fabrication, when the source of the data is unauthorized. The fabrication thus affects data authenticity, because an attacker may require forged or spoofed credentials.

Stallings (2014) adds a fifth:

5. Flow replay, when the data are duplicated and retransmitted at a later point in time. Replay attacks are a form of integrity violation.

The unique operational requirements of an ICS or SCADA network, which can include real-time or reliability requirements, distributed sensing, and closed-loop control, suggests two more scenarios:

6. Flow delay/acceleration, when the transit time of the data is altered. Depending on the context, this attack can affect either the availability or integrity of the data. For instance, many timing protocols (e.g. Precision Timing Protocol (PTP), etc.) rely on bidirectional consistency in transport latency times to

properly synchronize. An attacker can skew the clock value of a remote node by imposing a one-way flow delay, which can then affect real-time system performance.

7. Flow suppression, where the transit path of the data is degraded, resulting in unreliable service. Suppression reduces data availability.

Finally, due to the reliance of many ICS systems on networked sensor data, a threat analysis should consider the scenario of *sensor poisoning*, where sensor readings are modified through direct physical action.

3.3 Quantifying the Consequence of Loss

Four objective criteria help to quantify the consequence of losing a cyber service or one or more of its cyber dependencies:

1. Time before impact occurs

   An organization might not suffer immediately from the loss of a cyber service or its dependency. The impact might be delayed if the organization has a capacity to withstand the loss. Estimating the time before impact helps to quantify the consequence.

2. Extent of impact

   The extent of impact can vary. Potential threats, including adverse physical events, can cause short-term interruptions, degradation of quality or integrity, or failure of a cyber dependency. Estimating the extent of impact should aim to measure the percentage of normal cyber functions that would be lost or degraded due to the potential threats. If measuring the percentage of loss is not feasible, the analyst can estimate the extent of impact in ordinal terms such as "severely," "moderately," or "minimally impacted."

3. Degree of dependency

   The reliance on a cyber dependency can vary. If the failure of a dependency directly causes the failure of an important cyber service or the business objective, then the dependency is high. If the failure of the dependency requires the organization to initiate its contingency plan or alternative measures, then the dependency is medium. If the organization can tolerate the failure of the dependency and continue to operate, then the dependency is low.

4. Time to recover

   Consequence varies with the time to recover, which can be expressed as a Mean-Time-to-Recovery (MTTR) metric (Moore and Cho 2019). The more quickly an organization recovers, the lower the consequence. With hot-standby, failover redundancy, an organization can tolerate loss. Without readily available recovery methods, the consequence of loss will increase.

4 Methodologies Used to Assess the Risk of Cyber Dependencies

This section discusses four methods for assessing the risk of cyber dependencies; these methods are briefly summarized in Table 2.

4.1 *Expert-Weighted Survey Methodology*

As part of an overall assessment of cyber protection and resilience, the United States Department of Homeland Security (DHS) developed a cybersecurity assessment tool named the Cyber Survey Tool (CST) (Joyce et al. 2017). The C-IST evaluates cyber dependencies as part of an overall assessment of cyber protection and resilience. The methodology uses a survey to collect and inventory the attributes of an organization's cyber protection and resilience program. The answers to the survey are scored to create a Cyber Protection Resilience Index (CPRI) that indicates the effectiveness of an organization's cyber protection and resilience program.

The survey is a set of questions and answers that are designed to measure the attributes of cybersecurity. The attributes are organized by category and scored by expert opinion. There are five categories: (1) Cybersecurity Management, (2) Cybersecurity Forces, (3) Cybersecurity Controls, (4) Incident Response, and (5) Dependencies.

Expert opinion is elicited to measure the relative importance and contribution of each category and of each of the question sets within the categories. By combining the relative importance of all of the attributes of cybersecurity, the methodology creates a composite index that reflects the relative effectiveness of an organization's cyber protection and resilience program as compared to that of other organizations with similar critical cyber services (CCSs).

Table 2 Cyber dependency risk assessment methodologies

Expert-weighted survey	Facilitated survey identifies and quantifies the severity and duration of cyber dependency loss
Pathway analysis	Uses graph model of cyber dependencies to identify critical systems (nodes) and relationships (edges)
Model checking	Applies formal proofs to system model to verify the presence of a security or resilience property
System dynamics simulation	Characterizes the time-bound or causal effects of dependency disruption by modeling attack scenarios in a simulated OT deployment

4.1.1 Using Dependency Curves

Dependency curves provide a method of visualizing the impact of losing dependencies over time. As shown in Fig. 3, the dependency curve starts when the loss occurs, and the first shaded region (time before impact) shows how long before the impact has its effect. The next period, or the second shaded area (backup duration), depicts the reduction or degradation in service caused by the impact. During this period, the organization also relies upon backup operations, if available. If the backup operations fully replace the lost service, the period would remain steady at 0% loss of service.

The next shaded period (time before mandate to shutdown) shows the continued degradation of service until the organization is unable to continue, or is required to shut down, operations. The dependency curve would remain constant, and not decrease, during this period if the backup systems remain fully operational.

The curve next depicts the full outage (blank) and the restoration (fourth shaded region, or time for full resumption of operations) periods, followed by the actual resumption of full service.

4.2 *Pathway Analysis Methodology*

Pathway analysis encompasses a broad class of risk assessment techniques that model systems as node-and-edge graphs, then apply graph-theoretic algorithms to identify architectural vulnerabilities. While the scope of these techniques is limited, the simplicity, adaptability, and explainability of these models has made them popular subjects of research (O'Neill 2013; Srivastava et al. 2018; Umunnakwe et al. 2021; Wan et al. 2021).

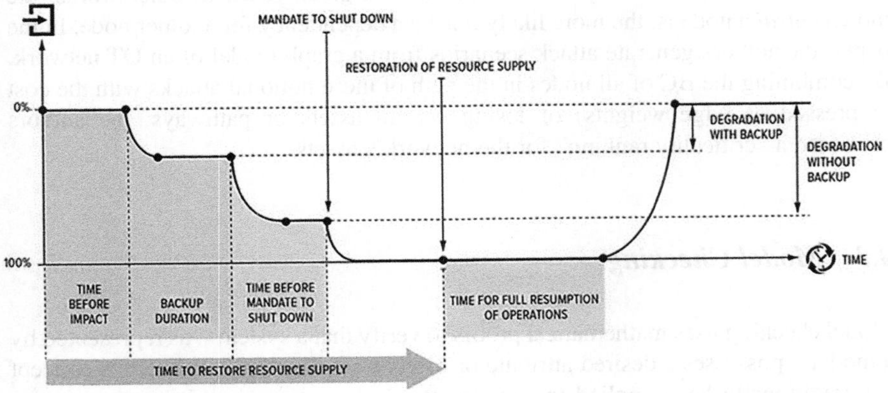

Fig. 3 Dependency curves

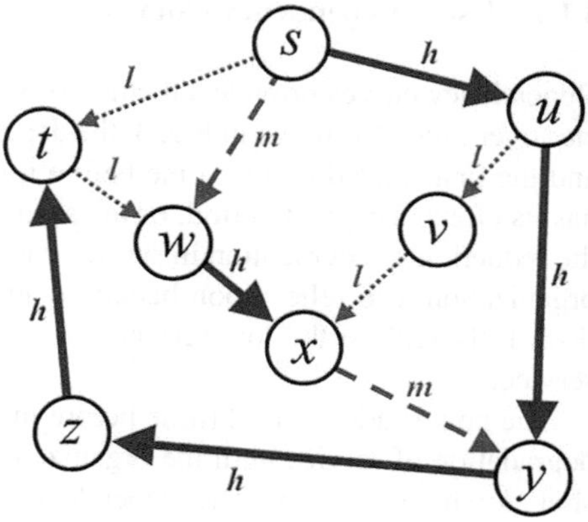

Fig. 4 This directed graph depicts devices as labeled circles and shows dependencies as edges whose weights indicate the degree of dependency (solid arrow ¼ high impact, dashed arrow ¼ medium impact, dotted arrow ¼ low impact). (O'Neill 2013)

In O'Neill (2013), the author uses links in a directed graph to model dependencies, then uses graph features to identify the paths of highest dependency, or the paths that pose the highest risk.

After eliciting expert opinion, the user constructs a directed graph by connecting assets through weighted links, which each represent the degree and direction of a dependency between two assets (see Fig. 4). This approach can accommodate multiple upstream and downstream links, and indirect dependencies, which are a series of two or more links, or a path, from one asset to other assets. Using the directed graph, an analyst can then evaluate the degree of dependency, if any, between any two asset nodes. The analyst can also estimate the severity of a fault at one or more nodes by computing the weight of the edges along the fault's propagation path or paths.

Other research has explored different graph metrics. In Umunnakwe et al. (2021), the researchers apply *betweenness centrality* (BC), which quantifies the likelihood that the shortest path between two nodes will traverse another node, and is a convenient indicator of a system's reliance on a given asset. In other words, the more *central* a node is, the more likely it is be a dependency for another node. In the paper, the authors generate attack scenarios from a graph model of an OT network. By combining the BC of all nodes in the path of these notional attacks with the cost (expressed as edge weights) of losing certain assets or pathways, the authors produced a "criticality ranking" for the network's assets.

4.3 Model Checking

Model checking uses mathematical proofs to verify that a system—as represented by a model—possesses a desired attribute or meets a set of criteria. While this concept has traditionally been applied to concurrent system verification, its usefulness for proving performance or qualification assertions has inspired broad adoption in

```
system SensorBoard

    features
        --input
        current_reading: in data port Base_Types::Integer;

        --output
        current_out: out data port Base_Types::String;
        millis_out: out data port Base_Types::String;

    annex verdict {**
        CyberRel "current_out_I" = current_reading:I => current_out:I;
        CyberRel "millis_out_I" = current_reading:I => millis_out:I;
    **};

end SensorBoard;
```

Fig. 5 A simple VERDICT system asset definition. The 'CyberRel' statements identify and describe the outbound cyber relationships ("I" identifies "integrity" as the chief concern)

academia and industry. For this chapter, we deal only with model checking as it applies to cyber dependencies in OT systems.

The strengths of model checking, besides its ability to provide provable guarantees, are its speed and its ability to pinpoint where an assertion fails. Its persistent weakness is the risk of combinatorial explosion, where the set of possible paths through a system grows faster than a user's computer can manage. To avoid this, model checkers limit the set of system interactions that they can represent, often leading to simple models in tightly constrained domains. The user must learn to translate model-checking results into useful real-world insight, and may struggle to apply the results from a model-checker built for one domain to a system in another.

A good example of a cyber dependency model checker is the VERDICT tool (Meng et al. 2021). Built on top of the AADL language (Feiler and Gluch 2012) and the Assume-Guarantee Reasoning Environment (AGREE) model checker (Cofer et al. 2018), VERDICT introduces cybersecurity requirements and properties (e.g. "insideTrustedBoundary", "deviceAuthentication", etc.) to AADL/AGREE's system asset and relationship descriptors, as well as a cost estimator that weighs attack impact costs against control costs (see Fig. 5).

The user must first describe their system's security requirements, current security controls, and topology with VERDICT's AADL language extensions, then provide cost estimates of both security failures and security upgrades. VERDICT builds an undirected node-and-edge system model, then evaluates whether the security controls are sufficient to prevent the worst attacks by checking each edge in the paths between nodes of interest. If any edge fails for a given requirement, VERDICT notifies the user about the location, which requirement wasn't met, and provides the difference between the cost of a successful attack and the cost of a security upgrade. (VERDICT will also notify the user about where they may be *overspending* on security.)

Like most formal modeling tools, VERDICT cannot provide fine-grained security guarantees, so it is left to the user to interpret the results. As such, this tool (and others like it) is best used *in combination with* other assessment methodologies, especially the Expert-Weighted Survey process.

4.4 System Dynamics Simulation

Some ICS or SCADA network cyber dependencies are too complex to assess with model checking, pathway analysis, or expert surveys alone, especially if the dependency has a temporal element. This is where simulation comes in. Simulation uses a representation of a component—often an equation, a state machine, or trained model—to predict that component's behavior when presented with one or more stimuli, the component's internal state, and the global state of the simulation environment (including other components). Networked control systems are complex, time-dependent compositions of interacting components in a physical environment, and so are natural candidates for the simulation approach.

Simulation is an empirical enterprise—it depends on running experiments on artificial systems. As such, a user must pay close attention to detail, and often consult experts, when constructing an accurate model of a physical system. Running the simulation is often computationally expensive, and sometimes involves the use of specialized software or equipment. The user must interpret simulation results in much the same way they would physical test results. When properly employed, however, simulation is extremely flexible, easy to instrument, portable, and second only to physical tests in terms of available resolution. Indeed, due to cost or ethics concerns, some system failure scenarios (e.g. safety subsystem faults, natural disasters, etc.) can only be investigated with simulations.

Due to heavy use in general system design and engineering, a number of major proprietary (MathWorks 2024; Modelica Association 2024) and open-source simulation tools (Dassault Systemes 2024) are available for use. However, the heterogeneity of the ICS domain and the rarity of support for explicit cybersecurity attributes means that custom solutions are often required. In Alves et al. (2018), the authors construct a modular SCADA testbed for cybersecurity research from unrelated open- and closed-source components (Fig. 6). They start by breaking a generic SCADA

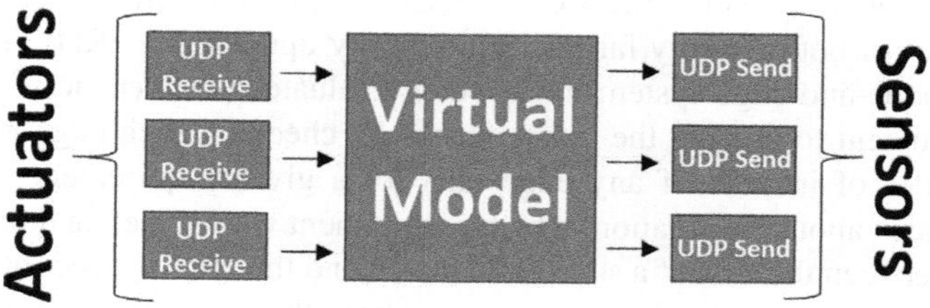

Fig. 6 High-level I/O diagram for a physical system simulation model. (Alves et al. 2018)

architecture into its constituent parts, then assigning a simulation or virtualization program to each (e.g. Simulink for physical simulation, OpenPLC for distributed control simulation, etc.). The researchers ensure interoperability by drawing module boundaries at protocol interfaces. A hypervisor manages the interconnect and concurrent processing for each module. The researchers then demonstrate the versatility of their approach with several case studies.

4.5 Discussion of Assessment Methodologies

The methodologies discussed in this section provide complementary approaches to identifying, characterizing, and quantifying the impact of cyber dependency disruption. The dependency curve analysis, for instance, helps an organization understand its resiliency to the loss of a dependency, including the temporal effects of delayed impact of the use of backup systems, but cannot efficiently characterize multipath systems or event trees. Pathway analysis methodologies, on the other hand, excel at characterizing networks of dependencies, which help an organization locate indirect dependencies and critical paths, but cannot represent temporal relationships.

Model checking and simulation, while more difficult to use than the previous two, offer analysis capabilities that certain ICS deployments may require. Model checking provides dependency *assurance* at the cost of *resolution*; it can directly pinpoint locations of weakness in a network, and provides strong formal guarantees about selected dependency attributes, but it struggles with randomness and complex concurrent systems. Simulation, on the other hand, provides *resolution* at the cost of *assurance*. A user with sufficient knowledge and computational resources can simulate detailed interactions between components—including time-bound and nondeterministic behavior—but will need to run many tests before the results are statistically significant.

Ideally, any attempt to understand ICS cyber dependencies would use all of these methodologies in parallel, but it is dependent on the user to identify how and when to use them. For instance, a user might begin with the facilitated survey technique to produce a detailed account of use cases, failure modes, contingency plans, and hardware models. They would then use this information to build pathway graphs and dependency curves, which would then be used to identify critical components and dependencies. The user could then use model checking to identify and verify large-scale architectural and policy changes, and simulation to understand and secure sensitive timing or control dependencies.

5 Managing the Risk of Cyber Dependencies

The risk assessment of cyber dependencies helps an organization determine the relative importance of managing the cyber elements upon which they depend. Although an organization will use standard security practices for managing cyber dependencies, some special attention or consideration might be necessary. Dependencies represent relationships between cyber elements. These relationships might require special configurations, such as software or hardware versions, network addressing, identification certificates, and permissions. This section reviews security practices that merit special attention for cyber elements used in cyber dependencies. The relationship formed by a dependency requires special attention because it might pose a trust-based security risk and it might be difficult to monitor. To permit the communication between devices required by the dependency, the organization might relax security controls or grant special permissions. Relaxing security controls might permit the introduction of threats. Granting special permissions forms a trust between the elements that might permit the transmission of threats. Organizations need to assess these security risks and determine whether they need to add controls to mitigate the risk posed by this trusted relationship.

The organization needs to monitor the dependency to detect threats and evaluate uptime. Because the dependency might create an attackable opening, the monitor needs to analyze the communication for signs of threats. Because the organization needs to use the dependency, the monitor needs to understand how to measure whether the dependency is operating as expected. If the monitor detects problems, it should provide information on the type and source of the problem.

When problems with cyber dependencies occur, the organization needs a contingency plan that addresses how to respond and recover. When developing the contingency plan, the organization should evaluate whether the recovery time (MTTR) of the dependencies meets the organization's recovery time requirements. If not, alternative recovery strategies should be considered, such as provisioning redundant systems. Redundant systems vary in cost and complexity, ranging from spares on a shelf to fully provisioned, hot-standby, redundant components with automated failover.

The cyber elements used in cyber dependencies might require special configurations or use specialized software and hardware. The plan should address how to acquire and configure replacement elements. Some specialized elements, such as control systems, might have reached end-of-life which will require the organization to develop a plan to find and configure suitable replacements.

The contingency plan should also address, if possible, the use of alternatives other than cyber elements. For example, some computerized control systems permit manual control.

In addition to preparing for the loss of internal dependencies, the contingency plan should address the potential loss of external dependencies, if required. The organization needs to ensure that its external partner has a recovery or restoration plan that meets its own requirement. If not, the organization needs to develop its own alternative.

Finally, the organization may consider transitioning all or part of its system to a zero trust architecture. As defined by the Department of Defense (DoD), a zero trust security model "eliminates the idea of trusted or untrusted networks, devices, personas, or processes, and shifts to multi-attribute-based confidence levels that enable authentication and authorization policies based on the concept of least privileged access" (DISA and NSA 2022). Core principles of zero trust are to verify explicitly, use least privileged access, and to assume breach. According to Microsoft's Zero Trust Adoption Report (Microsoft Security 2021), this security model has been gaining significant popularity in recent years, with 96% of surveyed cybersecurity professionals convinced that the adoption of zero trust is a critical improvement to the security posture of their organizations.

6 Conclusion

Listed below are the key steps in managing cyber dependencies:

1. Identify the critical cyber services (CCS) depended upon by the organization.
2. Identify the cyber dependencies of those CCS and their cyber components:
   (a) Assess the relative importance of the CCS and their cyber components.
3. Characterize each cyber dependency:
   (a) Determine whether the organization owns and controls the cyber dependency components or whether it depends upon an external organization.
   (b) If external, evaluate whether the external organization has sufficient protection and recovery procedures to maintain the dependency.
   (c) Quantify the consequence of its loss: the extent of impact and the time before impact.
   (d) Define its type and the threats it faces:
      (i) Flow interruption, interception, modification, fabrication, replay, delay/acceleration, and suppression
   (e) Evaluate its vulnerabilities.
4. Evaluate how the organization manages the risk of its cyber dependency components:
   (a) Assess whether the organization sufficiently monitors the performance of the cyber dependency.
   (b) Assess how quickly and effectively the organization can recover the lost cyber dependency.
   (c) Measure whether the organization can recover before serious harm occurs. Assess whether the organization can restore the cyber dependency.

References

Alves T, Das R, Werth A, Morris T (2018) Virtualization of SCADA testbeds for cybersecurity research: a modular approach. Comput Secur 77:531–546. https://doi.org/10.1016/j.cose.2018.05.002

Clemente D (2013) Cyber Security and global interdependence: what is critical? Chatham House, London

D. Cofer et al (2018) A Formal Approach to Constructing Secure Air Vehicle Software, in Computer, vol. 51, no. 11, pp. 14-23, Nov, doi: https://doi.org/10.1109/MC.2018.2876051

Dassault Systemes Xcos (2024) Scilab. In: Xcos. https://www.scilab.org/software/xcos. Accessed 6 May 2024

DISA, NSA (2022) Department of Defense Zero Trust Reference Architecture

Feiler PH, Gluch DP (2012) Model-based engineering with AADL: an introduction to the SAE architecture analysis & design language. Addison-Wesley, Upper Saddle River

International Communication Union (2004) Quality of service and network performance handbook. ITU-T telecommunication standardization of ITU. Geneva: ITU

Joyce A, Petit F, Nowak L, & Evans N (2017) Cyber protection and resilience index: An account of the elicitation process. Argonne: Argonne National Laboratory, Global Security Sciences. ANL/GSS-17/3

MathWorks (2024) Simulink – simulation and model-based design. In: Simulink. https://www.mathworks.com/products/simulink.html. Accessed 6 May 2024

Meng B, Larraz D, Siu K et al (2021) VERDICT: a language and framework for engineering cyber resilient and safe system. Systems 9:18. https://doi.org/10.3390/systems9010018

Microsoft Security (2021) Zero trust adoption report. Microsoft

Modelica Association (2024) In: Modelica. https://modelica.org/. Accessed 6 May 2024

Moore TJ, Cho J-H (2019) Applying percolation theory. In: Kott A, Linkov I (eds) Cyber resilience of systems and networks. Springer, Cham, pp 107–133

O'Neill P (2013) Protecting critical infrastructure by identifying pathways of exposure to risk. Technology Innovation Management Review 3:34–40

OECD (2012) Quality of service and network performance: handbook. Organisation for Economic Co-operation and Development, Paris

Pederson P, Dudenhoffer D, Hartley S, Permann M (2006) Critical infrastructure interdependency modeling: a survey of US and international research. Idaho National Laboratory, Idaho Falls

Petit F, Verner D, Brannegan D et al (2015) Analysis of critical infrastructure dependencies and interdependencies. Argonne National Lab. (ANL), Argonne

Rinaldi SM, Peerenboom JP, Kelly TK (2001) Identifying, understanding, and analyzing critical infrastructure interdependencies. IEEE Control Syst Mag 21:11–25. https://doi.org/10.1109/37.969131

Srivastava AK, Ernster TA, Liu R, Krishnan VG (2018) Graph-theoretic algorithms for cyber-physical vulnerability analysis of power grid with incomplete information. J Mod Power Syst Clean Energy 6:887–899. https://doi.org/10.1007/s40565-018-0448-7

Stallings W (2014) Cryptography and network security: principles and practices, 6th edn. Pearson Education, Boston

Umunnakwe A, Sahu A, Narimani MR et al (2021) Cyber-physical component ranking for risk sensitivity analysis using betweenness centrality. IET Cyber-Phys Syst Theory Appl 6:139–150. https://doi.org/10.1049/cps2.12010

Wan Z, Mahajan Y, Kang BW et al (2021) A survey on centrality metrics and their network resilience analysis. IEEE Access 9:104773–104819. https://doi.org/10.1109/ACCESS.2021.3094196

Quantifying and Reducing System Non-Resilience: Methodology, Metrics, and Case Study

Jose Luis Castanon Remy, Ekzhin Ear, and Shouhuai Xu

1 Introduction

The importance of assuring system resilience against attacks can never be overestimated. However, it is largely an open problem to quantify the resilience of systems in general and cyber resilience in particular (Cho et al. 2019; Xu 2021), partly because of the dynamic nature of cyber attacks and cyber attack-defense arms races that make it elusive to characterize the evolution of attack capabilities (see, e.g., Li et al. 2019, 2023, 2024, 2021a, b).

In this chapter, we investigate the other side of the same coin which to our knowledge has not been explored in the literature, namely: How should we define and quantify the non-resilience of cyber-enabled systems? Answering this question would deepen our understanding of the factors that cause system non-resilience, and thus help us in reducing system non-resilience and improving system resilience.

1.1 Our Contributions

This chapter makes three contributions. First, we initiate the study of system non-resilience, which is the other side of the resilience coin. This is a new perspective that aims to understand what makes a system non-resilient. To further appreciate the importance of studying non-resilience, let us consider one analogy in mathematics: Human kind first studied positive numbers, which is in a sense parallel to the

J. L. C. Remy · E. Ear · S. Xu (✉)
Department of Computer Science, University of Colorado Colorado Springs, Colorado Springs, CO, USA
e-mail: jcastano@uccs.edu; eear@uccs.edu; sxu@uccs.edu

© The Author(s), under exclusive license to Springer Nature Switzerland AG 2025

I. Linkov, A. Kott (eds.), *Cyber Resilience: Applied Perspectives*, Risk, Systems and Decisions, https://doi.org/10.1007/978-3-031-90109-6_4

notion of resilience; human kind later found it necessary to study negative numbers, which is in the same sense parallel to studying the notion of non-resilience.

Second, we propose a novel methodology for quantifying and analyzing system non-resilience. The methodology has the following features: (i) it provides a novel metric, dubbed shot, to study and quantify system non-resilience; (ii) it leverages the notion of attack flow (Zeien et al. 2024; Ear et al. 2023; Mireles et al. 2016) to describe attacks, including both real-world attacks and hypothetical attacks (for "what if" analysis); (iii) it allows us to identify the root cause of attacks and guide the reduction of system non-resilience. Moreover, the shot metric has the following features: (i) In the context of the present study, one shot represents a successful offensive cyber action against a system for the purpose of disrupting or compromising a functionality or service of the system and/or for the purpose of accomplishing another attacker objective. Nevertheless, it can be easily adapted to other application domains; for instance, in the earthquake domain, one shot can represent a seismic activity. (iii) It is intuitive to understand. For instance, the well-known concept of single point of failure corresponds to the special case of 1-shot, meaning that a single successful attack can disrupt or compromise a functionality or service of a system. (iii) It can be used to measure non-resilience at the system and sub-system levels. (iv) It can be extended to describe the consequences to a system or sub-system incurred by one or multiple shots. The preceding (ii)–(iv) show that the shot metric actually leads to a family of associated metrics. For instance, a system may only be able to withstand n-shots, where $n \geq 1$ (and $n = 1$ corresponds to the notion of single point of failure).

Third, we show the usefulness of the methodology and metric by applying them to conduct a case study on a space infrastructure and service, where the shot metric can be used to measure both real-world attacks and hypothetical attacks. In particular, we show how to apply them to identify and employ countermeasures to prevent attacks and potentially reduce system non-resilience.

1.2 Related Work

We divide related prior studies into three categories: studies on resilience, studies related to our methodology, and studies related to our case study.

In terms of research on resilience, we focus on resilience of cyber-enabled systems because it is the theme of this chapter. For instance, the National Institute of Standards and Technology (NIST) defines resilience as "The ability to prepare for and adapt to changing conditions and withstand and recover rapidly from disruptions. Resilience includes the ability to withstand and recover from deliberate attacks, accidents, or naturally occurring threats or incidents" (Ross 2018). However, documents of this nature do not provide specific resilience metrics, which play a fundamental role. Indeed, resilience metrics are little investigated, with three exceptions: the STRAM metrics framework (Cho et al. 2019) includes a pillar of resilience metrics; the SARR framework (Xu 2021) presents a foundational characterization on the relationships between security (e.g., confidentiality, integrity, and availability properties), agility, resilience, and risk metrics; and the agility metrics framework (Mireles et al. 2019) is closely related to resilience. Our study offers an innovative metric to describe non-resilience, n-shot, which can be incorporated into these frameworks.

Our analysis methodology is inspired by the Cybersecurity Dynamics framework (Xu 2019, 2020, 2014a), even though the result presented in the present chapter does not explicitly consider the time dimension yet (despite the notion of attack flows). It is an interesting future work to incorporate our methodology into the Cybersecurity Dynamics framework by explicitly considering the time dimension so as to leverage the rich analysis techniques and results in both first-principle cybersecurity modeling and analysis (Li et al. 2011; Xu et al. 2012a, b, 2014, 2015a, b; Xu and Xu 2012; Da et al. 2014; Xu 2014b; Zheng et al. 2015, 2018; Han et al. 2021; Lin et al. 2019; Zheng et al. 2018; Paarporn et al. 2023) and cybersecurity data analytics, especially cyber threats forecasting capabilities (Sun et al. 2023; Fang et al. 2019, 2021; Peng et al. 2017, 2018; Xu et al. 2017, 2018; Zhan et al. 2013, 2015), which would enable proactive defenses (Chen et al. 2021, 2022; Lu et al. 2013; Han et al. 2014), and are important means to achieving resilience. Moreover, our analysis methodology is somewhat reminiscent of the notion of attack tree (Schneier 1999), which is one approach to describing possible attacks. By contrast, we use the notion of attack flow to describe attacks. Furthermore, the notion of attack graph (Phillips and Swiler 1998) is remotely related to our methodology because we do not necessarily assume all the vulnerabilities in a system are known to the defender. Our analysis is centered on the novel n-shot metric.

In terms of our case study, we choose space cybersecurity because it is not adequately understood and benefits from more investigation. The case study is based on a real-world attack against a space infrastructure and service reported in (Ear et al. 2023), which presented an extrapolation methodology to deal with the problem of missing information that is often encountered when analyzing real-world attacks against space infrastructures. However, we go much beyond (Ear et al. 2023) by introducing and investigating the innovative notion of non-resilience.

1.3 Chapter Outline

The rest of the chapter is organized as follows. Section 2 defines the terms used in this chapter and provides a primer on space cybersecurity to help understand the case study. Section 3 presents our methodology for quantifying and reducing non-resilience, including the n-shot metric. Section 4 conducts a case study on applying our methodology to quantify and reduce the non-resilience of a satellite infrastructure and service. Section 5 discusses future research directions. Section 6 concludes the chapter.

2 Preliminaries

In this section we clarify the terms used in this chapter and briefly review the concepts that are used in space infrastructures and services.

2.1 Terminology

In this chapter, the term system is defined in a broad sense to include cyber-enabled critical infrastructures, cyber-enabled Operational Technology (OT) networks, Information Technology (IT) networks, and individual computing devices, where the term cyber-enabled means that cyber technology is integral to the functioning of the system. In particular, we treat space infrastructures as one kind of critical infrastructure and one kind of cyber-enabled OT network.

Designers and defenders of a system have to make assumptions about the attackers they deal with, in terms of a threat model, which includes attacker objectives and attacker capabilities. Attacker objectives are often geared towards the design objectives of a system, in terms of some functionality or service (e.g., assuring the confidentiality of the data stored in the system).

It would be ideal if we could assure that a system cannot be successfully attacked, which is however impossible in most if not all cases. An attack contains one or multiple offensive action(s) for achieving an attacker objective. An attack, both real (i.e., occurred in the real world) and hypothetical (i.e., can occur in principle but has not occurred in the real world) alike, can be described by one or multiple attack flow (s) (Zeien et al. 2024; Ear et al. 2023; Mireles et al. 2016). An attack flow describes how an attack progresses through a victim system, starting at an entry point whereby the attacker penetrates into the system from the outside and ends at an impact point where the damage is incurred.

Given an attack flow, one can identify the root cause that enabled the attack. A countermeasure remediates or overcomes an offensive action by addressing the root cause that enabled the offensive action. A defensive operation may employ one or multiple countermeasures to prevent attacks that would otherwise succeed.

2.2 A Primer on Space Cybersecurity

To help understand the case study, Fig. 1 presents a simplified space infrastructure model, which is inspired by (Ear et al. 2023). As shown, a space infrastructure is comprised of four segments: space, ground, user, and link, where each segment contains some modules that perform certain functionalities.

The space segment often contains satellites (or other spacecraft) in Earth orbits for conducting some missions, such as providing satellite-based communication and

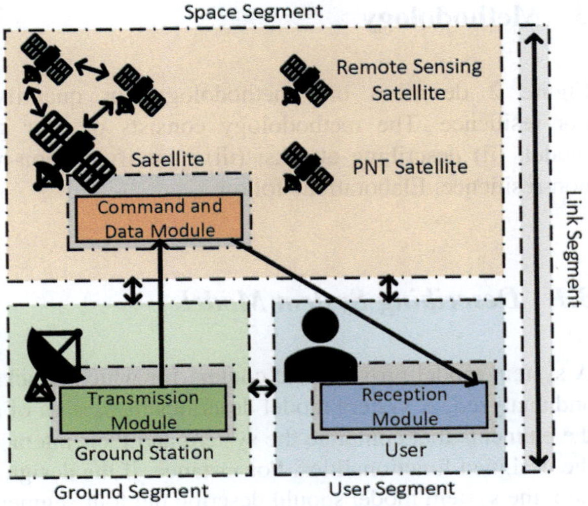

Fig. 1 A simplified space infrastructure model

remote sensing services and providing Position, Navigation, and Timing (PNT) services. For instance, the command and data module in the space segment is responsible for processing commands issued from the ground segment and data generated by the satellite.

The ground segment is a collection of cooperative terrestrial modules that communicate with the space segment (e.g., satellites) to command and control satellites and to receive, process, and analyze the health condition of satellites (i.e., telemetry) and mission data. For instance, the transmission module in the ground segment is responsible for sending data (e.g., commands) to the reception modules of satellites in the space segment.

The user segment describes the recipients of the services of satellites by receiving and processing the mission data either directly from satellites in the space segment or indirectly from the other modules in the ground segment. For example, the reception module in the user segment receives data from either the space segment or the ground segment.

The link segment describes connectivity across the space infrastructure between segments (e.g., between satellites in the space segment and modules in the ground segment) or within a segment (e.g., between two satellites in the space segment).

All four segments may be susceptible to and impacted by cyber attacks. This is evident from real-world incidents analyzed in (Ear et al. 2023) as well as theoretical attacks across the literature that were systematized in (Remy et al. 2024).

3 Methodology

Figure 2 describes our methodology for quantifying and reducing system non-resilience. The methodology consists of four steps: (i) describing system model; (ii) describing attacks; (iii) quantifying non-resilience; and (iv) reducing non-resilience. Elaborations follow.

3.1 *Describing System Model*

A system model provides the context for which attacks can be precisely described and analyzed. A system model describes the design of a system in detail, including the elements that comprise the system and their interactions in order to accomplish the designed functionalities. For instance, if the design concerns a space infrastructure, the system model should describe the four segments (i.e., space, ground, user, and link) and the modules in each segment. Moreover, the system model should clearly describe the condition under which we say an attack succeeds (e.g., what is compromised, exposed, or damaged).

Inspired by (Xu 2021), we advocate explicitly stating the assumptions that are made by the system model, dubbed system-model assumptions. This is important because these assumptions could have cybersecurity implications, meaning that an invalid system-model assumption could be exploited by attackers to wage successful attacks. For instance, if a system model assumes that each computer has a tamper-resistant hardware token for conducting certain tasks, then the absence of such a hardware token in one computer could undermine the cybersecurity properties of that computer and possibly for the entire network in question.

3.2 *Describing Attacks*

We propose describing attacks in the following three steps.

- **Describing threat models.** This is to clearly specify the threat model(s), including assumptions about attacker objectives and attacker capabilities, dubbed

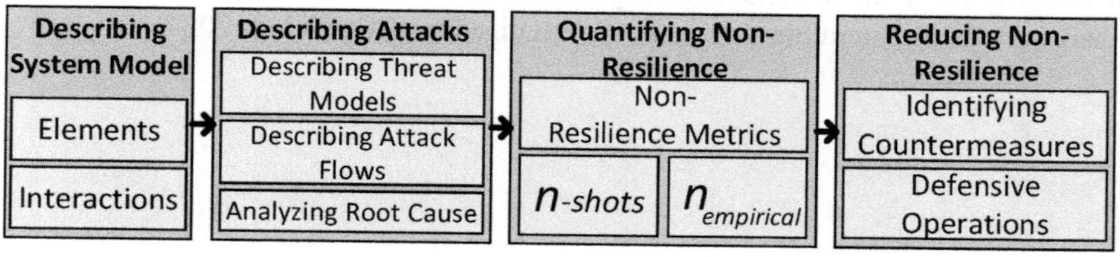

Fig. 2 Our methodology

threat-model assumptions. An attacker objective is often to disrupt one or multiple designed functionalities or services of a system. This is relevant because even for attacks that already occurred in the real world, we might not have all the information about the attacker objectives and/or attacker capabilities. Thus, an analyst would have to make assumptions, while leveraging domain expertise and the system model, to describe the threat model clearly.

This is equally applicable when speculating hypothetical attacks. We stress that multiple threat models can be considered when dealing with multiple kinds of attackers (e.g., criminals vs. nation-state) or when multiple threat models of interest formulate a partial order (i.e., the most powerful threat model does not exist).

- **Describing attack flows.** We propose precisely describing attacks, real and hypothetical alike, via their attack flows (Zeien et al. 2024; Ear et al. 2023; Mireles et al. 2016). An attack flow describes an actual or hypothetical attack scenario, from the entry point where the attacker penetrates into a system to the impact points where the damage(s) is/are incurred. An attack flow should be faithful to both the system-model assumptions and the threat-model assumptions, meaning attacker objectives and capabilities must not violate these assumptions. It would be ideal if we can enumerate all possible attack flows, which may not be possible or feasible in general.
- **Analyzing root causes.** It is important to understand the root cause or enabler of an attack, such as a software vulnerability or design flaw because this sheds light on identifying countermeasures to prevent the attack from succeeding, potentially reducing system non-resilience, and increasing system resilience.

3.3 *Quantifying Non-Resilience*

We propose using the shot metric to describe the non-resilience of a system.

***Definition 1** **(shot)** A shot is a successful offensive action by an attacker in realizing an attacker objective against a system.*

For instance, a shot can be the offensive action for compromising the password of a victim, invalidating a system-model assumption, or applying a capability (e.g., applying an exploit against a software vulnerability). We stress that a shot does not necessary correspond to an attack or attack flow. For example, an attack or attack flow may require two shots, one for penetrating into an entry point of a network, and the other for laterally moving to another computer in the network by exploiting a vulnerability in that computer.

***Definition 2** **(n-shots non-resilience)** If every attack (or attack flow) that can accomplish an attacker objective requires at least n shots, then we say the system is n-shots non-resilient.*

Definition 1 has several features. First, the higher the number n, the more resilient the system. Second, the metric degenerates to the well-known notion of single point of failure when $n = 1$. Third, a system functionality that cannot be compromised even if 1/3 nodes in a network of N nodes (e.g., Byzantine Fault-Tolerance or BFT (Lynch (1996)) is ($N/3 + 1$)-shots non-resilient.

Having defined the n-shots metric, a natural question is: How can we measure this metric? The ultimate solution is to develop a systematic theory for this purpose, which goes far beyond the present study. As a first attempt, we propose the following best-effort method in three steps.

The first step is to identify attacks and their attack flows. As mentioned above, it may not be possible or feasible to enumerate all possible attacks and all possible flows. However, this does not prevent analysts from leveraging their domain expertise and/or automated tools (e.g., artificial intelligence/machine learning tools) to identify relevant attacks and attack flows. Suppose ℓ attacks are identified, denoted by $A_1, \ldots, A_\ell$. Suppose for attack i ($1 \leq i \leq \ell$), z_i attacks flows are identified, denoted by $AF_{i,1}, \ldots, AF_{i,z_i}$, respectively. Note that $z_i > 1$ can happen because, for instance, when dealing with real-world attacks with missing information, which forces us to extrapolate the attack flows with respect to an incident (Ear et al. 2023).

The second step is to quantify the system non-resilience with respect to each of an attack flow and an attack, ideally also by designing tools to automatically compute this metric. For a given attack flow, we propose:

***Definition 3** **(empirical non-resilience with respect to an attack flow)** Given an attack, which is described by its attack flows, accomplishing the attack to achieve the attacker objective requires the attacker to wage n shots or successful offensive actions.*

Note that Definition 3 is defined with respect to a specific attack flow, whereas Definition 2 does not have to be based on explicitly enumerating all attacks or all attack flows. Given the z_i attack flows corresponding to attack i, Definition 3 leads to the measurements of $n_{i,1}, \ldots, n_{i,z_i}$ shots, respectively. Then, the empirical non-resilience with respect to at attack is given be:

***Definition 4** **(empirical non-resilience with respect to an attack)** Consider a given attack A_i ($1 \leq i \leq \ell$) with z_i attack flows $AF_{i,1}, \ldots, AF_{i,z_i}$, which respectively lead to $n_{i,1}, \ldots, n_{i,z_i}$ shots non-resilience. Then, the empirical non-resilience with respect to attack A_i is given by $n_i = \min(n_{i,1}, \ldots, n_{i,z_i})$.*

The third step is to use Definition 5 below to compute the empirical system non-resilience.

***Definition 5** **(empirical non-resilience)** Consider ℓ given attacks $A_1, \ldots, A_\ell$ against a system, where each attack is described by an attack flow and leads to an n_i-shots non-resilience for $i = 1, \ldots, \ell$. Then, the empirical non-resilience, denoted by $n_{empirical}$, is defined as $n_{empirical} = \min(n_1, \ldots, n_\ell)$.*

Note that Definition 5 can be equally used to compute the empirical non-resilience incurred by a single attack with multiple possible attack flows, which can happen when dealing with real-world attacks with missing information (Ear et al. 2023). This means that Definition 5 can be equally used at both the attack and attack-flow levels. Note also that $n_{empirical}$ reflects the smallest number of shots that need to be waged among the given set of ℓ attacks. If we can enumerate all possible attacks, then $n_{empirical}$ is the least number of offensive actions that must be conducted by the attacker to accomplish damage, namely the inherent and invariant n-shots non-resilience property of the system.

3.4 *Reducing Non-Resilience*

To demonstrate the usefulness of the n-shots metric, we show how to leverage it to guide the process of reducing non-resilience (e.g., hardening the system), by employing defensive operations. Intuitively, the goal is to disrupt attacks, namely their attack flows, such as thwarting an attacker capability (e.g., patching a vulnerability or employing an advance defense tool to disrupt a lateral movement). This can be done by identifying countermeasures and then conducting defensive operations to eliminate the root cause that enables an attack, make at least one shot no more viable, and/or invalidate an assumption that is made by an attacker.

Note that it is not necessarily true that one defensive operation can invalidate one attacker assumption; rather, invalidating one attacker assumption may require multiple defensive operations. Thus, the employment of defensive operations may eliminate some of the ℓ attacks or attack flows mentioned in Definition 5. However, employing defensive operations may or may not improve the resilience of a system because $\min(n_1, \ldots, n_\ell) \leq \min(n_{i \in I})$, where $I \subseteq \{1, \ldots, \ell\}$ is a subset of the attacks that are not removed by the employment of defensive operations, while letting $\min(n_{i \in I}) = \infty$ when $I = \emptyset$ for notational convenience (to deal with a corner case with a due cybersecurity significance, namely the absence of no known attacks).

4 Case Study: Non-Resilience of the Intelsat Communications Satellite Infrastructure

In this case study we investigate how a space infrastructure and its services were not designed to be resilient against signal hijacking attacks, causing 1-shot non-resilience or a single point of failure. We consider a real-world attack against the Intelsat infrastructure via four hypothetical but plausible attack flows. This includes one attack that has been publicly reported and two hypothetical attacks that could occur but have yet to be observed. We choose this case study because space resilience is little understood and benefits from more investigation.

4.1 Describing System Model

Intelsat is a multinational satellite communications service provider offering end-to-end data connectivity through a network of ground stations, or Teleports, fiber networks, and a constellation of 32 geosynchronous and 19 low-Earth orbit (LEO) satellites (Intelsat 2024).

Figure 3 illustrates the Intelsat space infrastructure, with its main service of offering users end-to-end data connectivity and communications capabilities. The infrastructure includes eight Intelsat Teleports located around the world. Intelsat users send and receive data from the closest Teleport through fiber networks. A Teleports' main function is to encode users' data into Radio Frequency (RF) signals that are relayed to the satellite through the Teleport's transmission module (i.e., uplink antenna). Teleports can also receive RF signals from the satellite through the Teleport's reception module (i.e., downlink antenna). A Teleport decodes the data from the received RF signals and relays the data to the user through fiber networks. In addition to receiving and transmitting through Teleports, Intelsat satellites can relay signals to other satellites (i.e., satellite-to-satellite transmissions).

The Intelsat constellation consists of two kinds of communication satellites (which are not illustrated in Fig. 3): (i) simple communications satellites, which only amplify the received signals and relay these signals to other satellites; and (ii) complex communications satellites, which may process the received signals before transmitting the processed data to other satellites, end users, and/or the Teleports. More specifically, a simple communications satellite receives a signal (from another satellite in the same constellation or from a Teleport) through the communications reception antenna, amplifies the received signal, shifts the frequency of signal transmission as necessary (e.g., uplink transmissions may not use the same frequency as downlink transmissions or satellite-to-satellite transmissions), and transmits the signal through a transmission antenna. A complex communications satellite receives a signal via the receiving antenna, demodulates the receiving signal,

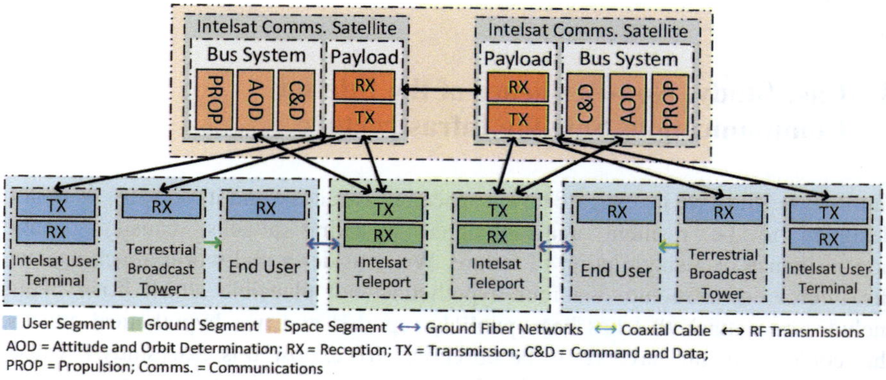

Fig. 3 System model of Intelsat space infrastructure and services

decodes the data in the signal, re-encodes the data to be transmitted, modulates the data into the transmission signal, and transmits the signal through the transmission antenna. We treat all Intelsat satellites as simple communications satellites because we find no evidence that Intelsat encoded their signals in or before 2005 (the year of the incident described later in the case study).

Example Use Cases of Intelsat One use case of Intelsat is to serve those users who have no access to fiber networks or are unable to transmit or receive data from the Teleport, either because of their remote location or because they are on the move. In these cases, the users must use the Intelsat terminal to transmit and receive data directly from the communications satellite (e.g., providing users with Internet connectivity). Intelsat user terminals comprise of an antenna and a modem that permit the user to modulate/demodulate data into RF signals, establish a connection with an Intelsat communications satellite, and transmit or receive data. The users can connect their computing devices to the Intelsat terminal via the modem using an Ethernet cable.

Another use case of Intelsat is to serve those users who may not have access to fiber networks or access to Intelsat user terminals. These users can receive satellite signals by connecting to a satellite-connected Terrestrial Broadcast Tower. Terrestrial broadcast towers work similarly to Intelsat user terminals but only have downlink capabilities (i.e., can only receive signals from satellites). The broadcast tower demodulates received signals and transmits the data to end users, who could be connected to the broadcast tower through a coaxial cable.

System-Model Assumptions There is no document on the system model assumptions made by Intelsat designers. It appears that a major system-model assumption that is inherent to the design of the Intelsat space infrastructure is that only legitimate user terminals will communicate with satellites, Teleports, and Terrestrial Broadcast Towers. This system-model assumption, if made, has led to the lack of defenses (e.g., authentication) against attackers' employment of rogue user terminals. Effectively, this unrealistic system-model assumption has been exploited by attackers to wage multiple attack flows against the Intelsat space infrastructure, as discussed below.

4.2 *Describing Attacks*

We consider three plausible attacks against the Intelsat space infrastructure: Attack 1 is extrapolated from a real world incident (extrapolation because of missing information) (Ear et al. 2023) and Attacks 2–3 are hypothetical.

4.2.1 Attack 1

In 2005, the Sri Lankan separatist group Liberation Tigers of Tamil Eelam (LTTE) was able to illegally broadcast their propaganda (television and radio signals) through Intelsat communications satellites across Europe and Asia (Fritz 2013; Reuters 2024). LTTE achieved this by identifying appropriate frequency channels within the Ku Band used by the Intelsat communications satellites in their footprint when they orbited over the Indian Ocean, where footprint of a satellite means the geographic area that the satellite can transmit signals to or receive signals from. Intelsat (i.e., the service provider) had no knowledge of the illegal broadcast until notified by the Sri Lankan government. Intelsat did not provide any details about how LTTE has achieved their impact (e.g., whether or not LTTE used Intelsat user terminals to gain access); nor did Intelsat provide the specific geographical locations from where LTTE transmitted the illegal signal (spacenews.com 2024). Given this limited information, we aim to recover highly probable attack flows so as to understand non-resilience of the Intelsat infrastructure.

Describing Threat Models Despite the lack of technical details in the incident, the attacker demonstrated the ability to hijack Intelsat satellite signals. This makes following threat-model assumptions plausible.

- Assumption 1.0 (Attacker Objective): The attacker's objective was to broadcast its own data to Intelsat service users, which thwarts the designed functionality that only authorized users can broadcast signals.
- Assumption 1.1 (Attacker Capability): This assumption provides the context of the attack. The attacker targeted Intelsat communications satellite #906 (IS-906) in Intelsat's constellation because: (i) IS-906 had a stationary orbit over the Indian Ocean, meaning that its orbit matched the Earth's rotation and gave it a constant position with respect to Earth's surface; (ii) IS-906's footprint covered Sri Lanka; and (iii) IS-906 was capable of transmitting over the Ku band. (Note that at the time of writing, IS-906 is not the only Intelsat satellite whose stationary orbit is over the Indian Ocean; the other Intelsat satellites with the same orbit include IS-22, IS-33e, IS-38, and IS-39, which were launched after 2005, meaning that these satellites could not have been targeted by the attacker simply because there were not yet launched into space when the attack occurred.)
- Assumption 1.2 (Attacker Capability): This assumption also provides the context of the attack. The attacker did not have access to fiber networks and thus lacked connectivity to any of the four Intelsat Teleports located in Asia. This assumption provides the context that the attacker cannot use a Teleport to transmit the illegal signal to IS-906.
- Assumption 1.3 (Attacker Capability): If the attacker used a rogue terminal, the attacker was able to identify an appropriate frequency in the Ku band. For example, the attacker could have monitored the radio frequency spectrum for an available frequency by measuring the RF noise levels of the channels within the target frequency range. If the noise readings of the channel are close to zero,

then no signals are being transmitted in that channel, meaning the channel can be used by the attacker's rogue terminal to transmit signals. This also means IS-906 was actively listening for transmissions in the Ku band.

- Assumption 1.4 (Attacker Capability): If the attacker used a rogue terminal, the attacker was able to modulate its data into an RF signal. This means the attacker knows the modulation that was used by the IS-906 reception module; otherwise, IS-906 would not correctly demodulate the attacker's signal, thus preventing the attacker's signal from being transmitted.
- Assumption 1.5 (Attacker Capability): If the attacker used a legitimate Intelsat user terminal, the attacker was capable of physically accessing the legitimate Intelsat user terminal and of injecting data into the terminal. This is plausible because Intelsat user terminals were available in the Sri Lankan region, having physical access to legitimate terminals greatly decreases the technical barrier for exploitation, and because we found no evidence for the ability to remotely employ the terminals (e.g., local area network). If the attacker used a legitimate terminal, the assumption that the attacker used a rogue user terminal should be correspondingly dismissed. That is, the attacker would not need to identify an available Ku band or the modulation used by IS-906 because a legitimate terminal can do these for the attacker.

Describing Attack Flows The preceding assumptions lead to two possible attack flows corresponding to the incident, owing to the lack of detailed information about the incident. The two attack flows differ in whether the attacker used a rogue user terminal or a legitimate Intelsat user terminal. Figure 4 highlights the two attack flows, which are elaborated below.

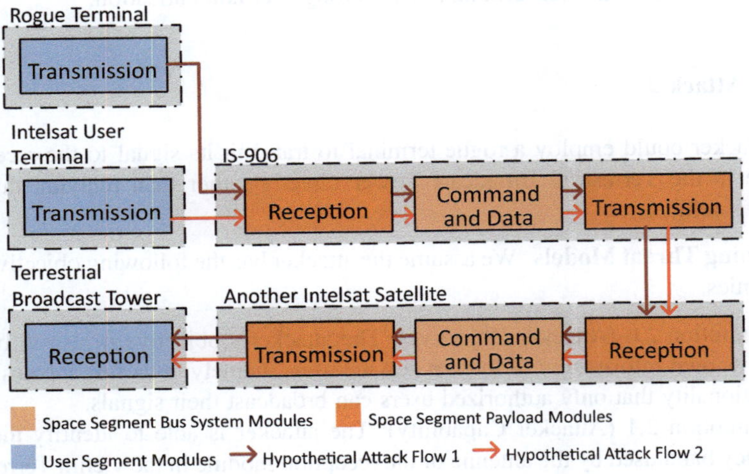

Fig. 4 Two possible attack flows of Attack 1

The first attack flow started with a rogue terminal and proceeded as follows. (i) The attacker constructed a rogue terminal to transmit the attacker's signal to the reception module in IS-906. Specifically, the transmission module in the rogue terminal modulated LTTE's data into an RF signal and transmitted the signal in an available Ku band frequency channel used by IS-906. (ii) The IS-906 reception module received the signal, demodulated it, and sent the data to the command and data module in the IS-906 bus system. (iii) The command and data module processed the data and sent it to the payload transmission module in IS-906. (iv) The payload transmission module modulated the data into a different RF signal based on the frequency used between IS-906 and another Intelsat communications satellite for relaying the signal. (v) The reception module in the relaying satellite received the signal from IS-906, demodulated the signal, and sent the signal to its bus system command and data module. (vi) The command and data module processed the data and sent it to the payload transmission module. (vii) The payload transmission module modulated the signal using the appropriate modulation and frequency used by the Terrestrial Broadcast Tower and transmitted the signal to a Tower. (viii) The reception module in the Terrestrial Broadcast Tower received the signal, demodulated the received signal, and transmitted the data to the victim systems.

The second attack flow is the same as the first attack flow, except that the attacker had access to a legitimate Intelsat user terminal and used the terminal to transmit the attacker's signal to the victim.

Analyzing Root Cause Having described the real-world attack against the Intelsat space infrastructure (via two possible attack flows), we now analyze its root cause. The first attack flow succeeded because the design of the space infrastructure (Fig. 4), more specifically the payload reception module in IS-906, lacks the capability to authenticate user terminals. The second attack flow succeeded because of the design of the space infrastructure (Fig. 4), more specifically the transmission module in the Intelsat user terminal, lacks strong user authentication.

4.2.2 Attack 2

The attacker could employ a rogue terminal to transmit its signal to the reception module in the Terrestrial Broadcast Tower directly rather than relaying through Intelsat satellites.

Describing Threat Models We assume the attacker has the following objective and capabilities.

- Assumption 2.0 (Attacker Objective): The attacker's objective is also to broadcast its own information to Intelsat service users, namely thwarting the designed functionality that only authorized users can broadcast their signals.
- Assumption 2.1 (Attacker Capability): The attacker is able to identify the frequency band used by the antenna of the reception module in the victim Terrestrial Broadcast Tower of interest as different Towers may use different frequency

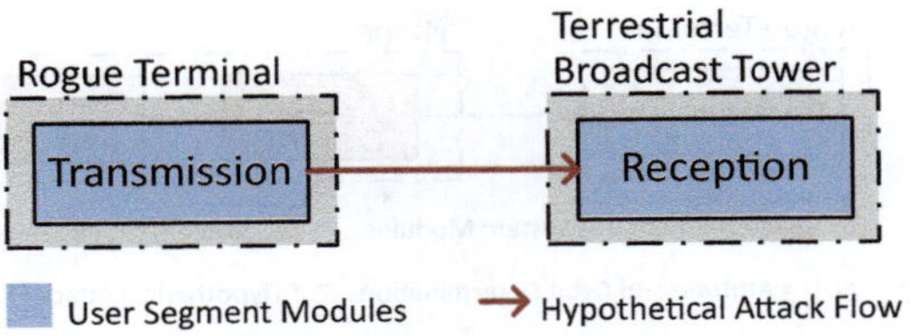

Fig. 5 Attack flow of hypothetical Attack 2

channels. To identify the frequency band used by the antenna of the reception module, the attacker is able to perform reconnaissance against the infrastructure providers (e.g., the company that built the antenna or the company that built the Terrestrial Broadcast Tower of interest).

- Assumption 2.2 (Attacker Capability): The attacker is able to identify an available channel within the frequency band used by the Terrestrial Broadcast Tower of interest. This may be achieved by monitoring frequency channels in the frequency band and measuring the noise levels of each channel. If the noise level of the channel is close to zero, then no signals are being transmitted through that channel and thus the attacker's rogue terminal can use it to transmit the attacker's signals. The attacker's success also requires that the Terrestrial Broadcast Tower is actively listening for transmissions in the frequency band.
- Assumption 2.3 (Attacker Capability): The attacker is able to modulate its data into an RF signal. To modulate the illegal signal, the attacker would need to know what modulation is used by the antenna in the reception module in the Terrestrial Broadcast Tower.

Describing Attack Flow Figure 5 describes the attack flow, which is elaborated below. (i) The attacker uses the transmission module in a rogue terminal to modulate its message into an RF signal and transmits the signal to a Terrestrial Broadcast Tower via an available frequency channel. (ii) The antenna of the reception module in the Terrestrial Broadcast Tower receives the signal from the transmission module of the rogue terminal, demodulates the received signal, and transmits the data to the victim user terminals.

Analyzing Root Cause The attack succeeds because the reception module in the Terrestrial Broadcast Tower lacks the capability to authenticate the terminal and the received signals. As a consequence, any signal received by the reception module will be treated as a legitimate signal and demodulated and broadcast.

4.2.3 Attack 3

The attacker can employ a rogue terminal to transmit a control command to satellite IS-906.

Describing Threat Models We make the following assumptions about the threat model.

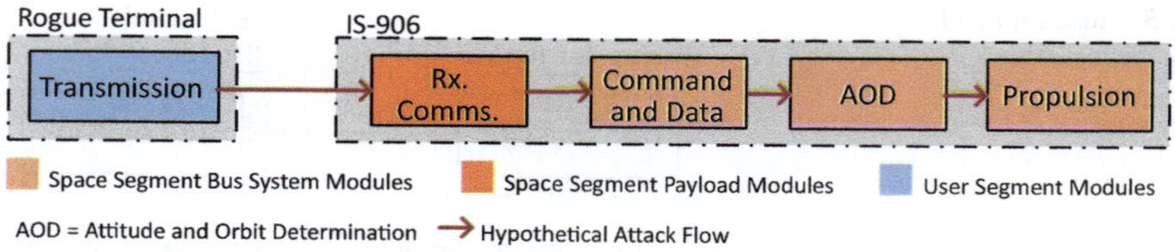

Fig. 6 Attack flow of hypothetical Attack 3

- Assumption 3.0 (Attacker Objective): The attacker's objective was to control IS-906 (e.g., causing IS-906 to stop processing RF signals and incur denial-of-service).
- Assumption 3.1 (Attacker Capability): The attacker is able to cause the command and data module in IS-906 to parse and execute the attacker's control commands. This means the attacker knows the protocols and syntax used by Intelsat to build a control command. This information may be attained, for example, by social engineering a knowledgeable operator. If the attacker is able to access the command and data module and issue the illegal command appropriately, the command and data module will parse and broadcast the command to the applicable the modules in IS-906 (e.g., the propulsion module for a control command).
- Assumption 3.2 (Attacker Capability): This is the same as Assumption 1.3 (i.e., the attacker is able to find an available Ku frequency channel to transmit the signal to IS-906).
- Assumption 3.3 (Attacker Capability): This is the same as Assumption 1.4 (i.e., the attacker is able to use the same modulation as IS-906).

Describing Attack Flow Figure 6 describes the attack flow, which is elaborated below. (i) The attacker employs a rogue terminal to modulate the attacker's control command into an RF signal and transmits it to the reception module of IS-906. (ii) The reception module in IS-906 receives, demodulates, and sends the received command to the bus system command and data module in IS-906. (iii) The command and data module in IS-906 processes and broadcasts the command to the applicable modules of the bus system and payload of IS-906 (e.g., a maneuver command to the attitude and orbit determination (AOD) module or a telemetry command to trigger a malware in the electrical power module). (iv) The receiving module executes the control instructions. For instance, the AOD module parses the maneuver command and transforms the instructions into electrical signals that are transmitted to the propulsion module. (v) The propulsion module executes the command by producing thrust to maneuver the satellite.

Analyzing Root Cause The attack can succeed because the command and control module in an Intelsat communications satellite (e.g., IS-906) lacks the capability to authenticate commands received from the Teleports. As a consequence, any valid command would be interpreted and executed by the satellite.

4.3 *Quantifying Non-Resilience*

Having described three attacks that have or could have occurred in the real world against the Intelsat space infrastructure, we now quantify the non-resilience of the Intelsat space infrastructure.

4.3.1 Quantifying Non-Resilience via Individual Attacks

We quantify non-resilience with respect to individual attacks as follows.

Quantifying Non-Resilience According to Attack 1 Since Attack 1 has two possibly attack flows, we can effectively treat them as two individual attacks according to Definition 5.

With respect to attack flow 1 of Attack 1, the attacker uses a rogue terminal to broadcast its own data to Intelsat users through Intelsat communications satellites. Thus, Assumptions 1.3 (the attacker is able to find an available frequency channel in the Ku band to transmit its signals to IS-906) and 1.4 (the attacker is able to use the same modulation as IS-906) must hold. To realize Assumption 1.3, the attacker must accomplish one successful offensive action, by identifying at least one available frequency channel to communicate with IS-906 (e.g., randomly selecting a channel and measuring its noise level). To realize Assumption 1.4, the attacker must accomplish another successful offensive action, by employing reconnaissance against publicly available information published by Intelsat to understand the modulation used by IS-906. These lead to a 2-shots non-resilience.

With respect to attack flow 2 of Attack 1, the attacker uses an Intelsat user terminal to broadcast its own data to Intelsat service users through Intelsat communications satellites. This means that Assumption 1.5 (i.e., the attacker has physically access to an Intelsat user terminal) must hold. To realize Assumption 1.5, the attacker needs to conduct one successful offensive action, by gaining physically access to a legitimate user terminal to inject the attacker's data into the Intelsat space infrastructure. This leads to a 1-shot non-resilience.

To accommodate the worst-case scenario, we can use Definition 5 to obtain that the infrastructure has a 1-shot non-resilience, namely that there is a single point of failure with respect to Attack 1, as $n_{empirical} = \min(2, 1) = 1$. Note that Assumptions 1.1 (the attacker targeting IS-906) and 1.2 (the attacker having no access to fiber networks or any Teleport) did not contribute to the analysis of non-resilience but are important as they provide context for Attack 1.

Quantifying Non-Resilience According to Attack 2 With respect to the attack flow of Attack 2, the attacker uses a rogue terminal to broadcast its own data to Intelsat users through a satellite-connected Terrestrial Broadcast Tower. This means the following Assumptions must hold: 2.1 (the attacker is able to identify the frequency band used by the Terrestrial Broadcast Tower), 2.2 (the attacker is able to find an available frequency channel to transmit the signal to the Terrestrial

Broadcast Tower), and 2.3 (the attacker is able to use the same modulation scheme as the Terrestrial Broadcast Tower). To realize Assumptions 2.1, 2.2, and 2.3, the attacker needs one successful offensive action, by employing reconnaissance of publicly available information published by Intelsat or the organization managing the Terrestrial Broadcast Tower to identify the band, frequency, and modulation used by the Terrestrial Broadcast Tower. This leads to a 1-shot non-resilience because the requirements of the three Assumptions are closely related to each other, highlighting that there is no one-to-one mapping between Assumptions and offensive actions.

Quantifying Non-Resilience According to Attack 3 With respect to the attack flow of Attack 3, the attacker uses a rogue terminal to control IS906. This means that the following Assumptions must hold: 3.1 (the attacker knows the command format used by IS-906), 3.2 (the attacker is able to find an available Ku frequency channel to transmit the signal to IS-906), and 3.3 (the attacker is able to use the same modulation as IS-906). To realize Assumption 3.1, the attacker needs one offensive action, such as targeting an Intelsat satellite operator through social engineering to understand the command and data module protocols in use. To realize Assumptions 3.2 and 3.3 (i.e., Assumptions 1.3 and 1.4), the attacker, respectively, needs one offensive action as mentioned above. These lead to a 3-shots non-resilience.

4.3.2 Quantifying Overall Non-Resilience

By applying Definition 5 with respect to the three attacks, we obtain the empirical non-resilience of the system as $n_{empirical} = \min(\min(2, 1), 1, 3) = 1$. This means that the attacker can inflict damage against the Intelsat space infrastructure and impact its services with a single shot, namely that the Intelsat space infrastructure suffers from a single point of failure.

4.4 Reducing Non-Resilience

Having showed that the Intelsat space infrastructure has a 1-shot non-resilience, now we show how to leverage the attack flows to identify countermeasures and leverage them to conduct defensive operations to reduce its non-resilience, namely increasing the $n_{empirical}$ value. The basic idea is to disrupt the attack flow(s) that incurred the smallest n-shots, effectively increasing $n_{empirical}$ according to Definition 5. Countermeasures should be applied first to disrupt Attack 1 (more specifically, its attack flow 2) and Attack 2 because both attack flows incurred 1-shot non-resilience. In order to further reduce non-resilience, countermeasures can be applied to disrupt attack flow 1 of Attack 1 and Attack 3.

4.4.1 Identifying Countermeasures

Figure 7 provides an overview of the identified countermeasures and how they can be applied to disrupt the attacks (more specifically, attack flows).

Countermeasure 1 (for invalidating Threat-Model Assumptions 1.4, 2.3, and 3.3) This countermeasure employs cryptographic techniques to authenticate user terminals, Terrestrial Broadcast Towers, Teleports, and satellite transmission and reception modules prior to establishing a communications connection, as well as to protect data in-transit. For example, each module can be associated with a pair of public and private keys, and the public key is certified by a certificate authority (e.g., the module manufacturer or vendor). This allows, for example, a satellite reception module to authenticate a user terminal transmission module with which the satellite is communicating. Another cryptographic approach is to use a symmetric-key authentication scheme, by sharing a secret key between modules where each module can authenticate the other via a message authentication scheme. However, this countermeasure encounters the scalability issue of key management (e.g., all modules need to share a secret key). This issue is further complicated when modules (e.g., of user terminals) are dynamically added to the space infrastructure.

We stress that in order to make the resulting system (i.e., the Intelsat infrastructure in this case study) resilient against, the cryptographic techniques must be made resilient. This is relevant because cryptographic techniques are not necessarily resilient against the compromise of cryptographic keys (Loscocco et al. 1998; Xu and Yung 2009); for instance, compromising a cryptographic key only requires one shot or successful offensive action as shown in (Harrison and Xu 2007). Thus, it is necessary to use more advanced techniques to enhance their resilience (e.g., Dodis et al. 2002, 2003, 2012; Dai et al. 2012; Xu et al. 2008, 2011). Another challenge is incurred when legitimate devices can be captured or obtained by attackers, which highlights the importance of cryptographic re-keying techniques to render the

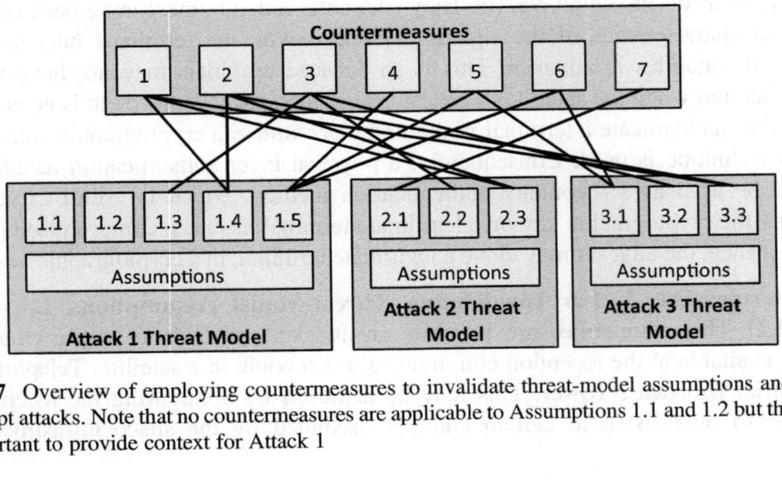

Fig. 7 Overview of employing countermeasures to invalidate threat-model assumptions and thus disrupt attacks. Note that no countermeasures are applicable to Assumptions 1.1 and 1.2 but they are important to provide context for Attack 1

devices obtained by attackers useless. Resilient re-keying techniques are challenging to design and validate, as demonstrate in (Xu 2007), showing how to attack elegantly designed cryptographic re-keying techniques (including Wong et al. 2000; Naor et al. 2001).

Countermeasure 2 (For Invalidating Threat-Model Assumptions 1.4, 2.3, and 3.3) This countermeasure employs a physical layer signal authentication scheme to authenticate user terminals, Terrestrial Broadcast Towers, Teleports, and satellite transmission and reception modules. Physical Layer Authentication (PLA) (Wesson et al. 2017; Rothmaier et al. 2021a, b, c; Gross et al. 2019; Rustamov et al. 2020; Jovanovic et al. 2014; Jansen et al. 2016; Schmidt et al. 2020; Humphreys 2013; Psiaki et al. 2013; Falletti et al. 2021; Ranganathan et al. 2016; Chen et al. 2021, 2022) leverages unique physical characteristics associated with RF signals. For instance, the reception module in a Terrestrial Broadcast Tower can deploy signal authentication defenses based on the Angle of Arrival (AoA) (Liu et al. 2021; Lo et al. 2018). The AoA measures the direction where the signal is transmitting from, which allows a reception module to authenticate the signal based on the AoA of the signal. The AoA of a legitimate signal transmitted by an Intelsat communications satellite would come from space. This method prevents the attacker from terrestrially transmitting its signals to the reception component of a Terrestrial Broadcast Tower. Another countermeasure is to use beamforming techniques whereby the operator of a satellite can modify the footprint of the satellite to prevent an attacker from using a rogue terminal to transmit signals to the satellite. One report (spacenews.com (n.d.) states that this was the initial reactive defense employed by Intelsat upon being notified of the attack.

Similarly, resilience of the resulting infrastructure requires these countermeasures to be resilient against attacks as well. These countermeasures has the following known weaknesses. (i) They cannot achieve the strong security guarantee offered by cryptographic authentication techniques, assuming cryptographic private keys are not compromised or their compromises are adequately mitigated by advanced techniques as discussed above. (ii) They encounter a deployment issue because the physical characteristics of the signals associated with the terminals may not be known to satellites in advance. This is so because terminals may not have been manufactured when the satellites were launched into space. Thus, there is no strong method to authenticate a terminal in the first place unless a cryptographic authentication technique is used. Consequently, a physical layer authentication technique should be used as a secondary authentication method, which is useful to verify whether the cryptographic key on a legitimate terminal has been compromised (as a consequence, the attacker may clone a legitimate terminal, in a cryptographic sense).

Countermeasure 3 (For Invalidating Threat-Model Assumptions 1.3, 2.2, and 3.2) This countermeasure prevents an attacker from identifying a channel that is available at the reception communications module in a satellite, Teleport, or Terrestrial Broadcast Tower. This may be achieved by configuring the reception module to only listen to certain channels assigned by the space infrastructure operator. However, this incurs a high management cost. This could also be achieved by deliberately introducing noise into the channel until a legitimate transmission needs to use the channel.

Countermeasure 4 (For Invalidating Threat-Model Assumption 1.5) This countermeasure employs a strong user authentication mechanism to prevent an unauthorized user from using a legitimate terminal to send signals. A strong user authentication mechanism can require a strong password and possibly multi-factor authentication. This countermeasure cannot cope with the situation that an operator is compromised, either as an insider threat or as a victim who is not even aware that its computer or device for operating the terminal has been compromised (e.g., via adversarial malware that can evade the employed malware detectors (Li et al. 2019, 2021a, b, 2023, 2024).

Countermeasure 5 (For Invalidating Threat-Model Assumptions 1.5 and 3.1) This countermeasure prevents an authorized terminal operator (or user) from becoming an insider threat. This can be achieved by leveraging a combination of technical and non-technical solutions. A technical solution can help detect anomalies in the behavior of an operator in cyberspace so as to recognize that the operator is suspicious as soon as possible. A non-technical solution can help detect anomalies in the behavior of an operator in the physical space so as to recognize that the operator may behave abnormally in their daily activities (Montañez et al. 2022).

Countermeasure 6 (For Invalidating Threat-Model Assumption 1.5 and 3.1) In the case that a terminal can be remotely accessed by an authorized operator, the attacker can compromise the operator's computer that is used to communicate with a terminal and thus exploit the compromised user to broadcast signals to satellites without being noticed by the victim operator. An attacker may achieve by waging cyber social engineering attacks or using evasive malware (Li et al. 2019, 2021a, b, 2023, 2024). We highlight that despite numerous studies, effective countermeasures against cyber social engineering attacks remain elusive. One main reason for their ineffectiveness is the mismatch between the attacker's heavy use of psychological elements and the defender's light use of psychological elements, known as Psychological Factors, Psychological Techniques, and Psychological Tactics (Longtchi et al. 2024; Longtchi and Xu 2024a, b; Montañez et al. 2020, 2023; Rodriguez and Xu 2022).

The relevance of cyber social engineering attacks to space cybersecurity is evidenced in (Ear et al. 2023), which showed that such attacks against the ground segment have been leveraged to attack satellites in a substantial number of cases. Although we have no access to information about whether the Intelsat incident was indeed enabled by a social engineering attack or not, this possibility cannot be ruled out.

Countermeasure 7 (For Invalidating Threat-Model Assumption 2.1) This countermeasure prevents an attacker from having access to the information about the Terrestrial Broadcast Towers that are targeted by the attacker, namely the frequency band used by the Terrestrial Broadcast Towers to communicate with Intelsat satellites.

4.4.2 Reducing Non-Resilience

Now we show how to employ countermeasures to disrupt individual attacks, whether effectively reducing non-resilience or not, and then how to select countermeasures to disrupt multiple attacks to effectively reduce system non-resilience.

Reducing Non-Resilience Incurred by Attack 1 (Flow 2) Attack 1 (more specifically its attack flow 2) can be disrupted by employing Countermeasures 4 (employing strong authentication to prevent an unauthorized user from using a legitimate terminal), 5 (a combination of technical and non-technical solutions to prevent an authorized user from becoming an insider threat), and 6 (employing countermeasures to prevent a user or an operator from being compromised by social engineering attacks and others). A defensive operation can employ all three countermeasures to invalidate Assumption 1.5 and thus the attack flow. The eliminating of attack flow 2 incurs, for notional convenience, an ∞-shot non-resilience against attack flow 2. However, attack flow 1 is viable, leading to a 2-shot non-resilience as $2 = \min(2, \infty)$. This means employing countermeasures 4–6 is only able to eliminate attack flow 2 of Attack 1, but not able to reduce the empirical system resilience as we still have $n_{empirical} = \min(2, 1, 3) = 1$.

Reducing Non-Resilience Incurred by Attack 1 (Flow 1) Attack flow 1 of Attack 1 can be disrupted by employing Countermeasures 1 (authenticating terminals to Intelsat communications satellites through cryptographic techniques), 2 (authenticating terminals to Intelsat communications satellites through PLA), 3 (preventing an attacker from identifying an available frequency channel), and 7 (preventing attacker from identifying the frequency band used by the Terrestrial Broadcast Towers). A defensive operation can employ (i) either Countermeasures 1 or 2 to invalidate Assumption 1.4 and (ii) Countermeasure 3 to invalidate Assumption 1.3, effectively disrupting the attack flow. Note that countermeasures 1 and 3, or 2 and 3, need to be taken together to disrupt the attack flow. This leads to an ∞-shot non-resilience with respect to attack flow 1. Without disrupting attack flow 2 of Attack 1, the non-resilience with respect to Attack 1 is still $1 = \min(\infty, 1)$, meaning that the system resilience is still $n_{empirical} = \min(1, 1, 3) = 1$.

Reducing Non-Resilience Incurred by Attack 2 Attack 2 can be disrupted by employing countermeasures 1 (authenticating Intelsat communications satellites to Terrestrial Broadcast Towers through cryptographic techniques), 2 (authenticating Intelsat communications satellites to Terrestrial Broadcast Towers through PLA), 3 (preventing an attacker from identifying an available frequency channel), and 7 (preventing attacker from identifying the frequency band used by the Terrestrial Broadcast Towers). A defensive operation can employ Countermeasures 1 or 2 to invalidate Assumption 2.3, employ Countermeasure 3 to invalidate Assumption 2.2, and Countermeasure 7 to invalidate Assumption 2.1. Note that countermeasures 1, 3, and 7, or 2, 3, and 7, need to be taken together to disrupt Attack 2. Eliminating Attack 2 leading to a ∞-shot non-resilience with respect to Attack 2. Without eliminating any (attack flow) of the other attacks, we have an empirical system non-resilience $n_{empirical} = \min(1, \infty, 3) = 1$.

Reducing Non-Resilience Incurred by Attack 3 Attack 3 can be disrupted by employing Countermeasures 1 (authenticating Teleports to Intelsat communications satellites), 2 (authenticating terminals to Intelsat communications satellites through PLA), 3 (preventing an attacker from identifying an available frequency channel), 5 (combining technical and nontechnical solutions to prevent an authorized user from becoming an insider threat), and 6 (protecting a user or operator from social engineering attacks). A defensive operation can employ Countermeasures 1 or 2 to invalidate Assumption 3.3, employ Countermeasure 3 to invalidate Assumption 3.2, and employ Countermeasures 5 and 6 to invalidate Assumption 3.1. Note that either Countermeasures 1, 3, 5, and 6, or Countermeasures 2, 3, 5, and 6, need to be taken together to disrupt the attack flow. This eliminates Attack 3 and leads to an ∞-shot non-resilience with respect to Attack 3. Without eliminating any (attack flow) of the other attacks, we have an empirical system non-resilience $n_{empirical} = \min(1, 1, \infty) = 1$.

Reducing Non-Resilience by Eliminating Multiple Attacks The preceding discussion shows that employing countermeasures to disrupt a single attack (or attack flow) may not be sufficient to effectively reduce non-resilience. This suggests that we may have to employ countermeasures to disrupt multiple attacks (or attack flows) to effectively reduce non-resilience. Recall that before employing any countermeasure we have $n_{empirical} = \min(\min(2, 1), 1, 3) = 1$. This means: (i) by employing countermeasures to disrupt Attack 1 via its attack flow 2 and Attack 2 leads to $n_{empirical} = \min(\min(2, \infty), \infty, 3) = 2$; (ii) by additionally employing countermeasures to disrupt Attack 1 via its attack flow 1 leads to$n_{empirical} = \min(\min(\infty, \infty), \infty, 3) = 3$; (iii) by additionally employing countermeasures to disrupt Attack 3 leads to $n_{empirical} = \min(\min(\infty, \infty), \infty, \infty) = \infty$. These are options for a decision-maker to choose.

5 Discussion

This study initiated a new direction in analyzing and reducing system non-resilience. Future research directions are abundant. First, it is important to build a systematic theory and design automated tools to compute the n-shots non-resilience of a system based on Definition 2. This may take many years. Thus, it is important to empirically measure non-resilience based on Definition 5. This means that these tools can compute the n-shots non-resilience with respect to each attack flow as described in Definition 3, ideally in an automated fashion. Thus, it is important to design automated tools to enumerate possible attacks via their attack flows. These attacks may include what already occurred in the real world and what could occur but have not occurred. The tools may or may not leverage human expertise, may leverage known vulnerabilities in the system (i.e., known to the defender), and my include various flavors of "what if" analysis such as the presence of hypothetical zero-day vulnerabilities (i.e., vulnerabilities that are known to the attacker but are unknown to the defender). These tools are different from existing tools that often aim to show, if not rigorously prove, that a system is secure (i.e., cannot be compromised). However, most if not all systems in the real world can be compromised in one way or another.

Second, the methodology presented above does not adequately accommodate the time dimension. This is true despite the use of attack flows, where a flow indicates the "precedence-success" relation of offensive cyber actions rather than the evolution of attack objective or attack capabilities. Thus, it is interesting to incorporate the analysis of non-resilience into the Cybersecurity Dynamics framework (Xu 2014a, 2019, 2020), which aims to analyze cybersecurity (broadly defined to include resilience Xu 2021) from a holistic perspective that includes the evolution of attacker capabilities and defender capabilities. This is important because resilience, to which non-resilience is a stepping-stone, is ultimately a time-dependent property (Xu 2021; Chen et al. 2021, 2022; Mireles et al. 2019; Pendleton et al. 2016; Cho et al. 2019; Chen et al. 2018; Du et al. 2018; Lin et al. 2019).

Third, we provide a strategy to apply countermeasures to reduce system non-resilience based on eliminating attacks that render a system less resilient. However, we do not consider the cost incurred by a counter measure. This matter is relevant because some countermeasures may be too costly for employment. Thus, future research needs to investigate cost-effective, if not optimal, countermeasure selection. Moreover, dynamic employment of countermeasures is key to resilience (Cho et al. 2019) and is an exciting research topic that has not received the due amount of attention.

Fourth, it is important to extend the methodology to accommodate the effect of human factors on cybersecurity (Longtchi and Xu 2024a, b; Longtchi et al. 2024; Montañez et al. 2020). This is because for instance, operators in the ground segment have been exploited to attack the space segment (Ear et al. 2023). In particular, it is interesting to quantify and differentiate the non-resilience incurred by human factors from the non-resilience incurred by technological factors.

Fifth, it is important to incorporate the results (e.g., refined methodology) into industrial standards, as shown by what we recently did (Ear et al. 2024a). This is particularly important as there is a dearth of methodologies and tools for analyzing space cybersecurity risks (Ear et al. 2024b, c), which are ongoing efforts and can incorporate the methodology proposed in the present chapter.

6 Conclusion

We have presented a methodology with accompanying metrics for quantifying system non-resilience, which can be seen as one stepping stone towards making systems resilient against attacks. We showed how to leverage the quantified non-resilience to select countermeasures to harden systems by reducing its non-resilience. We demonstrated the usefulness of the methodology through a case study on space non-resilience. Although the methodology, including the metrics and attack flows, is described in the context of cybersecurity, they can be adapted to other domains as demonstrated by the fact that one shot can represent, for instance, one earthquake. In terms of future research, Section 5 outlines a range of exciting research directions.

Acknowledgement This research was supported in part by NSF Grants #2308142 and #2115134 and Colorado State Bill 18-086.

References

Chen H, Cho J, Xu S (2018) Quantifying the security effectiveness of firewalls and DMZs. In: Proceeding of the HoTSoS'2018, pp 9:1–9:11

Chen H, Cam H, Xu S (2022) Quantifying cybersecurity effectiveness of dynamic network diversity. IEEE Trans Dependable Secure Comput

Chen Z, Li H, Ziheng Z, Lu M (2021) An approach to separate GNSS spoofing signals from authentic signals using relative positioning. In: Proceedings of the 34th international technical meeting of the Satellite Division of The Institute of Navigation (ION GNSS+ 2021), pp 3642–3652

Cho J-H, Xu S, Hurley PM, Mackay M, Benjamin T, Beaumont M (2019) Stram: measuring the trustworthiness of computer-based systems. ACM Comput Surv 51(6):1–47

Da G, Xu M, Xu S (2014) A new approach to modeling and analyzing security of networked systems. In: Proceeding of the HotSoS'14, pp 6:1–6:12

Dai W, Parker P, Jin H, Xu S (2012) Enhancing data trustworthiness via assured digital signing. IEEE TDSC 9(6):838–851

Dodis Y, Katz J, Xu S, Yung M (2002) Key-insulated public key cryptosystems. In: Proceeding of the EUROCRYPT, pp 65–82

Dodis Y, Katz J, Xu S, Yung M (2003) Strong key-insulated signature schemes. In: Public Key Cryptography (PKC'03), pp 130–144

Dodis Y, Luo W, Xu S, Yung M (2012) Key-insulated symmetric key cryptography and mitigating attacks against cryptographic cloud software. In: Proceeding of the ASIACCS'12

Du P, Sun Z, Chen H, Cho JH, Xu S (2018) Statistical estimation of malware detection metrics in the absence of ground truth. IEEE T-IFS 13(12):2965–2980

Ear E, Remy JL, Feffer A, Xu S (2023) Characterizing cyber attacks against space systems with missing data: Framework and case study. In: 2023 IEEE conference on Communications and Network Security (CNS). IEEE, pp 1–9

Ear E, Bailey B, Xu S (2024a) Towards principled risk scores for space cyber risk management. CoRR abs/2402.02635. https://doi.org/10.48550/arXiv.2402.02635

Ear E, Bailey B, Xu S (2024b) Towards principled space cyber risk management: Characterizing notional risk scores. In: Manuscript under review

Ear E, Bailey B, Xu S (2024c) Towards principled space cyber risk management: defining desired properties. In: Manuscript under review

Falletti E, Falco G, Nicola M et al (2021) Performance analysis of the dispersion of double differences algorithm to detect single-source GNSS spoofing. IEEE Trans Aerosp Electron Syst 57(5):2674–2688

Fang X, Xu M, Xu S, Zhao P (2019) A deep learning framework for predicting cyber attacks rates. EURASIP J Inf Secur 2019:5

Fang Z, Xu M, Xu S, Hu T (2021) A framework for predicting data breach risk: leveraging dependence to cope with sparsity. IEEE T-IFS 16:2186–2201

Fritz J (2013) Satellite hacking: a guide for the perplexed. Cult Mandala 10(1):5906

Gross JN, Kilic C, Humphreys TE (2019) Maximum-likelihood power-distortion monitoring for GNSS-signal authentication. IEEE Trans Aerosp Electron Syst 55(1):469–475

Han Y, Lu W, Xu S (2014) Characterizing the power of moving target defense via cyber epidemic dynamics. In: HotSoS, pp 1–12

Han Y, Lu W, Xu S (2021) Preventive and reactive cyber defense dynamics with ergodic time-dependent parameters is globally attractive. IEEE TNSE 8(3):2517–2532

Harrison K, Xu S (2007) Protecting cryptographic keys from memory disclosures. In: IEEE/IFIP DSN'07, pp 137–143

Humphreys TE (2013) Detection strategy for cryptographic GNSS antispoofing. IEEE Trans Aerosp Electron Syst 49(2):1073–1090

Intelsat (2024) Intelsat Coverage Map. URL: https://www.intelsat.com/fleetmaps/

Jansen K, Tippenhauer NO, Pöpper C (2016) Multi-receiver GPS spoofing detection: error models and realization. Association for Computing Machinery

Jovanovic A, Botteron C, Farin'e P-A (2014) Multi-test detection and protection algorithm against spoofing attacks on GNSS receivers. In: 2014 IEEE/ION PLANS 2014. IEEE

Li X, Parker P, Xu S (2011) A stochastic model for quantitative security analyses of networked systems. IEEE TDSC 8(1):28–43

Li D, Li Q, Ye Y, Xu S (2019) Enhancing robustness of deep neural networks against adversarial malware samples: principles, framework, and AICS'2019 challenge. In: AAAI-2019 Workshop on Artificial Intelligence for Cyber Security (AICS'2019)

Li D, Qiu T, Chen S, Li Q, Xu S (2021a) Can we leverage predictive uncertainty to detect dataset shift and adversarial examples in android malware detection? In: The 2021 Annual Computer Security Application Conference (ACSAC)

Li D, Li Q, Ye Y, Xu S (2021b) A framework for enhancing deep neural networks against adversarial malware. IEEE Trans Netw Sci Eng 8(1):736–750

Li D, Li Q, Ye YF, Xu S (2023) Arms race in adversarial malware detection: A survey. ACM Comput Surv 55(1):1

Li D, Cui S, Li Y, Xu J, Xiao F, Xu S (2024) Pad: towards principled adversarial malware detection against evasion attacks. IEEE Trans Dependable Secure Comput 21(2):920–936

Lin Z, Lu W, Xu S (2019) Unified preventive and reactive cyber defense dynamics is still globally convergent. IEEE/ACM ToN 27(3):1098–1111

Liu S, Cheng X, Yang H, Shu Y, Weng X, Guo P, Zeng KC, Wang G, Yang Y (2021) Stars can tell: a robust method to defend against {GPS} spoofing attacks using off-the-shelf chipset. In: 30th USENIX Security Symposium (USENIX Security 21)

Lo S, Chen YH, Jain H, Enge P (2018) Robust GNSS spoof detection using direction of arrival: Methods and practice. In: Proceedings of the 31st ION GNSS+ 2018

Longtchi T, Xu S (2024a) Characterizing the evolution of psychological factors exploited by malicious emails. In: Proceedings of international conference on Science of Cyber Security (SciSec'2024)

Longtchi T, Xu S (2024b) Characterizing the evolution of psychological tactics and techniques exploited by malicious emails. In: Proceedings of international conference on Science of Cyber Security (SciSec'2024)

Longtchi T, Rodriguez RM, Al-Shawaf L, Atyabi A, Xu S (2024) Internet-based social engineering psychology, attacks, and defenses: a survey. Proceeding of the IEEE 112(3):210–246

Loscocco P, Smalley S, Muckelbauer P, Taylor R, Turner S, Farrell J (1998) The inevitability of failure: The flawed assumption of security in modern computing environments. In: Proceedings of the 21st National Information Systems Security Conference (NISSC'98)

Lu W, Xu S, Yi X (2013) Optimizing active cyber defense dynamics. In: Proceeding of the GameSec'13, pp 206–225

Lynch N (1996) Distributed algorithms. Morgan Kaufmann

Mireles JD, Cho J, Xu S (2016) Extracting attack narratives from traffic datasets. In: International conference on Cyber Conflict (CyCon)

Mireles J, Ficke E, Cho J, Hurley P, Xu S (2019) Metrics towards measuring cyber agility. IEEE Trans Inf Forensics Secur 14(12):3217–3232

Montañez R, Golob E, Xu S (2020) Human cognition through the lens of social engineering cyberattacks. Front Psychol 11:1755

Montañez R, Atyabi A, Xu S (2022) Book chapter in "Cybersecurity and cognitive science", Elsevier, chapter Social engineering attacks and defenses in the physical world vs. cyberspace: a contrast study

Montañez R, Longtchi T, Gwartney K, Ear E, Azari D, Kelley C, Xu S (2023) Quantifying psychological sophistication of malicious emails. In: Proceedings of International Conference on Science of Cyber Security (SciSec'2023), pp 319–331

Naor D, Naor M, Lotspiech J (2001) Revocation and tracing schemes for stateless receivers. In: Kilian J (ed) Proceeding of the CRYPTO 2001, Lecture Notes in Computer Science, vol 2139. Springer, pp 41–62

Paarporn K, Brown PN, Xu S (2023) Analysis of contagion dynamics with active cyber defenders. In: 62nd IEEE conference on decision and control, CDC 2023, Singapore, December 13–15, 2023. IEEE, pp 4990–4995

Pendleton M, Garcia-Lebron R, Cho J-H, Xu S (2016) A survey on systems security metrics. ACM Comput Surv 49(4):62:1–62:35

Peng C, Xu M, Xu S, Hu T (2017) Modeling and predicting extreme cyber attack rates via marked point processes. J Appl Stat 44(14):2534–2563

Peng C, Xu M, Xu S, Hu T (2018) Modeling multivariate cybersecurity risks. J Appl Stat 0(0):1–23

Phillips C, Swiler L (1998) A graph-based system for network vulnerability analysis. In: Proceedings of the 1998 workshop on New Security Paradigms (NSPW'98). ACM Press, New York, pp 71–79

Psiaki ML, O'Hanlon BW, Bhatti JA, Shepard DP, Humphreys TE (2013) GPS spoofing detection via dual-receiver correlation of military signals. IEEE Trans Aerosp Electron Syst 49(4): 2250–2267

Ranganathan A, Olafsd'ottir H, Capkun S (2016) Spree: a spoofing' resistant GPS receiver. In: Proceedings of the 22nd annual international conference on Mobile Computing and Networking

Remy JL, Ear E, Xu S (2024) Manuscript to be submitted for review. In: Sok: space infrastructures model, missions, vulnerabilities, attacks, and defenses

Reuters (2024) Intelsat stops Tamil tiger satellite broadcasts. https://www.reuters.com/article/idUSN24434540/

Rodriguez RM, Xu S (2022) Cyber social engineering kill chain. In: Proceedings of international conference on Science of Cyber Security (SciSec'2022), pp 487–504

Ross RS (2018) Risk management framework for information systems and organizations: a system life cycle approach for security and privacy

Rothmaier F, Taleghani L, Chen Y-H, Lo S, Phelts E, Walter T (2021a) GNSS spoofing detection through metric combinations: Calibration and application of a general framework. In: Proceedings of the 34th international technical meeting of the Satellite Division of The Institute of Navigation (ION GNSS+ 2021)

Rothmaier F, Chen Y-H, Lo S, Walter T (2021b) A framework for GNSS spoofing detection through combinations of metrics. IEEE Trans Aerosp Electron Syst 57(6):3633

Rothmaier F, Chen Y-H, Lo S, Blanch J, Walter T (2021c) Providing continuity and integrity in the presence of GNSS spoofing. In: Proceedings of the 34th international technical meeting of the Satellite Division of The Institute of Navigation (ION GNSS+ 2021)
Rustamov A, Gogoi N, Minetto A, Dovis F (2020) GNSS anti-spoofing defense based on cooperative positioning. In: Proceedings of the 33rd international technical meeting of the Satellite Division of The Institute of Navigation (ION GNSS+ 2020)
Schmidt E, Gatsis N, Akopian D (2020) A GPS spoofing detection and classification correlator-based technique using the lasso. IEEE Trans Aerosp Electron Syst 56(6):4224–4237
Schneier B (1999) Attack trees: modeling security threats. Dr. Dobbs J
spacenews.com (n.d.) Intelsat Vows to Stop Piracy by Sri Lanka Separatist Group. https://spacenews.com/intelsat-vows-stop-piracy-sri-lankaseparatist-group/
Sun Z, Xu M, Schweitzer K, Bateman R, Kott A, Xu S (2023) Cyber attacks against enterprise networks: Characterization, modeling and forecasting. In: Proceeding of SciSec'2023
Wesson KD, Gross JN, Humphreys TE, Evans BL (2017) GNSS signal authentication via power and distortion monitoring. IEEE Trans Aerosp Electron Syst 55:469–475
Wong C, Gouda M, Lam S (2000) Secure group communication using key graphs. IEEE/ACM Trans Netw (Preliminary version in SIGCOMM 1998) 8:16–30
Xu S (2007) On the security of group communication schemes. J Comput Secur 15(1):129–169
Xu S (2014a) Cybersecurity dynamics. In: Proceedings of the symposium on the Science of Security (HotSoS'14), pp 14:1–14:2
Xu S (2014b) Emergent behavior in cybersecurity. In: Proceeding of the HotSoS, pp 13:1–13:2
Xu S (2019) Cybersecurity dynamics: a foundation for the science of cybersecurity. In: Proactive and dynamic network defense, vol 74. Springer, pp 1–31
Xu S (2020) The cybersecurity dynamics way of thinking and landscape (invited paper). In: ACM workshop on moving target defense
Xu S (2021) Sarr: a cybersecurity metrics and quantification framework. In: Third international conference on Science of Cyber Security (SciSec'2021), pp 3–17
Xu M, Xu S (2012) An extended stochastic model for quantitative security analysis of networked systems. Internet Math 8(3):288–320
Xu S, Yung M (2009) Expecting the unexpected: towards robust credential infrastructure. In: International Conference on Financial Cryptography and Data Security (FC). Springer, pp 201–221
Xu S, Li X, Parker P (2008) Exploiting social networks for threshold signing: attack-resilience-vs. availability. In: ACM Symposium on Information, Computer and Communications Security (ASIACCS'08), pp 325–336
Xu S, Li X, Parker T, Wang X (2011) Exploiting trust-based social networks for distributed protection of sensitive data. IEEE T-IFS 6(1):39–52
Xu S, Lu W, Xu L (2012a) Push- and pull-based epidemic spreading in networks: Thresholds and deeper insights. ACM Trans 7(3):1–26
Xu S, Lu W, Zhan Z (2012b) A stochastic model of multivirus dynamics. IEEE Trans Dependable Secure Comput 9(1):30–45
Xu S, Lu W, Xu L, Zhan Z (2014) Adaptive epidemic dynamics in networks: thresholds and control. ACM Trans 8(4):1–19
Xu M, Da G, Xu S (2015a) Cyber epidemic models with dependences. Internet Math 11(1):62–92
Xu S, Lu W, Li H (2015b) A stochastic model of active cyber defense dynamics. Internet Math 11(1):23–61
Xu M, Hua L, Xu S (2017) A vine copula model for predicting the effectiveness of cyber defense early-warning. Technometrics 59(4):508–520
Xu M, Schweitzer KM, Bateman RM, Xu S (2018) Modeling and predicting cyber hacking breaches. IEEE T-IFS 13(11):2856–2871
Zeien L, Chang C, Ear E, Xu S (2024) Characterizing advanced persistent threats through the lens of cyber attack flows. Military Cyber Affairs 7(1):5

Zhan Z, Xu M, Xu S (2013) Characterizing honeypot-captured cyber attacks: Statistical framework and case study. IEEE T-IFS 8(11):1775–1789

Zhan Z, Xu M, Xu S (2015) Predicting cyber attack rates with extreme values. IEEE Trans Inf Foren Secur 10(8):1666–1677

Zheng R, Lu W, Xu S (2015) Active cyber defense dynamics exhibiting rich phenomena. In: Proceedings of the HotSoS

Zheng R, Lu W, Xu S (2018) Preventive and reactive cyber defense dynamics is globally stable. IEEE TNSE 5(2):156–170

Cyber-Social and Cyber-Physical Interactions

Navigating Socio-technical Influences Upon Cyber Resilience Adoption

Travis D. Howard and José de Arimatéia da Cruz

1 Introduction: A Call for National Cyber Resilience

Cyberspace has become a battleground for nation-states and state-sanctioned aggressors. Intelligence testimony to the U.S. Congress in recent years continues to report on a rise in cyber-enabled espionage, intrusion, information theft, and even seeding malware into cyber-physical, industrial control, and critical infrastructure systems that the nation's citizenry depends on (Office of the Director of National Intelligence 2023).[1] Seemingly in step with the increase in the cyber threat landscape, strategic direction at the U.S. federal level has pivoted decisively toward cyber resiliency as a means to reduce the impact of an attack on critical infrastructure and impose a cost on adversaries attempting it (Office of the President of the United States 2023).[2]

The authors predict that this call for cyber resilience at a national level will continue and become the dominant strategy for dealing with nation-state threats. This strategic shift represents not a "shield" against attacks, stopping them outright, but rather a "parry and riposte" in which the defender deflects the part of the attack intending to do damage, leaving critical systems intact and leaving open the option for an effective counterattack. In naval combat parlance, legendary strategist and

[1] Office of the Director of National Intelligence. Feb 6, 2023. "Annual Threat Assessment of the U.S. Intelligence Community." https://www.intelligence.gov/annual-threat-assessment

[2] Office of the President of the United States. Mar 1, 2023. "National Cybersecurity Strategy." https://www.whitehouse.gov/wp-content/uploads/2023/03/National-Cybersecurity-Strategy-2023.pdf

T. D. Howard (✉)
United States Navy (Retired), Washington, DC, USA

J. de Arimatéia da Cruz
Georgia Southern University, Savannah, GA, USA

U.S. Army War College, Carlisle, PA, USA
e-mail: jose.a.dacruz.civ@army.mil

© The Author(s), under exclusive license to Springer Nature Switzerland AG 2025

I. Linkov, A. Kott (eds.), *Cyber Resilience: Applied Perspectives*, Risk, Systems and Decisions, https://doi.org/10.1007/978-3-031-90109-6_5

fleet tactics author Captain Wayne P. Hughes, U.S. Navy (retired), advocated for victory through "fire effectively first" (Vencill 2019).[3] In this case, it would appear that the national cyber strategy to harness resiliency while simultaneously wielding the combat power of the Department of Defense's U.S. Cyber Command would make Captain Hughes proud.

Systems can be engineered and configured for redundancy and rapid fail-over, with segmentation and all the trimmings of engineered resiliency. Yet, organizations and people must adopt resiliency differently through adaptation and leadership. Only when technical and social factors are combined and harmonized can organizations achieve the cyber resiliency needed to sustain operations in this ever-changing digital battleground upon which every internet-connected public and private sector organization stands.

This chapter presents recommendations for overcoming and adapting to leadership and socio-technical challenges in adopting cyber resiliency as an organizational innovation through a case study in recent history that highlights the socio-technical nature of adverse cyber events: the 2021 Colonial Pipeline ransomware event. The authors then present a 2023 study exploring limiting and enabling factors toward cyber resiliency adoption in organizations based on the Diffusion of Innovations Theory. The chapter concludes with recommendations for practitioners based on deductive lessons from the 2021 Colonial Pipeline cyber incident and findings from research.

2 Review of the Colonial Pipeline Cyber Event: A Case Study in Socio-technical Cyber Resiliency

On May 7th, 2021, the American public learned that the Colonial Pipeline—responsible for 45 percent of the fuel supply for the East Coast of the United States, had shut down due to a cyber-attack (Marinos and Gordon 2021).[4] Ransomware inserted by a cyber-crime group known as DarkSide disrupted the company's business network; the operational technology (OT) that ran the 5500-mile-long pipeline was unaffected. The disruption lasted for 6 days, and on May 13th, the pipeline resumed delivery to all affected markets, yet the damage was already apparent. The public panic created by fears of fuel shortages caused a supply rush, creating a market frenzy and proof that disruptions to critical infrastructure would yield these results in the future (Marinos and Gordon 2021).

---

[3]Taylor Vencill 2019. "In Memoriam: Legendary Strategist and Fleet Tactics Author Wayne P. Hughes." Naval Postgraduate School. https://nps.edu/-/in-memorium-legendary-strategist-and-fleet-tactics-author-wayne-p-hughes

[4]Nick Marinos and Leslie V. Gordon. 2021. "Colonial Pipeline Cyberattack Highlights Need for Better Federal and Private-Sector Preparedness (infographic)." U.S. Government Accountability Office, WatchBlog. https://www.gao.gov/blog/colonial-pipeline-cyberattack-highlights-need-better-federal-and-private-sector-preparedness-infographic

Two years later, the CEO of Southern Company (a major pipeline operator in the United States, but not the owner of Colonial Pipeline) and the director of the Cybersecurity and Infrastructure Security Agency (CISA)—an agency within the U.S. Department of Homeland Security—issued a joint publication on the progress made as well as a call to action on the work still before them. The CISA release by then-CISA Director Jen Easterly and Sothern Company CEO Tim Fanning (2023) described working groups and public-private partnership efforts that stood up after the incident, including the Joint Cyber Defense Collaborative (JCDC).[5] These groups and initiatives were created to "share insights and information in real-time to understand threats and drive down risk to the nation" (para 2). However, the most impactful statement regarding resiliency came at the end of the release: "We cannot completely prevent attacks from happening, but we can minimize their impact by building resilience into our infrastructure and our society."

Scholarly studies to date suggest such resilience continues to elude us. A 2023 study by Greubel, Andres, and Hennecke found articles and media surrounding the Colonial Pipeline incident lacking details and preventative advice that would help students, experts, and the general public understand the need for such resilience, let alone how to achieve it at a technical level.[6] Lubin (2023) analyzed legislative action following the incident and noted a need for more legislative response in the form of reform, regulation, or omnibus legislation that would improve critical infrastructure cyber resiliency.[7] Indeed, Lubin (2023) offered this stark analysis: "The kneejerk litigatory, regulator, and legislative response that the Colonial Pipeline ransomware generated were casebook examples of all wrong with our existing cybersecurity law and policy ecosystem."

The lack of technical details and post-mortem analysis, due in no small part to the pipeline's private-sector ownership, prevents cybersecurity professionals from using the attack as a case for improving resiliency within their corporate networks. Two years after the Colonial Pipeline attack, no correct publicly available data has been collected or disseminated to show that, as a society and a community of cybersecurity experts, we have made progress in building cyber resiliency into critical infrastructure beyond creating collaborative groups to talk about it. In short, lessons learned from the Colonial Pipeline incident are closed and held only for select groups within the public and private sectors. Has Colonial Pipeline improved its socio-technical cyber resiliency? We will have to take their word for it.

---

[5] Jen Easterly and Tom Fanning. 2023. "The Attack on Colonial Pipeline: What We've Learned & What We've Done Over the Past Two Years." Cybersecurity & Infrastructure Security Agency, Blog. https://www.cisa.gov/news-events/news/attack-colonial-pipeline-what-weve-learned-what-weve-done-over-past-two-years

[6] André Greubel, Daniela Andres, and Martin Hennecke. 2023. "Analyzing Reporting on Ransomware Incidents: A Case Study." Social Sciences (2076–0760) 12 (5): 265. doi:10.3390/socsci12050265.

[7] Asaf Lubin 2023. "Cyber Plungers: Colonial Pipeline and the Case for an Omnibus Cybersecurity Legislation." Georgia Law Review 57 (4): 1605–32. https://search-ebscohost-com.captechu.idm.oclc.org/login.aspx?direct=true&db=aph&AN=169978264&site=eds-live

3 Understanding Socio-technical Influence Factors in Cyber Resiliency Adoption

Howard (2023), studying on influence factors affecting cyber resiliency adoption, presented nine influences identified and categorized through an extensive literature review.[8] The nine identified influence categories are described in Table 1. The researcher posited that these categories represent socio-technical influences organizations face within the innovation-decision process described within the Diffusion of Innovation (DOI) Theory (Rodgers 2003).[9]

Emmett Rodgers, first in 1962 then refined through several revisions until 2003, developed the DOI Theory to explain how innovation spreads through a social system. It is best known for the "bell curve" model, which describes the momentum of innovation through consumer groups known as innovators, early adopters, early majority, late majority, and laggards. Perhaps lesser known, Rodgers (2003) also postulated vital elements and characteristics of an innovative organization and the process by which organizations decide whether to adopt or reject an innovation, known as the innovation-decision process model (IDPM).

The IDPM is a five-step decision-making process (Fig. 1). Ryan and Gross identified the concept in 1943 in their diffusion study. Rodgers's seminal work, *Diffusion of Innovations 5th edition* codified the concept into the literature. Before the first step, conditions influence the process, such as a felt need or problem to be solved, the organization's overall innovation characteristics, and the social system's norms. Then, the knowledge phase occurs when the organization gathers awareness and information about the innovation, informed and influenced by the prior conditions. The organization is then persuaded to analyze how the innovation would benefit the whole (persuasion phase). Once the decision has been made, it is either adopted or rejected by the innovation (decision phase). In the implementation phase, if adopted, the innovation is implemented, and a confirmation phase determines over time if the decision made was correct for the organization (Rodgers 2003).

The 2023 study found that technical factors—such as implementation frameworks, network topologies, software and hardware designs, and available architecture—enabled overall (85.51%) versus limited (78.26%) adoption. Organizational influences such as internal policies, procedures, and leadership were high in percentile and found near-equally enabling (84.05%) as limiting (85.51%). Resources and funding were, predictably, overall limiting (85.95%) vice enabling (73.91%). These findings show that technical advances enable cyber resiliency adoption in organizations, while a lack of resources limits adoption and sustainment. Internal policies and leadership were significant influences and could limit or enable cyber resiliency adoption depending on the organization's characteristics (Howard 2023).

[8] Travis D. Howard. 2023. "Understanding Influence Factors in Cyber Resiliency Adoption: A Mixed Methods Study of Cybersecurity Leaders." PhD diss. Capitol Technology University.

[9] Emmett M. Rodgers. 2003. "Diffusion of Innovations" (5th Ed.). Free Press.

Table 1 Socio-technical influence factors affecting cyber resiliency adoption

Influence factor	Examples
Technical	Network topology, implementation frameworks, system engineering and design concepts, system security architecture
Cultural	Workplace culture and norms, cybersecurity awareness and agency, cultural adaptability to change, innovativeness characteristics
Organizational policies and leadership	Governance models, leadership and executive level support, policies, procedures, standards, guidelines
Workforce and skills	Qualifications, engagement skills and subject matter expertise, roles and responsibilities, team structures
Knowledge management and information access	Access to information, ability to categorize, organize, and share knowledge
Industry and competitiveness	Competitive advantage, industry infrastructure and characteristics, business-to-business factors, business-to-consumer factors, public or private ownership
Vendor and third-party support	Managed cloud providers, external consultancies, staffing agencies, software and hardware vendors, patching and update support for installed products
Legal and regulatory	Privacy and civil liberty issues, industry specific laws and regulations, organizational legal concerns to prevent lawsuits
Resources	Access to: financial resources, human resources, technological and material resources, rapid access to tap resources for innovations, ability to sustain funded efforts

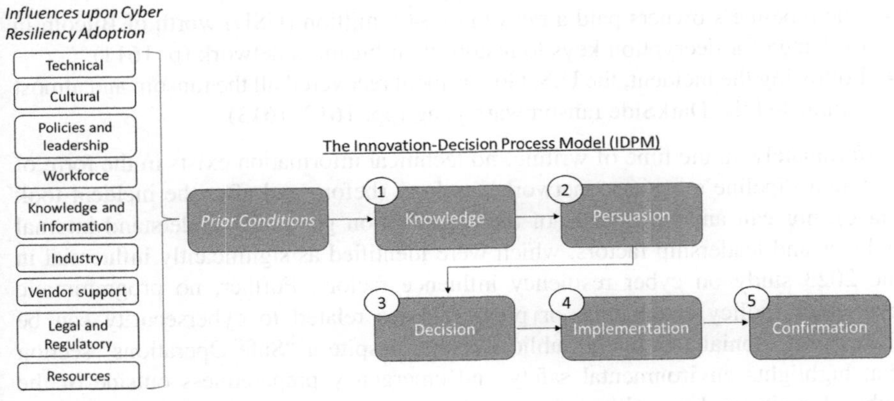

Fig. 1 The nine cyber resiliency influence categories feed the *prior conditions* pre-requisite to the IDPM (Left) and Socio-Technical Influences upon the IDPM for Cyber Resiliency Adoption (Center)

Participants in the 2023 study noted significant influences around workplace culture, organizational policies, practices, and leadership buy-in that led to either limiting or enabling factors towards cyber resiliency adoption or sustainment of the adopted capability as a part of cyber defense. Organizational cyber leaders, such as the Chief Information Security Officer or senior-level managers and directors of cybersecurity programs, noted that strong leadership on their part coupled with executive-level championship was essential to success. Cultural barriers around resistance to change were experienced, and some participants experienced headwinds in the form of internal politics and a lack of clear prioritization that led to funding shortfalls (Howard 2023).

4 Deconstructing the Influence Factors in the Colonial Pipeline Cyber Incident

While little actual data is publicly available to researchers in analyzing lessons learned from the 2021 Colonial Pipeline cyber incident, deductive reasoning combined with what information is available could allow practitioners and researchers alike to understand the possible socio-technical influence factors affecting cyber resiliency adoption in this case. One must first examine what is known, thanks to an excellent summation of events by the previously referenced Asaf Lubin (2023) in the *Georgia Law Review*:

1. The DarkSide ransomware attack affected the Colonial Pipeline business network and did not affect the pipeline's industrial control systems (ICS) (pp. 1609–1610).
2. The pipeline ceased operations due to an inability to perform business functions such as delivery tracking and billing (pp. 1609–1610).
3. The pipeline's owners paid a ransom of $4.5 million (USD) worth of Bitcoin in exchange for decryption keys to restore their business network (p. 1611).
4. Following the incident, the U.S. Government recovered all the ransom and almost eliminated the DarkSide ransomware gang (pp. 1612–1613).

Unfortunately, at the time of writing, no technical information exists in the form of Colonial Pipeline's business network topology (before and after the incident took place), nor can anyone outside of the organization genuinely understand internal policies and leadership factors, which were identified as significantly influential in the 2023 study on cyber resiliency influence factors. Further, no programmatic overviews, policy statements, or press releases related to cybersecurity can be found on Colonial Pipeline's public website, despite a "Safe Operations" section that highlights environmental safety and emergency preparedness outside of the cyber domain (such as spills and contamination).

Despite an unfortunate lack of publicly available data, several deductions can be made. Before the incident, many factors that influenced the adoption of cyber resiliency in Colonial Pipeline's business network were low to moderate. Technical aspects such as the suitability of the network topology to resiliency architecture and the general availability of resilient system security engineering would have appeared moderate. Likewise, the availability of knowledge and information on cyber resiliency as an accepted systems engineering practice and implementation frameworks

Table 2 Socio-technical factors affecting cyber resiliency adoption on Colonial Pipeline *before* the 2021 ransomware incident

Influence category	Possible level of influence favoring adoption
Technical	Moderate
Cultural	Low
Org Policies and Leadership	Low
Workforce	Low
Knowledge and Information (availability)	Moderate
Industry	Low
Vendor Support	Moderate
Legal and Regulatory	Low
Resources and Funding	Low

such as the U.S. National Institute for Standards and Technology (NIST) special publication (SP) 800–160 Volume 2 could be considered moderately influential. Finally, vendor support was available through commercial procurement to support cyber resiliency adoption, indicating at least a moderate level of influence favoring adoption. Based on the phenomenological results of the 2023 study on influences upon cyber resilience adoption, one might deduce a low level of influence favoring cyber resiliency adoption within Colonial Pipeline for cultural, organizational, industry, legal, and resource categories—indicating the demand signal for such adoption was not present before the incident; otherwise, at least rudimentary procedures would have been in place to operate the business side of Colonial Pipeline without networked software and hardware to support. Table 2 summarizes levels of influence favoring cyber resiliency adoption before the 2021 ransomware incident.

Post-incident, social pressures were exerted on Colonial Pipeline and the U.-S. Government to prevent such an incident. The 2021 ransomware incident sparked several journal, scholarly, and media articles on the subject, and the U.-S. Government formed several working groups and public-private collaboration initiatives as a result (Easterly and Fanning 2023). Technical, knowledge and information, and vendor support remained moderately influential as little had changed. Yet the social factors likely increased, particularly organizational policies and leadership (indicating a desire to change policies and procedures to prevent a recurrence), industry (pressure from other pipeline operators to understand what happened to mitigate their vulnerabilities), and regulatory influence in the form of a significant amount of government "attention" (despite any actionable legislation that resulted from it). Resources were made available to mitigate vulnerabilities to the business network and implement changes to the company's overall emergency preparedness for cybersecurity. Table 3 summarizes where influence factors changed post-incident.

Did Colonial Pipeline improve its cyber resiliency due to what it may have learned from this incident? Had a manual method to track pipeline deliveries and conduct billing been implemented, could the pipeline have remained operational through the incident response process? Only Colonial Pipeline can definitively

Table 3 Socio-technical factors affecting cyber resiliency adoption on Colonial Pipeline *after* the 2021 ransomware incident

Influence category	Possible level of influence favoring adoption
Technical	Moderate
Cultural	Moderate
Org policies and leadership	Moderate to High
Workforce	Low
Knowledge and information (availability)	Moderate
Industry	Moderate
Vendor support	Moderate
Legal and regulatory	Moderate to High
Resources and funding	Moderate

answer such questions; presented in this chapter are the authors' estimations and deductions based on limited publicly available data and by extrapolating results from the 2023 study on influences upon cyber resiliency adoption. Yet this thought experiment yields some interesting observations that many cybersecurity practitioners already know to be true: nothing spurs cybersecurity actions better than a significant cyber incident. One can see how the IDPM for Colonial Pipeline could have been influenced when mapped to socio-technical influence categories. The challenge to practitioners, therefore, must be to affect these influence factors in favor of resiliency adoption before such an incident occurs.

5 Recommendations for Practitioners

Deductive analysis of the 2021 Colonial Pipeline ransomware incident using the socio-technical influences upon the innovation-decision process model allows the authors to present recommendations along the probable five influence factors that may have increased the likelihood of resiliency adoption for Colonial Pipeline post-incident: cultural, organizational, industry vertical, legal, and regulatory, and resources.

Cultural Respondents in the 2023 study on influences upon cyber resiliency adoption indicated that cultural resistance to change was among the most limiting factors (Howard 2023). Adopting resiliency practices that change how the workforce operates—such as training on manual processes or running drills that involve suspending automation to test the effectiveness of resiliency protocols—will be a tough sell to a resistant workforce. Overcoming this influence factor as a limiter can be twofold. First, concentrate on automated resiliency methods, such as automated fail-over and network segmentation with single sign-on (SSO) capabilities, that make it easy for the workforce to adopt or bypass cultural adoption altogether. Secondly, when techniques and procedures must be changed to incorporate resiliency, do so over a more extended period, emphasizing over-communication to allow the workforce time to absorb information and adapt to change.

Organizational Policies and Leadership While understanding the technical aspects of their cyber defense programs is essential to the job, organizational cyber leaders must now shift to understanding the business at a deeper level to align with priorities and "tell the story" to executive leadership and the board of directors. Doing so generates executive buy-in essential for implementing new capabilities affecting the entire organization, such as cyber resiliency adoption. Leaders must communicate resiliency as a "whole of company" solution to vulnerabilities that, if or when exploited, would cause significant damage—not just a purely technical solution that can be implemented "behind the scenes." Respondents in the 2023 study on influences upon cyber resiliency adoption noted that executive championship was essential to successfully adopting cyber resilient capabilities (Howard 2023). Cyber leaders should identify and ally with other influential executives who can sway decision-making or help communicate the necessity of resiliency measures, such as the Chief Risk Officer or Chief Operating Officer. Daily revenue loss due to a cyber incident can be calculated, which can bring the Chief Financial Officer around to support these capabilities as well.

Industry Influences from the industry vertical or market segment can be complex to understand, let alone be predicted. Business strategies are crafted to compete within the market segment the organization finds itself within, or in the case of the public sector, better serve the population. As a cyber leader struggles to understand influence factors affecting the industry, they find that scholarship and networking are required. For the former, adopting an always-learning mindset enables leaders to challenge their thinking in favor of what innovations are possible in the market segment—not just technical but social to include different ways to examine problems or perform tasks. For the latter, many industries host conferences, summits, and working groups that include cybersecurity as a topic, and leaders should participate directly or send representation. Furthermore, cyber leader must also implement resilience-by-design approaches that "enhance the ability of critical infrastructure to prepare for, adapt to, and recover from changing conditions presented by new and emerging threats and hazards" (The White House 2024, p. 22).[10] This ensures that industry-affecting innovations are known and issues affecting the industry can be understood and acted upon before they become legal or regulatory headaches.

Legal and Regulatory Speaking of headaches, nothing brings one on faster than a new law or regulatory requirement that forces change upon an organization. While "compliance" is often regarded as a "dirty word," cyber leaders have learned to embrace it to generate the adoption of cybersecurity and cyber resiliency capabilities. As such, cyber leaders are advised to take full advantage of changes to laws and regulations that involve cyber resiliency to increase their adoption rate and generate

[10]The White House. April 30, 2024. "National Security Memorandum on Critical Infrastructure Security and Resilience," available at National Security Memorandum on Critical Infrastructure Security and Resilience | The White House. Accessed May 2, 2024.

resources in favor of implementing higher levels of resiliency. However, leaders are cautioned not to over-rely on such methods lest they backfire when laws change, or regulations are rolled back depending on the socio-political landscape.

Resources Consider again the case of the 2021 Colonial Pipeline cyber incident: while much government attention was received, it did not result in regulatory action or legislation (Lubin 2023). Suppose a leader were to rely on the threat of legal or regulatory action alone; in that case, they might experience a "rubber band" effect where early support for adoption is achievable only to wane in the "confirmation" phase of the IDPM. A measured approach is therefore recommended, balancing the call to action (and funding) that laws and regulations create with sustainable and affordable resiliency capabilities that can endure long after the initial surge of support has passed.

6 Conclusion

True organizational cyber resiliency considers not just technical but social factors, especially considering that fact the "in the 21st century, the United States will rely on new sources of energy, modes of transportation, and an increasingly interconnected and interdependent economy" (The White House 2024, p. 2). Socio-technical influences upon cyber resiliency adoption must be understood if such resiliency is to be achieved, particularly in critical infrastructure, as called for in national strategy and policy circles. This is not just a government problem to solve; every organization is affected by adverse cyber events, and when critical infrastructure is threatened, it affects society as a whole and—depending on severity—has repercussions in the global economy. It is up to cybersecurity leaders in every organization to take charge and ensure their organization can withstand, adapt to, and recover from adverse cyber events.

What ties these influence factors together is strong leadership on the part of the senior-most cybersecurity executive (Chief Information Security Officer-CISO) that communicates a clear vision and roadmap for adoption, provides direction to both internal staff and external stakeholders, but perhaps most importantly, inspires others to see what they see: an organization capable of not only defending against attacks but continuing essential functions despite them; to do so stop cyber attackers before they carry out their nefarious activities and achieve their ultimate goal.

References

Easterly J, Fanning T (2023) The attack on colonial pipeline: what we've learned & what we've done over the past two years. Cybersecurity & Infrastructure Security Agency, Blog. https://www.cisa.gov/news-events/news/attack-colonial-pipeline-what-weve-learned-what-weve-done-over-past-two-years

Greubel A, Andres D, Hennecke M (2023) Analyzing reporting on ransomware incidents: a case study. Social Sciences (2076–0760) 12 (5): 265. 2023. https://doi.org/10.3390/socsci12050265
Howard TD (2023) Understanding influence factors in cyber resiliency adoption: a mixed methods study of cybersecurity leaders. Ph.D. diss. Capitol Technology University
Lubin A (2023) Cyber plungers: colonial pipeline and the case for an omnibus cybersecurity legislation. Georgia Law Review 57(4):1605–1632. https://search-ebscohost-com.captechu.idm.oclc.org/login.aspx?direct=true&db=aph&AN=169978264&site=eds-live
Marinos N, Gordon LV (2021) Colonial pipeline cyberattack highlights need for better federal and private-sector preparedness (infographic). US Government Accountability Office, WatchBlog. https://www.gao.gov/blog/colonial-pipeline-cyberattack-highlights-need-better-federal-and-private-sector-preparedness-infographic
NIST Special Publication 800-160, Volume 2, Revision 1. Developing Cyber-Resilient Systems: A Systems Security Engineering Approach. Available at https://safe.menlosecurity.com/https://doi.org/10.6028/NIST.SP.800-160v2r1. Accessed 2 May 2024
Office of the Director of National Intelligence (2023) Annual threat assessment of the U.-S. Intelligence Community. Feb 6, 2023. https://www.intelligence.gov/annual-threat-assessment
Office of the President of the United States. National cybersecurity strategy. Mar 1, 2023. https://www.whitehouse.gov/wp-content/uploads/2023/03/National-Cybersecurity-Strategy-2023.pdf
Rodgers EM (2003) Diffusion of innovations, 5th edn. Free Press, New York
The White House (2024) National security memorandum on critical infrastructure security and resilience. April 30, 2003. Available at National Security Memorandum on Critical Infrastructure Security and Resilience | The White House. Accessed 2 May 2024
Vencill T (2019) In memoriam: legendary strategist and fleet tactics author Wayne P. Hughes. Naval Postgraduate School. Dec 11, 2019. https://nps.edu/-/in-memorium-legendary-strategist-and-fleet-tactics-author-wayne-p-hughes

Resilient Decision Making in Cyber Incident Response

Jelle Groenendaal and Ira Helsloot

1 Introduction

In the event of a cyber incident like a Distributed Denial of Service (DDoS) attack, widespread malware infection, or data breach, organisations must act swiftly to continue or restore operations and maintain stakeholders' trust. Typically well-prepared organizations activate their Cyber Incident Response Plan (CIRP). A key component of such a plan is the activation of the Cyber Incident Response Team (CIRT). CIRT's main duties include incident response (identification, containment, eradication, and recovery), forensics (collection and analysis of evidence post-incident), and incident management (coordinating responses and mitigating impacts across the organization). The composition of a CIRT can vary, including both in-house and external experts, depending on the specific demands of the incident. At its core, the team is generally formed of cybersecurity and IT personnel, with the possibility of incorporating additional expertise from other IT areas, business operations, risk management, communications, and legal teams as needed (Groenendaal et al. 2022).

Under high-pressure and uncertain conditions, CIRTs are tasked to make critical decisions that can significantly affect the organization. If the team fails to respond promptly or adequately, the cyber incident may escalate, magnifying the damage to the organisation. Conversely, making an ill-informed defensive decision, such as unnecessarily shutting down systems, could exacerbate the operational impact. CIRTs face complex decisions regarding the extent of investigations, selection of

J. Groenendaal (✉)
Crisislab, Renswoude, the Netherlands
e-mail: jelle@jellegroenendaal.nl

I. Helsloot
Crisislab, Renswoude, the Netherlands

Governance of Safety and Security, Radboud University Nijmegen, Nijmegen, the Netherlands

© The Author(s), under exclusive license to Springer Nature Switzerland AG 2025

I. Linkov, A. Kott (eds.), *Cyber Resilience: Applied Perspectives*, Risk, Systems and Decisions, https://doi.org/10.1007/978-3-031-90109-6_6

hypotheses to explore or disregard, and strategies for incident containment. Decisions also extend to measures for lessening the incident's impact, including the choice of external communication tactics. Moreover, CIRTs must navigate varying and sometimes opposing interests of stakeholders within the organization, balancing them carefully to effectively manage and resolve the cyber incident.

This chapter delves into the concept of resilience within the realm of decision-making in response to cyber incidents. Kott and Linkov (2021) define cyber resilience as a system's capacity to recover or restore its functionality following an unexpected event that causes a decline in performance. It may seems evident that CIRTs can significantly influence an organization's ability to rebound from adverse cyber events, thus bolstering its cyber resilience. When a team makes satisfying decisions promptly, the organization stands a better chance of swift recovery and potential damage mitigation. Conversely, suboptimal decisions can prolong recovery efforts and lead to avoidable harm. To delve deeper into cyber-resilient decision-making, we explore three interconnected concepts: resilient decision-making processes, resilient decisions themselves, and cyber-resilient performance. We then scrutinize the relationship between these elements. While it may seem logical that a resilient decision by a CIRT would enhance overall cyber resilience performance, we illustrate that this is not always the case; one doesn't necessarily hinge on the other. To illustrate these points, we discuss a real-world case study detailing how a global financial institution navigated emerging cyber threats during the COVID-19 pandemic.

2 Background on Cyber Incident Response Decision Making

In this section we introduce the concept of cyber incident response decision making using the framework of naturalistic decision making (NDM) (Zsambok and Klein 1997).

2.1 *Cyber Incident Response*

Cyber Incident Response is an organized approach to detect, contain, and eradicate cyber threats as well as to restore business operations to normal in a timely and cost-effective manner (Ahmad et al. 2021). In practice several approaches are being used yet they typically share a core set of phases, each with a distinct purpose (Groenendaal et al. 2022):

- *Preparation*: This initial phase is about establishing and maintaining the readiness to respond to cyber events. It involves developing policies, procedures, and setting up the necessary tools and response teams to handle potential threats effectively.

- *Detection and Analysis:* In this stage, the focus is on identifying and assessing potential cyber incidents. It includes monitoring systems for signs of a breach, analyzing anomalies to determine whether they constitute a security threat, and assessing the nature and scope of the identified incident.
- *Containment*: Once an incident is confirmed, the aim shifts to limiting its spread and impact. This step involves implementing measures to isolate affected systems or networks to prevent further damage while ensuring that business operations can continue with minimal disruption.
- *Eradication*: With the threat contained, efforts are then directed towards completely removing the cause of the incident. This may involve cleaning infected systems, closing security loopholes, and implementing fixes to vulnerabilities that were exploited by the attackers.
- *Recovery*: Following eradication, the next goal is to restore and return affected systems and services to their fully operational states. This phase includes verifying the integrity of systems and data, monitoring for any signs of residual issues, and gradually reintegrating systems back into the operational environment.
- *Post-Incident Activities*: The final phase focuses on learning from the incident to improve future response efforts. This involves conducting a detailed review of how the incident was managed, documenting lessons learned, updating policies and procedures as necessary, and implementing improvements to prevent recurrence or mitigate the impact of future incidents.

2.2 Cyber Incident Response Decision Making

Cyber incident response decision making can be defined as the process of making decisions about the response to, and management of, (potential) cyber incidents (Groenendaal et al. 2022). It involves the process of choosing or selecting a workable option among alternatives regarding preparation, detection, containment, eradication and recovery, often requiring judgment, analysis, and action.

During the 1980s and 1990s, extensive research was conducted on how professionals working under conditions of uncertainty and time-pressure make decisions in their respective working environments. This body of research is known as Naturalistic Decision Making (NDM) (Zsambok and Klein 1997). The primary conclusion drawn from these studies was that professionals working under challenging conditions typically make decisions that are in general satisfactory, albeit not always optimal, based on their experience. In rare situations their experience however may misguide them. Klein and colleagues encapsulated this finding in the 'Recognition-Primed Decision Making' (RPD) model (Klein et al. 2010; Klein 1993). According to this model, professionals working under time pressure and uncertainty can recognise new situations based on a set of indicators and then select an approach that has previously proven effective in similar circumstances (Klein 2008).

Klein (1993) identified three models within the RPD framework: (1) a simple match model, (2) an action strategy model, and (3) a complex RPD strategy model. In the first and simplest model, known as the simple match model, the decision maker recognises the situation and implements the most obvious response (Klein 1993). This model is primarily used when there is limited time available. The second model, developing a course of action, incorporates the simple match approach but also includes a conscious evaluation of the response, called mental simulation, to identify potential issues before implementation (Klein 1993). This model is more commonly used when decision makers have additional time. The third and most advanced model, known as the complex RPD strategy, is used when decision makers, after careful evaluation, determine that the situation does not align with any prior experiences. In this case, the decision maker will attempt to rationally analyze the new situation and determine the best course of action suited to it (Klein 1993).

Although originally grounded in research on American firefighting operations, the RPD framework has since been applied successfully to understand decision-making among medical doctors, pilots, chess players, police officers and other professionals (Klein 1993, 1998, 2008). Groenendaal et al. (2022) found indications that RPD can also be applied to cyber incident responders to understand and predict their decision making.

Despite its effectiveness in navigating challenging decision-making situations, RPD can sometimes lead to less than optimal outcomes under two main circumstances:

- Recognition bias may hinder judgment, where a seasoned professional might misinterpret a situation as typical, missing key (and possibly contradictory) signals (Orasanu and Martin 1998). Particularly under pressure, the rapidity of recognition can obstruct the acknowledgment of divergent evidence. For example, if an IT system fails during an update period, a cyber incident responder may prematurely attribute the fault to the update, disregarding the potential for an external attack (Groenendaal and Helsloot 2016).
- There may be a lack of recognition because the decision maker does not have the right experience and/or the (learning) environment does not provide reliable feedback. Without appropriate feedback, discerning the true effects of decision can become challenging (Kahneman and Klein 2009).

Furthermore, NDM research indicates that responders are susceptible to biases that impair their judgment, such as confirmation bias or the tendency to confirm a previously formed hypothesis by (a) interpreting all gathered information as such, (b) subsequently search only for signs that validate that hypothesis and (c) discount negative evidence (Ask and Granhag 2005). Another common bias is the sunk cost bias, where responders are driven to complete tasks nearing their final phase despite changing circumstances (Meij 2004).

Groenendaal and Helsloot (2016) have developed a model for professionals leading incident responders, from fire service to cyber incident response, to help them avoid decision-making pitfalls (see Table 1). Using NDM, they have outlined five generic steps of incident command and linked various insights from NDM and other human decision-making research to these steps, aiming to provide leaders with an understanding of the inherent limitations of professionals making decisions under time pressure, as well as offering them practical tools to identify and ensure a resilient decision making process.

Cyber incident response decision-making is closely related to, yet distinct from, cyber incident command. The first three steps of cyber incident command (fact finding, analysis and decision-making) focus on decision-making, while the subsequent steps (communication and monitoring) ensure that decisions are executed properly by CIRT members.

3 Three Perspectives on Resilience in Cyber Incident Response

In this section we introduce three perspectives on cyber incident response and explain why their relations are not trivial.

3.1 *Resilient Cyber Incident Response Decision Making Process*

A resilient cyber incident response decision making process means that cyber professionals or teams are able to overcome decision making biases in the face of unexpected events encountered during their work. For instance, a CIRT may spend numerous hours investigating a single hypothesis, which could make the CIRT susceptible to the sunk cost bias, i.e. the team's inclination to persist with a failing hypothesis due to the time already invested. Resilient decision-making would involve ceasing the investigation of the current hypothesis in a timely manner to explore an alternative.

Through education, training, and reflection on their experience, professionals can recognise and avoid decision-making pitfalls to some extent (Klein 1998). However, a significant insight from NDM research is that professionals working under time pressure and uncertainty are inherently limited in their decision-making ability. Therefore external adjustments may be necessary to ensure a resilient cyber decision-making process.

The NDM-model of cyber incident command thus fundamentally differs from the traditional rational choice and problem-solving models that are dominant within cyber incident response (Groenendaal et al. 2022). These rational models presume

Table 1 A model of incident command with potential pitfalls per phase and NDM insights to overcome these pitfalls (Groenendaal et al. 2022; Groenendaal and Helsloot 2021; Groenendaal 2015)

Phase	Description	Example of pitfall	Example of NDM insight
Fact finding	Collect relevant information regarding the incident to understand the scope and scale accurately. This involves gathering data on aspects such as the nature of the threat, affected systems, potential vulnerabilities exploited, and the business impact.	'Inaccurate situation awareness'—despite accurate knowledge and experience, cyber incident responders can still make wrong decisions if their perception of reality does not correspond to the actual situation.	Leaders should actively search for information and use this to validate their perception of reality in the light of the present situation. In addition, they must validate the accuracy of information they receive from incident responders and others outside the organisation
Analysis	Assess the gathered information to get an understanding of the situation that serves as a springboard for further action, which includes analysing the severity, potential impact, and immediate priorities.	'RPD'—fast, recognition-primed decision making (system 1) is dominant when working under time-pressure and uncertainty.	Leaders should consciously take time to engage their reasoning ability (system 2) to validate the decisions taken by incident responders
Decision-making	Decide on a workable response strategy based on the analysis. This includes selecting specific actions to contain, mitigate, or eradicate the threat while considering resource allocation and potential consequences.	'Group think'—Psychological phenomenon where the desire for harmony and conformity results in irrational or dysfunctional decision making	Leaders should proactively organise a devils advocate view by consulting experts who have not been previously involved in the decision-making process.
Communication	Communicate the decided action plan to the team and relevant stakeholders. This involves ensuring that everyone understands their roles and responsibilities and the expected outcomes of the response efforts.	'This is how we've always done it assumptions'—People make assumptions about how well they are understood by others	Leaders should communicate their intent and proactively verify if the receiver has the same understanding.
Monitoring	Oversee the implementation of the action plan to ensure it's executed as intended. Adjust the plan as necessary based on real-time feedback and the evolving situation to effectively mitigate the incident.	'Overconfidence in own influence'	Leaders should proactively check whether their orders are being carried out as agreed upon.

Table 2 Differences between recognition-primed decision making and classical choice model of decision making (Klein 1998, 2008)

Recognition primed decision making (RPD) in cyber response	Rational choice model (RCM) in cyber response
Context: Time pressure, high stakes, uncertainty	Context: Limited time pressure, sufficient information available,
First option usually workable	Random generation and selective retention
Serial generation of options	Concurrent evaluation of options
Satisficing	Optimising
Evaluation through mental simulation	Evaluation through decision analysis or statistics
Focus on elaborating and improving options	Focus on choosing between options
Focus on situation assessment	Focus on decision events
Decision maker primed to act	Decision maker primed to analyse

that cyber decision-makers follow a rational decision-making process, where they gather information, formulate a range of options, and systematically evaluate each option to arrive at the optimal decision. However, Simon (1973, 1992) demonstrated long ago that this is not a natural decision-making method for people. Specifically in frontline work, Weick (1995) argued that it is impossible for decision-makers to form a complete and accurate understanding of reality and retain it. Furthermore, Weick (1995) noted that professionals working under challenging conditions cannot focus on all individual aspects of complex problems. Cyber incident command is no different.

Enhancing a resilient cyber incident response decision process thus calls for a decision support system that helps cyber incident command professionals to 'escape' from the left side of Table 2 to the right side.

3.2 Cyber Resilient Decisions

A cyber resilient decision aims to enhance an organisation's ability to adapt its functions before, during, or after specific events, such as changes, disruptions, or opportunities (Hollnagel 2011). Hollnagel (2011, 2017) notes that adaptations can occur reactively, in response to feedback after an event, like organising an incident debrief with the incident response team and other internal teams. However, adaptations can also occur proactively, based on assumptions about the future. In the context of cyber incident response, an example of proactive adaptation might involve an organisation putting extra cyber incident responders on standby when a threat of an imminent DDoS attack is identified. Another example could be organising a large-scale exercise to prepare for the increased risk of a ransomware attack.

In practice, organizations likely engage in both types of adjustments. Hollnagel (2017) underscores that while reactive adjustments are more common, they don't ensure organizational continuity. For instance, businesses operating with narrow profit margins and failing to anticipate economic downturns struggle to navigate recessions effectively. Proactive adjustment involves pre-emptively altering the system to meet anticipated future demands, including transitioning from normal operations to a heightened state of readiness before an event occurs (Hollnagel 2017). In this heightened state, resources are allocated to align with anticipated needs, and specialized functions may be activated.

According to Hollnagel (2011) and others (Leveson 2016; Woods 2015), cyber resilient decisions can be categorised into decisions aimed at enhancing the organisation's ability to anticipate, monitor, respond or learn:

- Anticipation involves foreseeing future developments, including potential disruptions, emerging requirements, constraints, or opportunities (Hollnagel 2017). According to Chuang et al. (2020), the ability to anticipate future events and conditions confers a competitive advantage.
- Monitoring pertains to the organization's ability to detect changes in work conditions and utilize indicators to track internal and external environments (Chuang et al. 2020). Hollnagel (2017) emphasizes the importance of monitoring organizational performance and external factors.
- Response entails knowing how to act and promptly responding to unfolding events (Hollnagel 2017). This includes reacting to regular and irregular changes, disruptions, and opportunities by activating prepared responses, adjusting operations, or devising new approaches (Chuang et al. 2020).
- Learning involves reflecting on experiences and extracting insights (Hollnagel 2011). Hollnagel (2017) underscores the importance of learning from both specific incidents and underlying causes, through single-loop and double-loop learning processes (Hollnagel 2011; Chuang et al. 2020).

3.3 Cyber Resilient Performance

Cyber resilient performance can be defined as the ability of a system (i.e. organisation) to recover or regenerate its performance following an unexpected impact that results in a performance degradation (Kott and Linkov 2021). According to Kott and Linkov (2021), the central idea behind cyber resilience involves acknowledging the possibility of cyber compromise and the consequent organizational impact. Cyber resilience characterizes what happens after an adverse cyber event, and requires preparedness for both known and unknown threats (ibid).

In the line of Kott & Linkov, a cyber resilient performance evaluates the impact of decisions on the strategic objectives of the organisation. This view focuses on the tangible results of the decisions, particularly how they contribute to the system's recovery or enhancement following a (cyber) incident. In the context of cyber incident response, this could involve the decision or recommendation to pay or not to pay a ransom in the event of a ransomware attack. Suppose the advice is not to pay the ransom, with the expectation that backups can be restored within a reasonable time frame. If this recommendation proves effective and the organisation quickly resumes its primary operations, the decision made by the CIRT significantly contributes to the organisation's cyber-resilient performance.

However, in practice, it is often not so simple to determine the influence a CIRT has on an organisation's cyber-resilient performance. Firstly, because the CIRT is not the only team making decisions during a cyber incident. As we know from extensive research on incidents, decision-making is fragmented across the entire organisation, despite the common assumption that the CIRT coordinates and tightly manages everything. Secondly, it is often challenging to determine the extent to which a CIRT decision contributes to the organisation's cyber-resilient performance. This difficulty arises because the impact of alternative options (those not chosen) is unknown. Thirdly, there can be a difference between the short- and long-term effects. For instance, a CIRT may decide not to shut down potentially infected systems to maintain operations and resilience. However, this decision could prove harmful in the long term if a much larger number of systems need to be shut down a few weeks later.

The relationship between a cyber resilient decision making process, cyber resilient decisions and resilient performance is not tirival. Resilient decision-making and resilient decisions are predictors of resilient performance, but not a necessary prerequisite. Hollnagel (2017), for instance, recognises that making resilient decisions does not necessarily guarantee resilient performance during incidents. He states that organisations making resilient decisions are more likely to demonstrate resilient behaviour than those that do not. Conversely, organisations that fail to make resilient decisions are poorly equipped to achieve resilient performance (Hollnagel 2017).

The same applies to resilient decision-making. A decision-making process can be completely free of biases without contributing to the organisation's cyber-resilient performance. For instance, consider a CIRT that, based on an accurate understanding of the situation, decides not to escalate a threat, which ultimately leads to a disruption in an operational process. The opposite is also possible. A CIRT's decision-making can be negatively influenced by biases, such as when a particular decision is made based on incorrect assumptions. However, the decision itself can still positively impact the organisation's cyber-resilient performance. For instance, consider a CIRT that develops tunnel vision regarding a specific threat that could lead to a data breach. To counter this, the CIRT decides to have additional personnel on standby over the weekend to monitor systems. If this decision results in detecting another, less vocal threat over the weekend, it could rightly be considered a fortunate coincidence rather than a cyber-resilient decision and process.

Our treatment of the three concepts will have made clear that the relation between them is not straightforward. The case study in the next section will demonstrate this in practice.

4 Case Study

Our study focuses on a global financial institution (GFI) case,[1] which offers a wide range of financial services and products, including insurance, to a global clientele. The financial industry, including our case study subject, is undergoing a transformation, increasingly integrating technology into its core operations. Digital interactions have become the primary mode of engagement between the GFI and its customers, elevating the significance of cyber security to a strategic level within the institution.

The methodology included semi-structured interviews with eleven key individuals across the strategic, operational, and tactical levels of the organization (Table 3). The positions of these key cyber response decision makers within the organization give a comprehensive and valid insight in the way the GFI handled cyber threats and digitized fraud. This approach ensures that the study captures both the direct experiences at the front lines and the broader policy implications. The selection was aimed at understanding the resilience of the organization in the face of the COVID-19 pandemic from a cyber security standpoint, drawing on the premise that these selected respondents would be most knowledgeable about the institution's cyber resilience capabilities.

The questions posed during the interviews focused on the participants' interpretations of cyber resilience, their approach to responding to cyber incidents, and their process for making decisions that enhance cyber resilience. Other queries were inspired by examples and concepts introduced by Hollnagel in Hollnagel 2017.

The interviews were conducted over a period from April to October 2020, with each session lasting approximately 30–45 min. Key observations and notable quotations from these discussions were meticulously documented. A few of the participants were engaged in a second round of interviews, typically after an interval of 2–3 weeks, to address any subsequent questions or to delve deeper into specific points raised during the initial discussion.

Observation 1: Considering GFI's incident statistics as well as the perception of interviewees, the organization performed cyber resiliently during the COVID pandemic

A significant number of participants highlighted that the GFI faced heightened vulnerability to cyber threats with the onset of the COVID-19 pandemic crisis. One notable factor, as pointed out by ISM1, was the rapid transition to remote working, which expanded the potential for cyber-attacks. ISM2 further noted the challenge of monitoring endpoints outside the corporate network, increasing the risk of attacks leveraging employee endpoints as access points to the corporate network or for data theft or tampering.

[1] This case was initially examined in Groenendaal and Helsloot (2021), where we laid the groundwork for our current investigation.

Table 3 Roles of respondent, organizational level and label

#	Role of respondent	Category	Label
1	Deputy chief information security officer	Strategic	DCISO
2	Chief cyber security operations	Strategic	CCOPS
3	Information security manager 1	Tactical	ISM1
4	Information security manager 2	Tactical	ISM2
5	Head of cyber incident response team (lead CIRT)	Tactical	HIR
6	Business continuity manager	Tactical	BCM1
7	Digital fraud investigator 1 (member CIRT)	Operational	FI1
8	Digital fraud investigator 2 (member CIRT)	Operational	FI2
9	Cyber threat analyst (member CIRT)	Operational	CTA
10	Cyber intelligence analyst 1 (member CIRT)	Operational	CIA1
11	Cyber intelligence analyst 2 (member CIRT)	Operational	CIA2

Furthermore, FI1, FI2, CIA1, and CIA2 observed that large-scale social disruptions like the COVID-19 pandemic create fertile ground for more effective phishing attacks, as individuals may be more susceptible to deceptive communications under such circumstances. Moreover, the shift to remote work has heightened the risk of insider threats, as malicious activities by employees are harder to detect away from the office environment, a point emphasized by CIA1 through the observation that such threats are often uncovered by vigilant co-workers, a safeguard weakened by remote working conditions.

Despite these increased vulnerabilities, DCISO and CCOPS highlighted that the actual incidence and impact of cyber threats have not escalated as might be expected. While phishing attempts have risen, linked to the pandemic theme, the effectiveness of these campaigns has seen only a marginal increase. The GFI's standard operational procedures have been largely effective in managing these incidents. The institution experienced a slight uptick in cyber incidents compared to the previous year, but the management of these incidents was not significantly disrupted by the pandemic, thanks in part to the adaptability of cyber incident responders to remote work. In instances where incidents had a potentially larger impact, key decision-makers, including DCISO, CCOPS, and HIR, opted for physical meetings in a designated crisis control room, recognizing the value of face-to-face interaction for optimal coordination and communication among IT, security, and business units.

In conclusion, the impact of cyber incidents was not higher than in previous years, indicating that cyber-resilient performance was not significantly affected by the COVID pandemic.

Observation 2: Cyber resilient decisions were made prior and during the COVID-19 pandemic, but rarely resulted in anticipatory adjustment

The discussions with all participants revealed that the GFI made several resilient decisions focused on enhancing key resilience capabilities: the ability to anticipate, respond, monitor, and learn. However, despite these capabilities being developed, they rarely resulted in proactive adjustments before the crisis unfolded.

4.1 Anticipation Capability

The GFI's approach to anticipating cyber threats included a dedicated unit for cyber threat intelligence, which operated separately from the CIRT. This unit was tasked with creating quarterly threat landscape reports and conducting daily analyses of cyber threat intelligence. Despite these efforts, CIA1, CIA2, and DCISO acknowledged that the pandemic's specific threats were unforeseen and, consequently, not included in these reports. Additionally, the organization's Business Continuity Management (BCM) capability, responsible for crafting an annual corporate threat report, relied on sources such as the Business Continuity Institute (BCI) Horizon Scan Reports and the World Economic Forum (WEF) Global Risks Reports. Yet, none of the corporate threat reports from the previous 5 years had identified the risk of a major infectious disease outbreak, including extensive government-imposed lockdowns. BCM1 noted that while some business units and departments had infectious disease plans focusing on staff sickness and basic hygiene, these were inadequate for addressing the severe restrictions imposed during the COVID-19 crisis.

The interviewed members of the CIRT indicated that in some cases, specific threat information from the cyber threat intelligence unit was used in their work. In the case of the COVID pandemic, the threats, such as an increased risk of attacks on remote working capabilities, were so evident that the information from the cyber threat intelligence unit proved of little value.

4.2 Monitoring Capability

The GFI's commitment to monitoring cyber threats and incidents was evident through its investment in internal reporting mechanisms. Key performance indicators were established, with consensus from senior management, to guide the daily, weekly, or monthly reports that inform decision-makers across operational, tactical, and strategic levels. Beyond the CISO department's reporting, the IT risk department maintains a distinct line of reporting on IT security risks to both senior management and regulatory bodies, ensuring a comprehensive view of the organization's cyber threat landscape.

4.3 Response Capability

Far before the pandemic hit, the Global Financial Institution (GFI) has established robust mechanisms for managing incidents, including a comprehensive cyber incident response organization. Operating round-the-clock on a follow-the-sun model, the Security Operations Center (SOC) stands at the forefront of cyber incident response. This capability is seamlessly integrated with the IT incident management team, tasked with addressing IT failures that have a tangible impact on business operations. In the event of a significant cybersecurity incident, this IT incident management framework enables swift action, such as applying patches to vulnerable systems, altering configuration settings, or implementing temporary solutions to mitigate risk. Additionally, the GFI has developed a Business Continuity Management (BCM) function dedicated to ensuring the continuity of service delivery during disruptions. According to BCM1, this involves having both contingency plans and a crisis management structure in place, facilitating quick recovery of services and enabling senior management to make rapid, informed decisions during crises.

4.4 *Learning Capability*

The process of learning from cyber incidents is embedded within the GFI's operational practices. HIR highlighted that the evaluation of most cyber incidents occurs within the meetings of the cyber incident response team. For more significant incidents, a structured incident learning procedure is enacted, encompassing a thorough evaluation meeting with all relevant stakeholders. This process is designed to foster the development of comprehensive lessons learned, including a detailed root cause analysis and the formulation of an action plan, all subject to approval by senior management. BCM1 noted that certain departments within the GFI have previously navigated challenges posed by infectious diseases, such as the swine flu pandemic in 2009. These past experiences have been documented and incorporated into the organization's pandemic preparedness plans, underscoring a commitment to continuous learning and adaptation.

In conclusion, none of the interviewees could cite specific instances of the organization adapting in advance of the pandemic crisis. The discussions revealed a predominantly reactive stance by the Global Financial Institution (GFI) in response to the evolving situation. Early awareness was evident, as DCISO and BCM1 noted the organization began tracking the outbreak in China by mid-January 2020, initiating daily briefings to share information about the pandemic's progress. Despite these efforts, early actions were limited to implementing travel bans to affected regions. Significant strategic measures, including a review of existing pandemic plans, were only considered after the first local infections were confirmed and the imminent impact on neighboring countries became apparent, revealing that the pre-existing plans were largely outdated or impractical. Furthermore, the adaptation to enhance remote working capabilities and its cybersecurity considerations were prompted only by the imposition of government lockdowns, as mentioned by DCISO.

Observation 3: The response capacity outside the CIRT had the greatest positive impact on the GFI's cyber resilient performance

Despite the initial reactive approach, the GFI demonstrated a commendable capacity to respond to the crisis. DCISO, CCOPS, HIR, ISM1, ISM2, and BCM1 shared examples of the organization's effective adaptation to the pandemic, notably the swift expansion of remote working capabilities. Within a few weeks, the GFI managed to significantly transform its infrastructure and processes, enabling a majority of its employees to work from home securely. This achievement, described by DCISO as an "unprecedented success" for such a large entity, was attributed to the maturity of the GFI's IT and security infrastructure and solid partnerships with third-party suppliers.

Although the organization's anticipatory and monitoring capabilities were utilized during the crisis, it's evident that these functions were activated in a reactive mode. For instance, CCOPS and HIR highlighted that critical cybersecurity services remained uninterrupted, thanks in part to pre-emptive measures like establishing fixed teams to prevent the virus's spread among staff. Additionally, FI1 and FI2 pointed out the formation of a special task force focused on monitoring and countering specific fraud threats against the organization and its clients. These measures, including adjustments to email configurations and the launch of targeted awareness campaigns, exemplify the GFI's dynamic response once in crisis mode, showcasing its resilience and adaptability amidst unprecedented challenges.

Observation 4: The organization responded cyber resiliently despite flaws in the decision-making process

During the interviews, attention was also paid to the decision-making process of the CIRT during the coronavirus pandemic. Interviewees indicated that the decision-making process was sometimes chaotic. At the beginning of the pandemic, a daily call was set up between operational cyber incident responders to exchange signals with each other. Notably, not everyone always showed up, and representatives from some countries sometimes didn't respond for days. Additionally, higher-level managers, who were not part of the core team, regularly joined the calls and actively involved themselves in mapping out the situation and setting actions. As the pandemic continued, the composition of the core team changed, and more senior managers joined. Consequently, there was no organised opposition from managers and professionals who were not directly involved in decision-making and could therefore provide objective and more detached feedback.

In another example, the Deputy Chief Information Security Officer (DCISO) recounted an experience during a DDoS attack that occurred amidst the COVID pandemic. The majority of senior leadership was present in the war room, closely monitoring the situation and preparing to make decisions as needed. Some senior leaders stayed the entire day without considering who would take over during the night and subsequent days. Although the attack ended during the night, the DCISO highlighted that several leaders identified this oversight as a key lesson learned, as they believed that staying in a war room for too long does not positively contribute to the quality of decision-making.

5 Conclusion

In this chapter we introduced three perspectives on resilience in cyber incident response:

- a resilient cyber decision-making process,
- a resilient cyber decision,
- and a cyber resilient performance.

As we have shown these three concepts are not simply causally connected. Clearly the goal is a cyber resilient performance. Resilient cyber decisions should facilitate that goal. A resilient cyber decision-making process should result in those resilient cyber decisions.

NDM-theory tells us that guaranteeing a resilient cyber decision-making process is extremely different. Much research on cyber incident response is focused on this, often starting from a too optimistic rational choice model.

Our case study of a global financial institution's response to emerging cyber threats during the COVID-19 pandemic illustrates that even in the absence of both a resilient cyber decision-making process and resilient cyber decisions made during the crisis the GFI was exhibiting a cyber resilient performance. This was made possible by a priori resilient cyber decisions.

The practical implications of these insights suggest that enhancing a priori cyber resilience should be prioritized by far example effective resource management (Groenendaal and Helsloot 2021). NDM research clearly makes the point for investing in expertise up front, as highly experienced professionals are less likely to err when making decisions under time-pressure and uncertainty (Klein 1998). Based on a review of the empirical literature on organizational resilience, Barasa et al. (2018) found that the availability of resources is considered a key enabler of organizational resilience. When material resources are used strategically, the authors state, organizations can overcome disruption. The authors emphasize that financial resources are also considered necessary to mobilize other needed resources during the crisis. In a similar vein, what we have uncovered in our case study reflects an analogy for companies of a well-known fact when it comes to resilience of individuals; in the end, having money at hand to be able to respond is a factor of greater importance than having put much effort in preparation (Helsloot and Ruitenberg 2004).

The design of decision support systems capable of counteracting suboptimal decision-making processes in cyber incident response organisations remains at present an elusive goal. Van der Kleij et al. (2022) developed a critical-thinking memory aid that follows typical incident response process steps, but adds big picture elements and critical thinking steps. This approach is supposed to make cybersecurity threat and incident managers more aware of the broader operational implications of threats and incidents while keeping a critical mindset. However, the authors did not investigate whether and how the support tool would be used in practice, and whether it can be successful to overcome decision-making flaws.

To conclude, future research on cyber resilience that makes a clear distinction between these concepts is needed to come up with both theoretical and practical insights that can be used to further improve the important domain of cyber incident response.

References

Ahmad A, Maynard SB, Desouza KC, Kotsias J, Whitty M, Baskerville RL (2021) How can organizations develop situation awareness for incident response? A case study of management practice. Comput Secur 101:1–15

Ask K, Granhag PA (2005) Motivational sources of confirmation bias in criminal investigations: the need for cognitive closure. J Investig Psychol Offender Profiling 2(1):43–63

Barasa E, Mbau R, Gilson L (2018) What is resilience and how can it be nurtured? A systematic review of empirical literature on organizational resilience. Int J Health Policy Manag 7(6): 491–503

Chuang S, Ou JC, Hollnagel E, Hou SK (2020) Measurement of resilience potential - development of a resilience assessment grid for emergency departments. PLoS One 15(9)9): e0239472. https://doi.org/10.1371/journal.pone.0239472

Groenendaal J (2015) Frontline command: reflections on practice and research. Eleven International Publishing, Den Haag

Groenendaal J, Helsloot I (2016) The application of naturalistic decision making (NDM) and other research: lessons for frontline commanders. J Manag Organ 22(2):173–185

Groenendaal J, Helsloot I (2020) Organisational resilience: shifting from planning-driven business continuity management to anticipated improvisation. J Bus Contin Emer Plan 14(2):102–109

Groenendaal J, Helsloot I (2021) Cyber resilience during the COVID-19 pandemic crisis: a case study. J Conting Crisis Manag 29(4):439–444

Groenendaal J, Helsloot I, Reuter C (2022) Towards more insight into cyber incident response decision making and its implications for cyber crisis management. In: Proceedings of the 19th ISCRAM conference. Tarbes, France May 2022

Helsloot I, Ruitenberg A (2004) Citizen response to disasters: a survey of literature and some practical implications. J Conting Crisis Manag 12(3):98–111

Hollnagel E (2011) RAG-the resilience analysis grid. In resilience engineering in practice. A guidebook. Ashgate, Farnham

Hollnagel E (2017) Safety-II in practice: developing the resilience potentials. Taylor & Francis

Kahneman D, Klein G (2009) Conditions for intuitive expertise: a failure to disagree. Am Psychol 64(6):515–526

Klein GA (1993) A recognition-primed decision (RPD) model of rapid decision making. Decision making in action: Models and methods 5(4):138–147

Klein G (1998) Sources of power: how people make decisions. MIT Press, Cambridge, MA

Klein G (2008) Naturalistic decision making. Hum Factors 50(3):456–460

Klein G, Calderwood D, Clinton-Cirocco A (2010) Rapid decision making on the fire ground: the original study plus a postscript. J Cognit Eng Decis Making 4:186–209

Kott A, Linkov I (2021) To improve cyber resilience, measure it. arXiv preprint arXiv:2102.09455

Leveson NG (2016) Engineering a safer world: systems thinking applied to safety. The MIT Press, p 560

Meij G (2004) Sticking to plans: capacity limitation or decision-making bias?. Dissertation. University of Amsterdam, Faculty of Social and Behavioral Sciences, Amsterdam

Orasanu J, Martin L (1998) Errors in aviation decision making: a factor in accidents and incidents. In: Proceedings of the workshop on human error, safety, and systems development, pp 100–107

Simon HA (1973) The structure of ill-structured problems. Artif Intell 4(3):181–201
Simon HA (1992) What is an "explanation" of behavior? Psychol Sci 3(3):150–161
Van der Kleij R, Schraagen JM, Cadet B, Young H (2022) Developing decision support for cybersecurity threat and incident managers. Comput Secur 113:102535
Weick KE (1995) Sensemaking. Sage, Thousand Oaks
Woods DD (2015) Four concepts for resilience and the implications for the future of resilience engineering. Reliab Eng Syst Saf 141:5–9
Zsambok CE, Klein G (1997) Naturalistic decision making. Lawrence Erlbaum Associates, Mahwah

Rule-Making for Insider Threat Mitigation

Igor Linkov, Kelsey Poinsatte-Jones, Benjamin D. Trump, Alexander A. Ganin, and Jeremy Kepner

1 Introduction

As businesses and workplaces become more interconnected in the digital age, digital systems are becoming increasingly vulnerable to man-made attacks. Referred to as cybersecurity threats, man-made attacks to digital systems can occur both from a place of trust within an organization and from a nefarious party outside of the protected digital system. Insider threats abuse a placement of trust within an organization and include current or former employees, contractors, and business partners who intentionally use their authorized access to an organization in a manner that intentionally negatively affects the organization. Insider threats to an organization exclude those individuals who unknowingly contribute to cyber-attacks, as well as those who are incompetent or considered whistleblowers (Kepner et al. 2014; Silowesh and Nicoll 2013). External threats comprise those individuals or organized bodies that directly or passively exploit system vulnerabilities through the use of malware system attacks. Beyond the internal and external threats to a system, system users can also unintentionally contribute to cyber-attacks and/or become a direct or passive target of a cyber-attack themselves (von Solms and van Niekerk 2013).

This chapter subsequently focuses on the largest driver of cyber threats—the human factor. Specifically, this chapter focuses on rulemaking from a cognitive

I. Linkov (✉) · K. Poinsatte-Jones · B. D. Trump
Engineering Research and Development Center, U.S. Army Corps of Engineers, Concord, MA, USA
e-mail: igorlinkov@usace.army.mil

A. A. Ganin
Department of Systems and Information Engineering, University of Virginia, Charlottesville, VA, USA

J. Kepner
MIT Lincoln Laboratory, Cambridge, MA, USA

© The Author(s), under exclusive license to Springer Nature Switzerland AG 2025
I. Linkov, A. Kott (eds.), *Cyber Resilience: Applied Perspectives*, Risk, Systems and Decisions, https://doi.org/10.1007/978-3-031-90109-6_7

science perspective to mitigate system vulnerabilities and reduce the consequences of cyber threats. United States agencies such as the National Institute of Standards and Technology and Department of Homeland Security apply traditional risk assessment paradigms to cybersecurity to help identify computer system threats, reduce cyber vulnerabilities, and negate subsequent financial and political consequences. While these commonly employed strategies such as rulemaking can be advantageous at minimizing and preventing vulnerabilities and risk, risk assessments exclude overall evaluations of system performance, including the ability of systems to recover and adapt following an attack (Linkov et al. 2016). Resilience, on the other hand, is used to identify the critical functionality of a system and to evaluate the recovery of critical functions in response to adverse events (Linkov and Trump, 2019). Resilience measures a system's overall performance for its ability to perform four functions with respect to adverse events: (1) planning and preparation, (2) absorption, (3) recovery, and (4) adaptation, as defined by the National Academy of Sciences (NAS) and is used in response to both anticipated and unanticipated adverse events (Linkov et al. 2013; NRC 2012). Furthermore, resilience can be modeled by mapping the four functions of a system against the four domains of system, including the physical, information, cognitive, and social aspects of a system and assigning metrics of performance to each cell (Linkov et al. 2013). Each of these domains within a digital system can be defined in the following ways (Linkov et al. 2013; DiMase et al. 2015):

- **Physical:** sensors, facilities, equipment, system states and capabilities.
- **Information:** creation, manipulation, and storage of data.
- **Cognitive:** understanding, mental models, preconceptions, biases, and values.
- **Social:** interaction, collaboration and self-synchronization between individuals and entities.

This chapter utilizes the rules-based paradigm to explore minimization of cyber threats, and offers a new take on the cybersecurity field from a behavioral science lens as compared to the computer science lens through which cybersecurity initiatives are normally established. Furthermore, this chapter extends the discussion of rulemaking and vulnerability reduction of a system into the cognitive domains of resilience, where interventions are optimized across the four functions of resilience within the cognitive science domain and consider both top-down and bottom-up approaches to improve such resilience (Linkov et al. 2014a).

1.1 Examining the Rules-Risk-Resilience Relationship

The implementation of work-related rules is intended to decrease the risks related to external threats. However, rules implementation can also increase the risks that arise from employee disregard for rules, if they become too restrictive and employees become more likely to disregard the rules and subsequently make mistakes (CESG 2016). Furthermore, the more often employees disregard the rules both intentionally

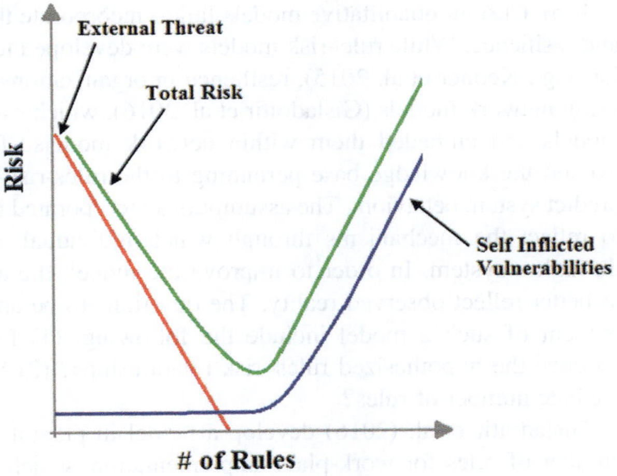

Fig. 1 System risk is a function of the number of rules. The red line represents the risk from external threats; the blue line represents the risk from internal vulnerabilities; and the green line represents the combined risk from external threats and internal vulnerabilities. (Gisladottir et al. 2016)

and unintentionally, the more likely insiders are able to observe and mimic employee behavior (Kepner et al. 2015; Williams and Adams 2012; Saji 2003; Ogus 2002). In-between the two extremes of 'too many rules' and 'not enough rules' lies the intermediate level of rules that both minimize external threats and internal vulnerabilities that arise related to employee behavior (Fig. 1). Ultimately, social behaviors and interactions have direct impact upon infrastructure and information systems, and must be accounted for to determine a system's operational capacity to achieve a given mission as well as its various facets of resilience or brittleness (Linkov et al. 2012; Trump et al. 2017).

Beyond the effects on success of risk reduction, overregulation may also affect a system's resilience pertaining to insider threats. According to the National Academy of Sciences, resilience is defined as the ability of a system to absorb, recover, and adapt to both known and unknown threats (Linkov and Florin 2016; NRC 2012). In the event that an attack exploits human-related vulnerabilities within a system, too many rules can hinder system recovery time and resilience (Zhu and Basar 2011). Evidence shows that too many rules employed within a technological system such as railways and nuclear power plants can lead to the loss of rule effectiveness. Specifically, overregulation can falsely signal that an environment is less risky and can lead to support against costly future rules, as well as the potential for underrepresentation of parties that bear the cost of the regulation (Aizenman 2009). Understanding the role of risk and resilience enables managers to better implement management strategies that regulate system functionality.

Compared to risk assessments, which start with the identification of hypothetical threat scenarios, resilience assessments start with the identification of a system's critical functions and their temporal patterns (Ganin et al. 2016; Connelly et al. 2017). Rules that are implemented to address specific system vulnerabilities might overlook the temporal nature of resilience, in comparison to rules that respond to the collective effect of a system's performance (Williams and Adams 2012; Linkov et al.

2014b). Current quantitative models fail to incorporate the ideas related to both risk and resilience. While rule–risk models were developed to analyze individual behavior (e.g., Kepner et al. 2015), resilience in organizations was traditionally analyzed using network models (Gisladottir et al. 2016), which extended individual rule–risk models and embeded them within network models. Thus, the model sought to expand the knowledge base pertaining to the rules-risk-resilience relationship and predict system behaviors. The assumptions incorporated in such models are intended to reflect the mechanisms through which individual actions and risk propagate through a system. In order to improve the model, the assumptions can be refined to better reflect observed reality. The questions to be answered through the development of such a model include the following: (1) Does the model adequately produce the hypothesized rules–risk relationships? (2) What are appropriate intermediate number of rules?

Gisladottir et al. (2016) develop a model in pursuit of determining an optimal number of rules for work-place implementation, which is further discussed in the methodologies section below. Specifically, the authors develop a model that shows the ways in which rules affect an organization's resilience with respect to the diffusion of a threat across its network of business relationships. Next, they model an example attack in which a virus is delivered via email to one employee's computer. The authors assume that the employee unintentionally propagates the virus to their collaborators. Thus, the model represents the diffusion of this virus among business colleagues at an organization. Within the network model, each node represents an employee and the links between the nodes represent the collaborations between employees on various projects. The model assumes that employees are connected to one another based on their business relationships. The acronyms, parameters and corresponding values used in the case study, and variables of the model are provided in Table 1.

The network model is assumed to be scale free in that its degree distributions follow a power law. The assumption takes into account the heavy-tailed degree distribution frequently found in communication systems (Hein et al. 2006). The authors use a *configuration algorithm* to construct the network and the algorithm starts with the generation of N power law distributed integers corresponding to the nodes' degrees, where N is the number of nodes in the network. Links are placed between the nodes in accordance with the nodes' degrees. In this study, the network consisted of 10,000 nodes and the authors assumed an average degree of 5 with a power law exponent of the degree distribution of 2.25, based on the power law exponents and degree distributions found within the (Barabasi 1999). The authors excluded nodes in this network that do not have a degree of at least 2 and, therefore, do not affect the network diffusion process. In addition to assumptions regarding business relationships, the model assumes that each employee has to follow a certain number of rules, which is a simplified representation of human behavior. The authors assume that employees with more connections have more expertise and therefore have greater ability or latitude to make work-related decisions. The maximal degree of a node was 88.

Table 1 Acronyms, parameters, and variables of the model

Main acronyms		
DL		Users' decision latitude
CF		Critical functionality of the network
R		Resilience of the network
Parameters and variables		
Name	Values	Description
b	1…50	Number of rules (barriers placed on the unit segment)
K		Critical functionality of the system
k		Degree of a node in the network model
k_j		Degree of node j in the network model
L_i		Amount of DL between barriers i and $i-1$
L_{min}	0.2, 0.25, 0.33, 0.5	Minimal DL required by users to perform normal activities in the 1D model
L^k_{min}		L_{min} for a node of degree k in the network model
L_T	0.5	Amount of DL required for a threat to succeed in an attack against a node
N	10,000	Total number of nodes
N_{in}		Number of infected nodes
N_{ns}		Number of not-strict barriers the threat has to breach
N_s		Number of strict barriers the threat has to breach
p_{det}	0.01	Probability that the user detects the threat at each time step
p_{new}		Probability that the threat succeeds in an attack against a node
p_{ns}	1	Probability that the threat breaches a not-strict barrier
p_s	0.1	Probability that the threat breaches a strict barrier
T_C	540	Control time
$T_{latency}$	10	Threat latency time
V		System vulnerability (same as L_{min})

Gisladottir et al. (2016)

In the modeled scenario, it is assumed that the employees disregard rules within the organization, if they are determined by the employee to inhibit their work-related tasks. The diffusion process begins when a malicious insider sends the infected email to their neighbors in the network. At this point, the degree to which the email is distributed depends on the level of barrier presented by the regulatory environment of the organization and the degree to which the employee follows the rules, described as strict and non-strict barriers. The authors note that organizations differ on implemented regulatory boundaries, access, and trust, particularly as it pertains to the position level of the employee. Barriers or rules are considered strict when employees are able to perform their work functions on a regular basis without crossing the regulatory barriers established by the organization. Nonstrict barriers, on the other hand, represent the rules that are occasionally circumvented. The authors assume in this scenario that an insider is able to cross some non-strict barriers without raising any red flags.

1.2 Applying the Rules-Risk-Resilience Relationship to Insider Threats

Within the workplace, rules establish the boundaries through which individuals can make work-related decisions, commonly referred to as decision latitude (DL) (Fagerlind et al. 2013). Low DL can lead to workplace stress and result in negative health consequences due to rules hindering individuals' ability to make workplace decisions. The measurement of DL or space between an individual action and an acceptable boundary is largely a subjective evaluation. For example, asking individuals in a specific environment about the desirability (or undesirability) of a specific hypothetical action can be used to establish the quantitative distance that action is from an acceptable boundary.

Breaking a rule requires that an acceptable boundary level is crossed. Employees might cross boundary levels for the benefit of the organization or individuals within the firm. Klieman (1996) examined the motivation for employees to cross rule boundaries and found that 2000 executive assistants were likely to destroy or remove damaging information (10%), write documents that contain misleading or false information (6.5%), or falsify vouchers or expense accounts (5.1%). When employees cross rule boundaries frequently, rules become less effective and insider threats become more likely to increase their latitude within an organization (Silowesh and Nicoll 2013).

Insider threats accomplish their goals of increasing DL and crossing rule boundaries by observing the boundaries that are being crossed by employees and orienting themselves to cross the same boundaries for the purposes of achieving new latitude. The behavior of an insider not only elevates their privileges such as through acquisition of co-worker passwords but also orients them to bypass security measures intended to prevent, detect, and react to unauthorized access (Claycomb et al. 2013; Myers et al. 2009). In an under-regulated environment, insiders are able to violate a security policy using legitimate access. In an over-regulated environment, frequent boundary crossing by individuals within an organization can lead to increased DL for insiders who wish to observe and mimic the same behavior.

Kepner et al. (2015) provide models for estimating the optimal number of workplace rules in the context of an insider threat. Specifically, Kepner et al. (2015) *employ a two-step approach to estimating workplace rules in the context of insider threats. The first approach comprises exploration of the Toy Model, which is based on a 1D concept of latitude and* allows for simple estimates of the impact of rules on latitude. Furthermore, the Toy Model illustrates the key concepts of over-regulation, under-regulation, and a tipping-point between these two regimes.

The Toy Model produces behaviors that are consistent with intuition: under-regulation, tipping-point, and over-regulation. However, it is not realistic to assume that complex human activities can be captured with such a simple model (at least not without more evidence). A more realistic model would likely have additional dimensions and more complex interactions among dimensions. Conducting a detailed analysis of the implications of adding dimensions and more complex interactions would be an overwhelming undertaking. Fortunately, the theoretical physics community has already made this investment. By changing the terminology of the Toy Model and converting it to a Percolation Model, it is possible to tap into this prior investment in theoretical work.

Thus, Kepner et al. (2015) change the terminology of the Toy Model to correspond to a Percolation Model in the second approach. The Percolation Model is widely used in theoretical physics for understanding a wide range of phenomena and allows for rapid generalization of the Toy Model into more complex, well studied models, including workplace cybersecurity measures. Further, the behavior of the Percolation Model is similar to many other complex models and use of the Toy Model captures the essence of behavior in these more complex models. Kepner et al. (2015) point out that more complex models may be quantitatively more accurate but that the qualitative behavior including workplace behavior is likely to remain the same.

2 Methodologies

2.1 *Toy Model*

The Toy Model represents the rule based environment as a 1D linear domain of points X where $0 < X < 1$ (Fig. 2). Hard boundaries are placed at $X = 0$ and $X = 1$ and the average employee performs their required activities within this domain. The latitude is defined as the total size of the domain in which an employee is able to perform their activities. In the initial configuration, the number of rules is $N = 0$ and the latitude of a normal individual in this environment is L_{normal} $(N = 0) = 1$ (Fig. 2a).

As the number of rules increases, new boundaries are created at random at a new location in the domain. The number of rules N reflect the number of internal boundaries N (Fig. 2b, c). Sorting the boundary locations in increasing order results in a boundary at X_1, another boundary at X_2, such that

$$X_1 < X_2 < \ldots < X_i < \ldots < X_N \tag{1}$$

A calculation of the difference between neighboring boundaries results in a sequence of N + 1 latitudes

$$\begin{aligned} L_1 &= X_1 \\ &\ldots \\ L_i &= X_i - X_{i-1} \\ &\ldots \\ L_{N+1} &= 1 - X_N \end{aligned} \tag{2}$$

Where the sum of the latitudes is always $\sum L_i = 1$.

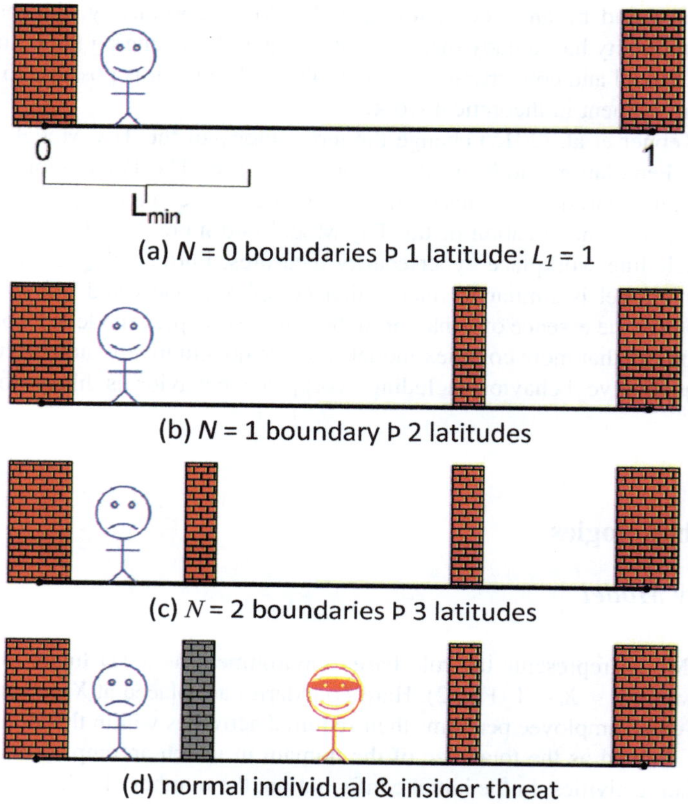

Fig. 2 Toy Model of an environment with a normal individual and an insider threat. The environment has boundaries at coordinates 0 and 1. L_{min} is the minimal latitude required for normal activity. From bottom to top: (**a**) no internal boundaries ($N = 0$) implies a single latitude $L_1 = 1$; (**b**) one internal boundary ($N = 1$) implies two latitudes; (**c**) two internal boundaries ($N = 2$) implies three latitudes, one of which is less than L_{min}; (**d**) the latitude less than L_{min} is eliminated for the adversary. (Kepner et al. 2015)

In any workplace environment, the average latitude (L_{normal}) for an employee is calculated as

$$L_{normal} = 1/_{N+1} \tag{3}$$

As more rules are added and more boundaries are created the average latitude of the normal individual decreases.

Within a workplace, the minimum latitude (L_{min}) needed for an employee to perform their required activities must be greater than the distance between any two boundaries. Alternatively, if any two boundaries are less than L_{min} apart, one of the boundaries is likely to be crossed by one of the employees. The average employee will most likely try to remain within the established workplace boundaries, thereby leaving their average latitude (L_{normal}) unchanged. In the case that any two boundaries are less than L_{min} apart, one of the rules will likely be broken by the average employee and the rules will most likely be completely ineffective at guarding against an insider threat. Therefore, in this case, the average latitude will not only be increased for the average employee but will also be increased for the insider threat.

In order to best understand the likelihood of insider threat success, one must calculate the number of insider threat boundaries N_{threat} as equal to the number of latitudes satisfying $L_i > L_{min.}$ As with the average employee, the number of threat latitudes must sum to 1 so that the average threat latitude is given by

$$L_{threat} = {}^{1}\!/_{N_{threat}+1} \tag{4}$$

2.2 Percolation Model

Percolation models define systems as set of vertices or sites that are connected to neighboring sites via edges. The most widely studied percolation models use a regular lattice. Each site is assigned a value. In many percolation models that site value can be -1 (indicating the site is "empty") or +1 (indicating the site is "occupied"). The probability that a site is -1 or +1 is then given by a probability P. The structure of the systems is then studied for given values of P. For example, in a 2D percolation model, it is common to compute the ability for a "drop" to "percolate" from the top to the bottom of the system by flowing through connected occupied sites.

The simplest of these models is 1D percolation. The Toy Model can be transformed into the Percolation Model in two steps. First, the latitudes L_i are "rotated" so that they are sitting on "sites" (see Fig. 3) and when $L_i < L_{min}$ the site is said to be "occupied." Second, the latitudes are turned into probabilities that the site is occupied. The latitudes L_i are sorted differences whose cumulative probability distribution is given by an exponential (see Fig. 4)

$$P(N, L) = 1 - e^{-(N+1)L} \tag{5}$$

This completes the transformation. Part of power of the Percolation Model is that it is effectively defined by only two parameters: the number of dimensions (in this case 1D) and the probability that a site is occupied (in this case P (N, L)).

This probability also provides a mechanism for computing the exact solution of the Toy Model since

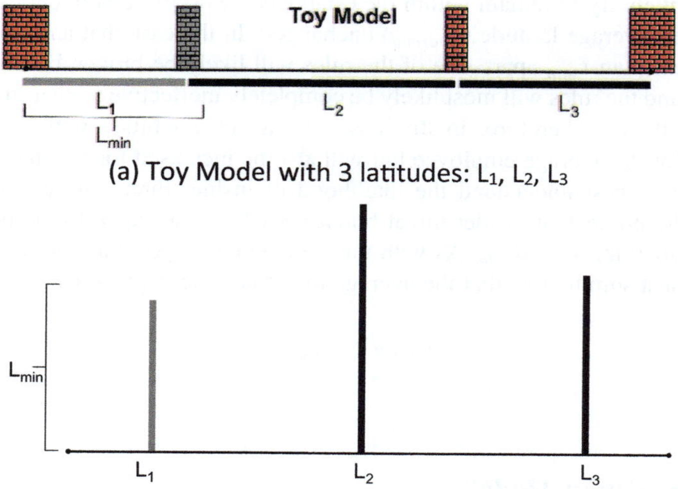

Fig. 3 Toy Model and Percolation Model. (**a**) Toy Model with three latitudes where $L_1 < L_{min}$. (**b**) Percolation Model that is equivalent to the Toy Model in (**a**). Site 1 is said to be occupied because $L_1 < L_{min}$. (Kepner et al. 2015)

Fig. 4 Probability of a percolation site being occupied versus latitude. Latitudes are sorted differences that are well modeled by an exponential distribution. (Kepner et al. 2015)

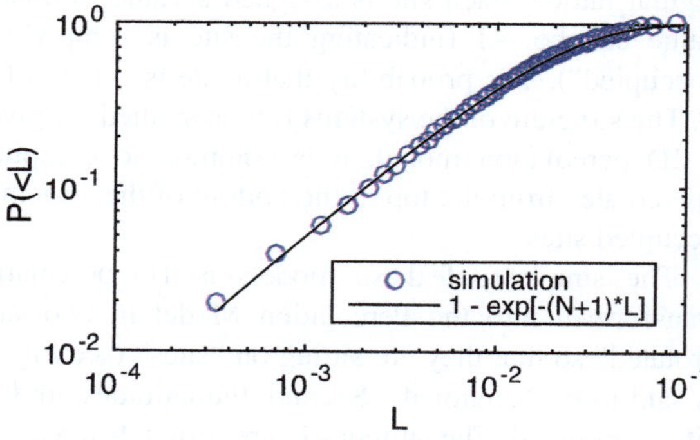

$$N_{threat} = (1 - P(N, L_{min})N) \tag{6}$$

resulting in

$$L_{exact}(N, L_{min}) = 1/(1 - P(N, L_{min}))N + 1 = 1/Ne^{-(N+1)L_{min}} + 1 \tag{7}$$

The above expression is minimized when $N_{min} = 1/L_{min}$ and results in

$$L_{exact}(N_{min}, L_{min}) = 1/N_{min}e^{-(L_{min}+1)} + 1 \tag{8}$$

that for $L_{min} \ll 1$ and $N_{min} \gg 1$ is

$$L_{exact}(N_{min}, L_{min}) \approx eL_{min} \tag{9}$$

Transforming the Toy Model into the Percolation Model connects it with the extensive theoretical work on percolation. One of the primary areas of study in percolation theory is the probability of clusters of occupied sites (i.e., areas of large latitude). The average size of a cluster in an infinite site 1D Percolation Model is (Stauffer and Aharony 1991)

$$S(P) = {}^{(P_c+P)}\!/_{(P_c-P)} \tag{10}$$

where P_c is the percolation threshold. For an infinite site 1D percolation $P_c = 1$. In the Percolation Model of an insider threat the average latitude of a site is

$$L_{avg} = {}^{1}\!/_{N+1} \tag{11}$$

Combining the formulas for L_{avg} and $S(P)$ gives the threat latitude in an infinite site 1D Percolation Model

$$L_{threat} = L_{avg}S(P(N, L_{min})) \tag{12}$$

2.3 *Modeling Exercise for Organizational Resilience*

Gisladottir et al. (2016) use the concept of DL to evaluate a node's susceptibility to a threat. Specifically, DL is defined as the measure of an employee's ability to make work-related decisions. Kepner et al. (2015) represent an employee's DL in Fig. 5 and show that the regulatory environment of a node comprises a set of barriers X_1, …, X_b that are placed on a unit segment of a straight line, where X_i is the position of the *i*th barrier. The amount of DL between barriers is therefore calculated as $L_i = X_i - X_{i-1}$. As described by Kepner et al. (2015), the sum of all DL amounts equals unity.

Gisladottir et al. (2016) assign the same unit segment of a possible rules implementation to the entire network, in that each employee is assumed to have the same set of rules. Within an actual organization, individual employees might differ in their levels of access and trust but for now this is ignored in the initial modeling approach. The barrier positions are drawn within a uniform distribution range (0; 1). The authors note that the modeled regulatory environment depends both on the number

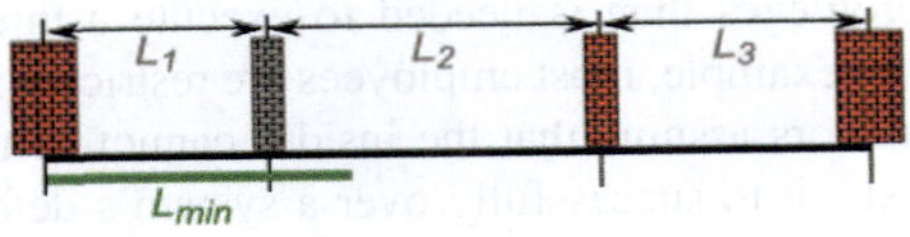

Fig. 5 One-dimensional users' DL model. (Gisladottir et al. 2016)

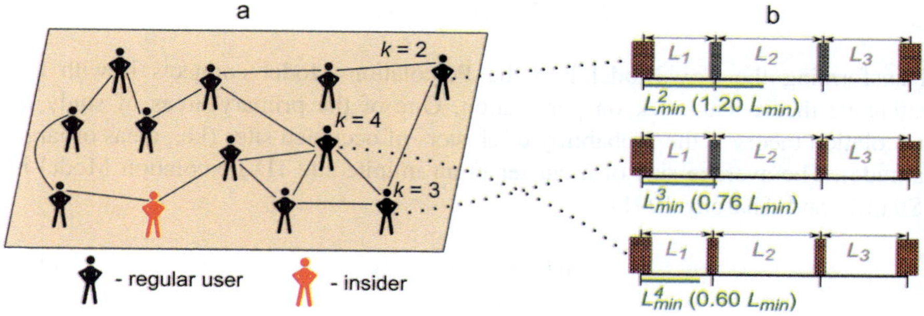

Fig. 6 Model of the business relationships network (**a**) and its nodes' minimum DLs (**b**). Notably, users of higher degrees have a smaller value of the DL, a higher number of strict barriers (red), and, hence, are less vulnerable. (Gisladottir et al. 2016)

of barriers (rules) and their stochastic placement. The authors determine the model placements by running a set of 10,000 Monte Carlo simulations for each data point. The variable L_{min} denotes the minimum DL value that employees need to perform their work-related tasks. In the original model, this value is a constant parameter. However, the authors note that employees with more connections (higher node degree) have more expertise and therefore need less DL to accomplish their business-related tasks. The authors incorporate the expertise level of individual employees by determining the minimum latitude (L_{min}) based on the node's degree and making the minimum latitude (L_{min}) inversely proportional to the natural logarithm of that degree. In the model, k_j represents the degree of node j. If the values of L_{min} are dependent on only the nodes' degrees and not on any other topological parameters or properties of the network, then the L_{min} of all nodes with degree k (L_{min}^{k}) are described as follows (Fig. 6):

$$L_{min}^{k} = \frac{L_{min}}{\ln(k)} N / \sum_{j=1}^{N} \frac{1}{\ln(k_j)} \tag{13}$$

In Eq. 1, the multiplication by $N/\sum_{j=1}^{N} \frac{1}{\ln(k_j)}$ takes into account the average value of a node's L_{min} across the whole network. Specifically, this formulation ensures that the smallest degree in the network is 2. Within the model, barrier strictness is accounted for if the amount of DL between barriers (L_i) is greater than the $L_{min.}$ The rules within the organization are nonstrict if this is not the case.

Within the model, L_T represents the minimum latitude (L_{min}) needed to execute a threat by an insider. As stated by the authors, employees typically have fewer privileges than is needed to execute a threat, therefore implying that $L_T > L_{\text{min.}}$ For example, most employees are restricted in access to sensitive corporate data. The authors assume that the insider cannot control the number of rules that they must exploit to successfully over a system's defense and that the threat latitude extends from 0 to L_T. Thus, the authors find that the number of strict and nonstrict barriers that the insider has to cross, represented as N_s and N_{ns}, are represented as the number of strict and nonstrict barriers within the range (0; L_T).

When the attack is initiated and the propagation of the attack occurs throughout the network, the probability of each new node becoming breached is represented in the following way (Eq. 2):

$$p_{new} = p_{ns}^{N_{ns}} p_s^{N_S} \tag{14}$$

Where the probability of each new breach is dependent on the probability of breaching of each strict and nonstrict barrier. Once the authors have modeled the extent to which threats propagate throughout a given system, the next step is to calculate the system's resilience when under attack. An organization's resilience is defined as the ability of an organization to prepare for, absorb, recover from, and adapt to adverse events (Linkov and Florin 2016; NRC 2012). In this article, Gisladottir et al. (2016) model only the absorption and recovery phases of resilience according to the methodology proposed by Ganin et al. (2016) and evaluate the system's *critical functionality* (CF), defined as a system's performance function of time, during the absorption and recovery phases. Specifically, Gisladottir et al. (2016) assume that the preparation and adaptation phases of resilience do not take place in the case of an attack, and, therefore, are not relevant in the case of a simulation. CF (K) is defined as the ratio of uninfected nodes to total number of nodes at each time step in a given simulation. If N_{in} represents the number of infected nodes, then (CF) K is found as follows:

$$K = 1 - \frac{N_{in}}{N} \tag{15}$$

In this exercise, resilience is calculated as taking place within the absorption and recovery phases, which take place within control time T_C. Specifically, the absorption phase begins when the initial nodes are breached and the recovery phase ends when organizational employees detect the breach. Thus, the resilience value R is calculated by integrating the (CF) K over the time interval [0, T_C] (Ganin et al. 2016):

$$R = \frac{1}{T_C} \int_0^{T_C} K \, dt \tag{16}$$

The authors model the system's evolution in discrete time steps, as visualized in Fig. 4. Therefore, Eq. 4 can be re-written to represent resilience (R) in the following way (Fig. 5):

$$R \approx 1 - \frac{1}{T_C} \sum_{t=0}^{T_C} \frac{N_{in}(t)}{N} \tag{17}$$

3 Results

3.1 Toy Model

A simulation of the employee and insider threat behavior stemming from the Toy Model is shown in Fig. 7, in which the average latitude (L_{normal}) for an employee decreases as the number of rules in a workplace increases. As shown in Fig. 7, the insider threat latitude first decreases with the number rules, reaches a minimum, and then increases as the number of rules increases. Thus, the average latitude for an insider threat starts to increase at the point in which the average employee reaches minimum latitude, as calculated

$$N_{min} = {}^{1}\!/_{L_{min}} \tag{18}$$

As expected, if $N > N_{min}$, then threat latitude increases.

The ratio of the threat latitude to the normal latitude is shown in Fig. 8. For modest numbers of rules the threat latitude is comparable to the normal latitudes. As N approaches N_{min} the ratio starts to increase. In this particular instance, the threat latitudes is ~3 times the normal latitude at $N = N_{min}$. The ratio increases dramatically as N surpasses N_{min}.

Figures 7 and 8 illustrate four possible workplace scenarios, including under-regulated, possibly optimal, tipping point, and over-regulated workplace environments, that are described in Fig. 9. In the under-regulated regime, the threat has a large latitude that is very similar to that of the normal individual. At the tipping-point, the threat latitude is minimized and the normal latitude is near the L_{min}. In the over-regulated regime, the threat latitude is increasing and the normal latitude is below L_{min}. Kepner et al. (2015) speculated that a possibly optimal regime exists between the under-regulated regime and the tipping-point.

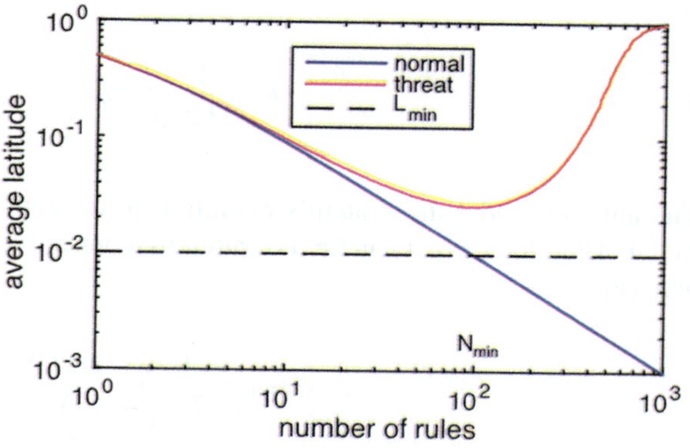

Fig. 7 Toy Model average latitude versus number of rules for a normal individual and an insider threat averaged over 100 simulations. Insider threat latitude reaches a minimum when the number rules is $N_{min} = 1/L_{min}$. (Kepner et al. 2015)

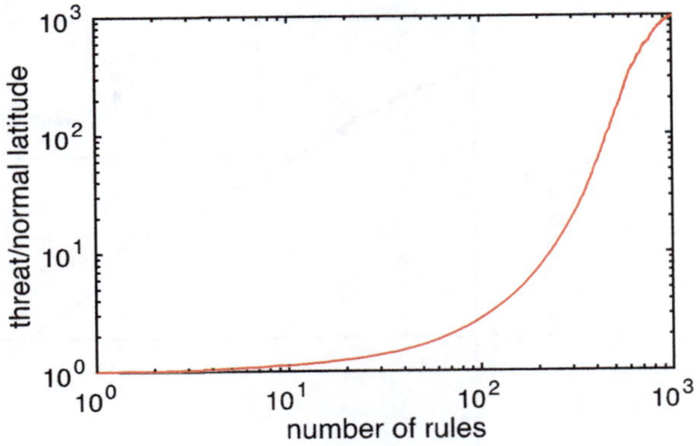

Fig. 8 Toy Model ratio of relative latitude of insider threat to normal individual. For modest numbers of rules the threat latitude is comparable to the normal latitudes. As N approaches N_{min} the ratio starts to increase. In this particular instance, the threat latitudes is ~3 times the normal latitude at $N = N_{min}$. The ratio increases dramatically as N surpasses N_{min}. (Kepner et al. 2015)

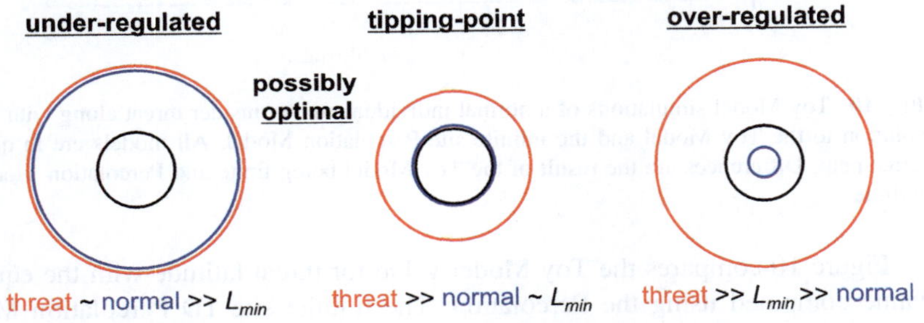

	under-regulated	tipping-point	over-regulated
pros	• normal latitude greatest • threat latitude comparable to normal	• threat latitude least	
cons	• normal latitude much higher than L_{min}	• threat latitude much greater than normal • normal latitude near L_{min}	• threat latitude greatest • normal latitude below L_{min}

Fig. 9 Toy Model regimes. In the under-regulated regime, the threat has a large latitude that very similar to that of the normal individual. At the tipping-point, the threat latitude is minimized and the normal latitude is near the L_{min}. In the over-regulated regime, the threat latitude is increasing and the normal latitude is below L_{min}. The authors speculated that a possibly optimal regime exists between the under-regulated regime and the tipping-point. (Kepner et al. 2015)

3.2 *Percolation Model*

It is not realistic to assume that complex human activities can be precisely captured with the simple Toy Model, however. A more realistic model requires representing environments with hierarchies of rules, complex connections between the domains of individuals, and variations in individual behaviors. However, it is plausible that the phenomena of under-regulation and over-regulation could be captured by a simple model. One way to rapidly increase the complexity of a system is to increase its dimensionality. If the qualitative behavior of a system is remains the same when the dimensionality is increased, then it may remain similar when other complexities are added.

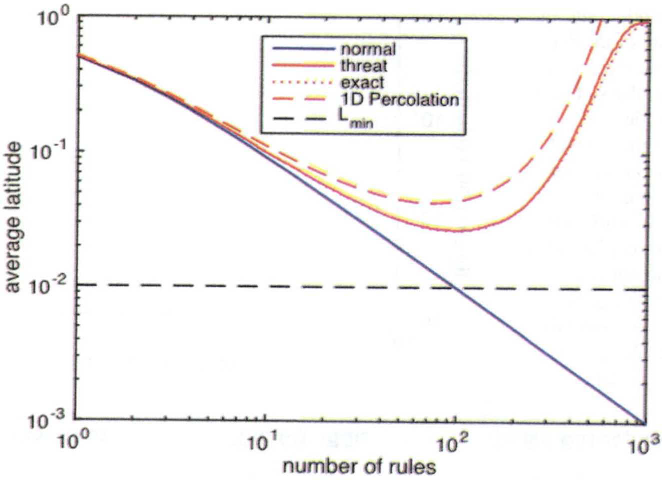

Fig. 10 Toy Model simulations of a normal individual and an insider threat along with the exact solution to the Toy Model and the infinite site Percolation Model. All models are in qualitative agreement. Differences are the result of the Toy Model being finite and Percolation Model being infinite

Figure 10 compares the Toy Model value for threat latitude with the equivalent value computed using the Percolation. The infinite site 1D Percolation Model is qualitatively equivalent to the Toy Model simulations. Small differences between the models are due to finite site simulations versus an infinite site model. Finite percolation models can also be calculated analytically, but they are beyond the scope of this work.

The qualitative agreement between the Toy Model and the Percolation Model allows the Toy Model to be connected with far more complex models. Percolation has been studied in many systems (see Fig. 8) and all exhibit the same qualitative behavior that is determined by a single parameter P. This suggests the qualitative behavior of the 1D insider threat model may also be similar to higher fidelity insider threat models. For example, in more complex percolation models the percolation threshold P_c gets smaller as the model gets more complex (see Fig. 11). This would mean that N_{min} becomes smaller in more complex systems and that insider threat latitude starts increasing with a smaller number of rules.

That the qualitative behavior of the Toy Model would be similar across a wide range of more complex models is consistent with the observation that under-regulation and over-regulation exist in a wide range of immensely complex system.

3.3 Modeling Exercise for Organizational Resilience

Figure 12 plots critical functionality as a function of time and the risk and resilience as functions of number of rules. Each line in Fig. 12b, c represents varying levels of system vulnerability (V). The graphs show that risk is high and resilience is low at

Dimension	Lattice	Site Neighbors	P_c
1d	linear	2	1
2d	honeycomb	3	0.696
2d	square	4	0.593
2d	triangular	6	1/2
3d	diamond	4	0.430
3d	dimple cubic	6	0.312
3d	body centered cubic	8	0.246
3d	face centered cubic	12	0.198
4d	hypercubic	8	0.197
5d	hypercubic	10	0.141
6d	hypercubic	12	0.107
7d	hypercubic	14	0.089
	Bethe lattice (Cayley graph)	z	1/(z-1)

Fig. 11 Percolation has been studied in a wide range of systems. The table shows a select few with varying number of dimensions, geometry and numbers of neighbors. P_c is the probability at which the system is likely to become fully connected. (Kepner et al. 2015)

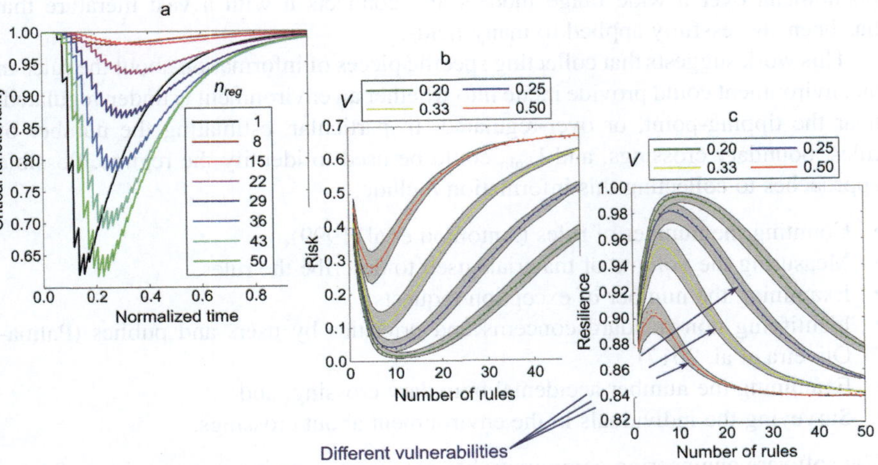

Fig. 12 **(a)** The CF of the system is plotted as a function of time for varying levels of rules, and **(b)** risk and **(c)** resilience are plotted as functions of rules for varying vulnerability levels. The shaded areas in panels **(b)** and **(c)** correspond to the values of risk and resilience for threat latitudes (LT) of 4.5 and 5.5, while the lines reflect the case of $LT = 0.5$. (Gisladottir et al. 2016)

very high levels of vulnerability, regardless of the number of rules implemented at an organization. Thus, at highly vulnerable organizations, risk and resilience cannot be effectively managed and, as a result, vulnerability within an organization must be dealt with first. Alternatively, when organizations possess lower levels of vulnerability, risk declines to a minimum under an increasing number of rules and then steadily increases as the number of rules increase. Resilience undergoes a similar trend, except in the opposite directionality. Thus, resilience increases as the number of rules increases and eventually declines under an increasing number of rules. Despite the level of vulnerability at an organization, the authors find that the maximum level of resilience falls between a small number of rules (between 5 and 15). System designers can therefore rely on this given estimate in the case that model assumptions and parameter values are not available.

4 Conclusions

Rules are the primary means of coordinating the activities of individuals who have a variety of goals and purposes. Over-regulation and under-regulation can both be exploited by insider threats. The Toy Model exhibits this regulation phenomena quantitatively. Adapting the Toy Model to the Percolation Model confirms this phenomena over a wide range models and connects it with a vast literature that has been successfully applied to many fields.

This work suggests that collecting specific pieces of information about the rules in an environment could provide incite into whether an environment is under-regulated, near the tipping-point, or over-regulated. In particular, estimating the number of rules, boundary crossings, and L_{min} could be used to identify the regime. Possible approaches to collecting this information include:

- Counting the number of rules (Antoniou et al. 1999),
- Measuring the volume of materials used to describe the rules,
- Examining the number of exception requests,
- Identifying non-standard concerns and priorities by users and publics (Palma-Oliveira et al. 2017)
- Examining the number accidental boundary crossing, and
- Surveying the individuals in the environment about crossings.

The software engineering community has extensive experience measuring software volume and some of these techniques are currently being applied to measure the U.S. Legal Code (Li et al. 2015). As with software engineering, care must be taken to distinguish active rules (e.g., source code) from inactive rules (e.g., commentary).

In the face of insider threats and cyber-attacks, system designers and administrators should aim to find the amount of organizational regulation that both decreases system risk and increases system resilience. Currently, insider risks are managed qualitatively with a focus on determining the appropriate number of employees to maintain security (Carroll 2004; Bunn 2004; Sagan 2004; Ghaffarzadegan 2008).

(Gisladottir et al. 2016) calls for a framework that systematically evaluated the rules, risks, and resilience of digital systems through the inclusion of behavioral science. Specifically, Gisladottir et al. (2016) argues for a framework that utilizes data from existing systems and establishes best practices based on an optimal number of rules. The benefits of optimization include expenditure of minimal resources and focus of security efforts on cardinal threats. Another benefit is that research models that incorporate resiliency of a cybersystem and an employee's ability to keep the security rules intact are often overlooked. However, employee behavior is a key component of the cybersecurity system and the critical systems and infrastructure of an organization. Gisladottir et al. (2016) note that the cybersecurity field would be benefit from increased management of resilience, with lesser focus on the implementation of new rules based on reactions to past consequences.

Optimization of rules and regulations to increase system resilience and performance requires a systematic framework. The results from the model described in this article demonstrate the hypothetical u-shaped system risk as a function of system behavior. Knowledge of the relationship between rules and risk allow system designers to move toward a more accurate predictive model of system behavior. Furthermore, the practical application of the model developed in this study involves estimation of the minimum DL (*L*min) using cognitive psychology methods where employees are asked a series of questions about their likelihood to follow a certain number of rules during the execution of particular tasks (Wood et al. 2017). Other methodologies exist to quantify the DL of specific employees. Once a more optimal number of rules are implemented, analytical techniques will be needed to evaluate the effect of new rules on the security of specific systems.

References

Aizenman J (2009) Financial crisis and the paradox of under- and over-regulation. National Bureau of Economic Research, Cambridge

Antoniou G, Billington D, Maher M (1999) On the analysis of regulations using defeasible rules. HISCC-32 proceedings of the international conference on systems sciences

Barabasi AL (1999) Emergence of scaling in random networks. Science 286(5439):509–512

Bunn M (2004) Thinking about how many guards will do the job. Risk Anal 24(4):949–953

Carroll JS (2004) Redundancy as a design principle and an operating principle. Risk Anal 24(4): 955–957

CESG (2016) Password guidance: simplifying your approach. Communications-Electronics Security Group, Government Communications Headquarters, Cheltenham. At: https://www.ncsc.gov.uk/guidance/password-guidance-simplifying-your-approach. Accessed 25 July 2017

Claycomb W, Huth C, Phillips B, Flynn L, McIntire D (2013) Identifying indicators of insider threats: Insider IT sabotage. 47th international Carnahan conference on security technology (ICSST), October

Connelly EB, Allen CR, Hatfield K, Palma-Oliveira JM, Woods DD, Linkov I (2017) Features of resilience. Environ Syst Decis 37(1):46–50

DiMase D, Collier ZA, Heffner K, Linkov I (2015) Systems engineering framework for cyber physical security and resilience. Environ Syst Decis 35(2):291–300

Fagerlind AC, Gustavsson M, Johansson G, Ekberg K (2013) Experience of work-related flow: does high DL enhance benefits gained from job resources? J Vocat Behav, (83), 2, 161–170. At: doi:https://doi.org/10.1016/j.jvb.2013.03.010. Accessed 25 July 2017

Ganin AA, Massaro E, Gutfraind A, Steen N, Keisler JM, Kott A, Mangoubi R, Linkov I (2016) Operational resilience: concepts, design and analysis. Sci Rep 6:19540

Ghaffarzadegan N (2008) How a system backfires: dynamics of redundancy problems in security. Risk Anal 28(6):1669–1687

Gisladottir V, Ganin AA, Keisler JM, Kepner J, Linkov I (2016) Resilience of cyber systems with over-and underregulation. Risk Anal 37(9):1644–1651

Hein O, Schwind M, Konig W (2006) Scale-free networks: the impact of fat trailed degree distribution on diffusion and communication processes. Wirtschaftsinformatik 48(4):267–275

Kepner J, Gadepally V, Michaleas P, Schear N, Varia M, Yerukhimovich A, Cunningham R (2014) Computing on masked data: a high performance method for improving big data veracity. IEEE HPEC

Kepner J, Gadepally V, Michaleas P (2015) Percolation model of insider threats to assess the optimum number of rules. Environ Syst Decis 35(4):504–510

Klieman C (1996), Secretaries often face ethical dilemma on job: many follow their bosses' orders to lie or violate rules, fearing for their jobs if they refuse. Orange County Register

Li W, Azar P, Larochelle D, Hill P, Lo AW (2015) Law is code: a software engineering approach to analyzing the United States code. J Bus Technol Law 10(2):Article 6

Linkov I, Florin MV (eds) (2016) International risk governance council (IRGC) resource guide on resilience. EPFL International Risk Governance Center, Lausanne. Available at: http://www.irgc.org/irgc-resource-guideon-resilience/

Linkov I, Rosoff H, Valverde LJ, Bates ME, Trump B, Friedman D et al (2012) Civilian response corps force review: the application of multi-criteria decision analysis to prioritize skills required for future diplomatic missions. J Multi-Criteria Decis Anal 19(3–4):155–168

Linkov I, Eisenberg DA, Bates ME, Chang D, Convertino M, Allen JH, Flynn SE, Seager TP (2013) Measurable resilience for actionable policy. Environ Sci Technol. Available at: http://pubs.acs.org/doi/abs/10.1021/es403443n

Linkov I, Anklam E, Collier ZA, DiMase D, Renn O (2014a) Risk-based standards: integrating top–down and bottom–up approaches. Environ Syst Decis 34(1):134–137

Linkov I, Bridges T, Creutzig F, Decker J, Fox-Lent C, Kroger W, Lambert JH, Levermann A, Montreuil B, Nathwani J, Nyer R, Renn O, Scharte B, Scheffler A, Schreurs M et al (2014b) Changing the resilience paradigm. Nat Clim Chang 4(6):407–409

Linkov I, Trump BD, Fot-Lent C (2016) Resilience: approaches to risk analysis and governance. In: Linkov I, Florin M-V (eds) IRGC resource guide on resilience. Available at: https://www.irgc.org/risk-governance/resilience/

Linkov, I., & Trump, B. D. (2019). The Science and Practice of Resilience. Springer, Amsterdam

Myers J, Grimaila M, Mills F (2009) Towards insider threat detection using web server logs. ACM proceedings of the 5th annual workshop on cyber security and information intelligence research: cyber security and information intelligence challenges and strategies

National Research Council (NRC) (2012) Disaster resilience: a National Imperative. The National Academies Press, Washington, DC, p 2012

Ogus A (2002) Comparing regulatory systems: institutions processes and legal forms in industrialised countries, Report No.: 35/2002. University of Manchester, Manchester. At: http://purl.umn.edu/30609. Accessed 25 July 2017

Palma-Oliveira JM, Trump BD, Wood MD, Linkov I (2017) Community-driven hypothesis testing: a solution for the tragedy of the anticommons. Risk Anal 38(3):620–634

Sagan SD (2004) The problem of redundancy problem: why more nuclear security forces may produce less nuclear security. Risk Anal 24(4):935–946

Saji G (2003) Safety goals in "risk-informed, performance-based" regulation. Reliab Eng Syst Saf 80(2):163–172

Silowesh G, Nicoll A (2013) Best practices and controls for mitigating insider threats. CMU Software Engineering Institute, DTIC, Pittsburgh

Stauffer D, Aharony A (1991) Introduction to percolation theory. Taylor and Francis, Boca Raton

Trump BD, Poinsatte-Jones K, Elran M, Allen C, Srdjevic B, Merad M et al (2017) Social resilience and critical infrastructure systems. In: Resilience and risk. Springer, Dordrecht, pp 289–299

von Solms R, van Niekerk J (2013) From information security to cyber security. Comput Secur. At: https://doi.org/10.1016/j.cose.2013.04.004. Accessed 31 July 2017

Williams R, Adams M (2012). Regulatory overload, Report no.: 103. Mercatus Center at George Mason University, Arlington, 4p. At: https://www.mercatus.org/system/files/Regulatory_Overload_WilliamsAdams_MOP103.pdf. Accessed 25 July 2017

Wood MD, Thorne S, Kovacs D, Butte G, Linkov I (2017) Mental modeling approach: risk management application case studies. Springer, New York, 251p

Zhu Q, Basar T (2011) Robust and resilient control design for cyber-physical systems with an application to power systems. IEEE, Piscataway, pp 4066–4071

Resilience in the Cloud-to-Things Continuum

David Perez Abreu, Karima Velasquez, Bruno Faria, Marilia Curado, and Edmundo Monteiro

1 Introduction

The Internet of Things (IoT) has opened up new possibilities for services and applications such as Augmented Reality (AR), smart homes and cities, self-driving cars, and eHealth, which have become an integral part of our daily lives. The IoT is a network of physical objects, like sensors and actuators, that collect and exchange data. These devices are typically resource-constrained and rely on batteries. To provide resources for IoT-based services, Cloud computing has emerged as a solution that offers a centralised pool of storage and processing power, providing scalability and reliability. However, this approach also poses challenges, as data has to travel across the network infrastructure to the core for processing and then come back to the Edge. This leads to increased delays for end-users, unnecessary network traffic and additional data privacy risks. Therefore, while the Cloud has proven to be an effective solution for many applications, it may not always be suitable for the various data generated by IoT devices and the diverse application requirements (Bittencourt et al. 2018). The Edge computing paradigm emerged as an extension of the Cloud, offering the very needed resources for IoT networks in their vicinity (not necessarily at IoT level), minimising the load of the Cloud and offering shorter response time to end users. The Edge provides a computational hierarchy that allows the collection, aggregation, and processing of data from/to IoT devices. The Edge (and related paradigms as Fog and Mobile Edge Computing (MEC)) are seen as more appropriate for handling the large volume of security-critical and time-sensitive data being produced and consumed by IoT devices (Yousefpour et al. 2019).

D. P. Abreu (✉) · K. Velasquez · B. Faria · M. Curado · E. Monteiro
CISUC, Department of Informatics Engineering, University of Coimbra, Coimbra, Portugal
e-mail: dabreu@dei.uc.pt

© The Author(s), under exclusive license to Springer Nature Switzerland AG 2025

I. Linkov, A. Kott (eds.), *Cyber Resilience: Applied Perspectives*, Risk, Systems and Decisions, https://doi.org/10.1007/978-3-031-90109-6_8

The combination of these paradigms (i.e., Cloud, Fog, Edge, and IoT) creates a new scenario known as the *Cloud-to-Things continuum*, an extension of resources and computational capabilities that provide analysis, processing, storage, and data generation capabilities (Moreschini et al. 2022). The Cloud-to-Things continuum is a highly dense, heterogeneous, distributed, and complex environment that is particularly prone to failures at lower levels of the infrastructure. The battery dependency of the IoT devices makes them even more liable.

The Cloud-to-Things continuum is a new approach that is transforming the way that resources are orchestrated in the context of IoT and Edge computing. While the Cloud has been reliable for many applications in the past, it struggles to handle the diverse and time-sensitive data generated by IoT devices. The integration of Cloud, Fog, Edge, and IoT paradigms offers a promising alternative, providing localised resources and reducing latency for end-users. However, this convergence also presents new challenges, particularly in terms of fault tolerance and resilience in the event of infrastructure failures at lower levels. As we explore this complex and dynamic ecosystem, it is crucial to develop innovative solutions that not only address these challenges but also leverage the unique capabilities of each paradigm to maximise the potential of the Cloud-to-Things continuum. The tendency is to automate most of the management procedures, which reduces operational costs and complexity while improving network resilience, performance, efficiency, and user experience. The rest of this chapter will unwrap key enabling technologies to improve resilience in the Cloud-to-Things approach, as well as discuss some open issues in this ecosystem and how to deal with them.

2 The Cloud-to-Things Continuum

Cloud computing is a paradigm that offers on-demand access to computing resources, making it easy to deploy, manage, and scale (distributed) applications. For many years, the Cloud represented the solution for services and applications that could not rely on local devices for execution. The Cloud computing paradigm adoption by network operators and service providers has been massive, given its benefits in cost-savings, enhancement in work and management response, business agility, and Quality of Service (QoS) (Sunyaev 2020). Despite the initial success of Cloud adoption, in recent years, there has been a tendency shift to bring computational resources and services towards the Edge of the network to fulfil the requirements of emerging paradigms such as the IoT. In this scenario, there is a larger scale of heterogeneous devices and lower latency that could represent a significant challenge for the traditional Cloud environments (Luckow et al. 2021). This tendency shift brought new paradigms, including Fog computing.

The Fog is an environment with a myriad of heterogeneous devices that work in a ubiquitous and decentralised manner, communicating and cooperating among themselves (Sabireen and Neelanarayanan 2021). Thus, the Fog emerges as an extension of the Cloud paradigm escalating from the core of the network towards its Edge; it relies on a heavily virtualised approach that makes it able to perform storage, processing, and networking activities between the Cloud servers and the end devices (Abdali et al. 2021). The Fog comes to support novel applications and services that are not completely fit for the Cloud, granting them ubiquity, high resilience, low latency, decentralised management, and cooperation (Mann 2022). This way, the Fog unburdens the load of the Cloud by hosting some services allowing them faster response times and improved data privacy.

One step further from the Fog is Edge computing, which takes place at the Edge of the Fog closer to the IoT level, not necessarily at the end device but at a one-hop distance. Hence, the Edge represents the entry point to the network for the end user, e.g., IoT device, allowing more computational resources closer to the data generation/consumption point. The Edge is characterised by even lower latency than the Fog. Fog and Edge nodes are notoriously more limited regarding their computational, storage, and networking capabilities when compared to the Cloud; however, they bring advantages that include minimising costs in data transfer, processing, and storage while increasing privacy levels.

Given the limitations in resource capacity of the devices at the Edge and Fog, some applications and services will necessarily be deployed in the Cloud. Thus, Fog and Edge computing emerge only as an extension of the Cloud, creating a continuum of resources.

The definition of the continuum has been studied in the past. The Cloud-to-Things continuum approach adds new tiers (Fog, Edge, and IoT) and challenges; thus, a new definition was proposed (Moreschini et al. 2022):

> "Cloud continuum is an extension of the traditional Cloud towards multiple entities (e.g., Fog, Edge, IoT) that provide analysis, processing, storage, and data generation capabilities."

A representation of the Cloud-to-Things continuum is depicted in Fig. 1. As we move up in the continuum, there are more computational resources (e.g., processing, storage) but also less privacy and more latency. These conditions reverse as we move closer to the end user. If we were to compare both ends, there would be considerable differences. However, these differences are blurred among neighbour nodes inside the continuum.

The impact of the Cloud-to-Things continuum in the development of novel services and applications will depend in large part on the efficiency of the orchestration and security mechanisms specifically designed to manage such a complex scenario. Accordingly, it is necessary to pay special attention to maximising the availability of services and applications, even in case of failures or vulnerabilities.

3 Resilience in the Cloud-to-Things Continuum

Despite the literature on network resilience being relatively broad, there is almost no consensus regarding resilience in the network management context. Some of the definitions available for resilience in this field include (1) the ability to come back to

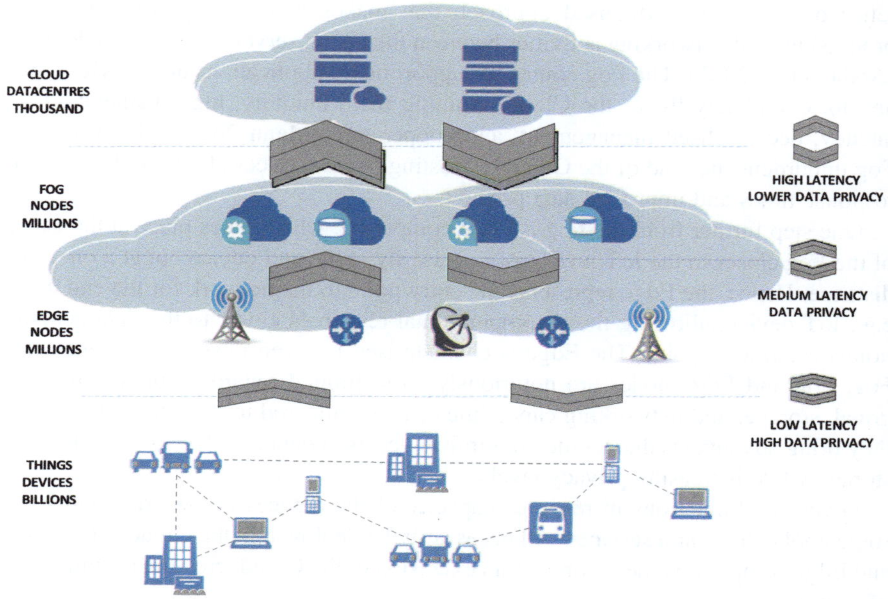

Fig. 1 Cloud-to-Things continuum

normal conditions after the occurrence of a disruptive event, (2) the capability of a system to maintain its functions in the face of internal and external changes, (3) the ability of a system to absorb unpredictable change and still keep its essential functions, (4) capacity of a system to withstand and recover from security threats, maintaining essential functions and protecting data integrity and availability, and (5) the aptitude of the system to withstand a significant disruption within admissible degradation parameters and to recover within an acceptable time and cost (Ahmadian et al. 2020). For this reason, and before moving forward with the discussion, it is necessary to specify the resilience definition to adopt for this chapter.

Starting from the principle that the main aspect of resilience lies in the ability of an entity to recover from an external disruptive event, in this chapter, resilience in the Cloud-to-Things continuum is defined as *a set of mechanisms to ensure service/application and networking robustness, by guaranteeing that resources and components are restored in case of failures or attacks*. For the restoration, it is possible to use a proactive approach or protection (actions before an anomaly) and/or a reactive approach or restoration strategy (actions after an anomaly) to improve the availability of nodes and links of a communication infrastructure (Amiri et al. 2023).

Reactive techniques do not pre-allocate resources for backup. Instead, they deal with anomalies once they take place. Using this approach might lead to a slower reaction time for recovery but would be less demanding with regard to resource consumption from the service provider's perspective. Contrarily, proactive solutions pre-allocate backup resources to guarantee fast recovery of services/applications in case of anomalies of components in the communication infrastructure (Amiri et al. 2023).

From the viewpoint of node failure (from unforeseen causes or via attacks), proactive approaches use backups or replicas of services that are allocated since the beginning of the lifetime of the application, while reactive approaches instantiate new replicas of services affected by failures after said anomaly occurs. For link failures, resilience is focused on the path concept. A path is a logical communication provision identified by source and destination end components, which is associated with a physical interface (Chiesa et al. 2021). Thus, the proactive approach or path protection is understood as the reservation of resources at the time the flow on the path is set up. In path restoration (reactive approach), when a path is broken, a Management and Orchestrator (MANO) module starts the restoration process by first calculating the backup path using the available resources and then re-establishing the flow.

Despite the kind of recovery technique used, it is necessary to perform the following tasks (Papadimitriou and Mannie 2006): (1) *Detection*, where an anomaly situation is identified; (2) *Localisation*, during this phase the point of anomaly is determined; (3) *Notification*, where the anomaly is informed to the appropriate entity that has to trigger the recovery process; and (4) finally, the actions to manage the problems and restore the normal operation are executed in the *Recovery Switching* phase. The MANO module in the Cloud-to-Things continuum is in charge of executing the previous tasks to trigger the proper actions to guarantee the resilience of services/applications in case of any disruption while achieving the end-users requirements.

Taking into consideration the previous discussion, resilience (*Res*) denotes a proportionality of the availability (*Ava*) and recovery (*Rec*) (Perez Abreu 2021), as per Eq. 1.

$$Res = Ava * Rec \tag{1}$$

The main focus of resilience is to provide mechanisms to support the continuity of systems and infrastructures (applications and networks). However, when different resilience techniques are available, it is important to make a conscious decision about the proper technique regarding a particular set of requirements. In order to make a choice in this context and also to compare the performance of mechanisms that enhance resilience, it is necessary to define metrics to measure the behaviour and results of the mechanisms that deal with resilience.

3.1 Resilience Metrics

To assess the best mechanisms to improve the resilience of services and applications in the Cloud-to-Things continuum, resilience metrics are defined in this subsection. An important concept that has to be defined in this context is the Quality of

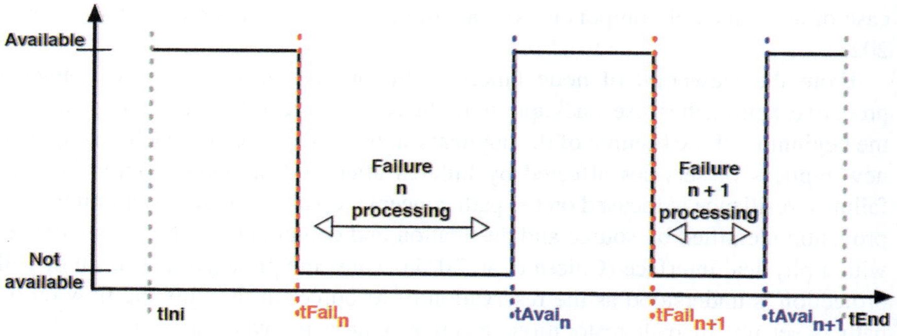

Fig. 2 ITU-T E.800 Availability Model. (Source ITU-T 2009)

Resilience (QoR). QoR is defined as the features of a communication infrastructure that affect the QoS perceived by users and are related to resilience (Chiesa et al. 2021). Considering this definition, it is possible to divide the metrics that measure resilience according to their *reliability attributes*. A discussion about the *reliability attributes* is presented below (Chiesa et al. 2021).

Continuity is the first reliability attribute and denotes the length of a period of time during which a service is not interrupted as a consequence of a failure. The Mean Time To Failure (MTTF) could be used to measure the continuity. MTTF is defined as the average duration of time from when a service request was received, assuming that the service was up until the service failed for the first time. In many systems, this metric is estimated or approximated using the notion of Mean Up Time (MUT). Figure 2 shows a binary scenario of availability originally proposed by the ITU-T (2009), where the system is considered available ($state = 1$) or unavailable ($state = 0$). The figure also describes the variables used in the model. Using this model, MUT corresponds to the moments where the system is in an available state, namely, where the model's function reaches the upper bound. Equation 2 represents a generic case of the MUT with n failures. In the absence of failures, Eq. 3 is used since $tFail_n = 0$ and $tAvai_n = 1$.

$$MUT = (tFail_1 - tIni) + \sum_{i=2}^{n} (tFail_i - tAvai_{i-1}) + (tEnd - tAvai_n) \quad (2)$$

$$MUT = tEnd - tIni \quad (3)$$

Downtime is the second reliability attribute and is related to the time period in which a service is inaccessible because of a failure in the communication infrastructure. The length of the interruption may be determined using Mean Down Time (MDT). MDT is the mean time duration from a service failure to the point when the service is recovered. In Fig. 2, MDT is represented by the moment when the service is unavailable or reaches the lower bound of the model's function. In a general case,

MDT is determined by Eq. 4. In the absence of failures, the MDT is equal to zero ($MDT = 0$).

$$MDT = \sum_{i=1}^{n} (\mathrm{tAvai_i} - \mathrm{tFail_i}) \tag{4}$$

With the definitions of MUT and MDT, it is possible to describe the **availability** (*Ava*) of a system as a function of these two metrics. The ITU-T (2009) proposes Eq. 5 to relate the availability to the MUT and MDT. In the computing and communication environment, it is most useful the concept of availability as a steady state, which may be understood as the probability of finding an item (i.e., object, device, network, and connection) in an operating state at any time that its service is required (Netes 2020).

$$Ava = \frac{MUT}{MUT + MDT} \tag{5}$$

The definitions of resilience discussed so far are general enough to be applied with different methods and mechanisms with the respective adjustments. The metrics exposed (*Ava*, MUT, MDT) can be used to determine the performance of mechanisms aimed at improving the resilience of the Cloud-to-Things continuum, which provides the communication infrastructure to end-user services/applications. Some trending approaches to infuse intelligence to achieve resilience in the Cloud-to-Things continuum are described below.

4 Intelligence in the Cloud-to-Things Continuum

The distributed and heterogeneous nature of the Cloud-to-Things continuum has necessitated the automation of network and service management operations (Benzaid and Taleb 2020). Various paradigms have emerged in pursuit of this objective.

Self-Organising Networks (SONs) seek to provide a cost-effective solution to both users and operators of communication networks by incorporating intelligence and autonomous adaptivity into network management. SONs have three main characteristics: (1) self-organisation; (2) self-optimisation; and (3) self-healing. Cognitive approaches can be used by management modules that gather knowledge based on current and past observations; thus, Machine Learning (ML) algorithms can allow the collection, exchange, and usage of data to enhance the learning process leading to a faster convergence incurring in lower costs (Moysen and Giupponi 2018). One fundamental area in the *Self-X* paradigm is auto-recovery or self-healing. The idea of guaranteeing this characteristic on SONs is to restore infrastructure components automatically and intelligently to the best possible status with a set of equipment, algorithms, and communication technologies. Thus, successful implementation of Self-healing depends on the comprehensive deployment of anomaly location, isolation, and service restoration, which requires data analysis & data engineering techniques. Intent-Based Network (IBN) is another approach for automating network management. The term intent is adopted as an evolution of the term policy for policy-driven management systems. Intents are defined as a set of specific policy types written in high-level operational and business objectives. The main idea is to meet system requirements without detailing how to achieve these objectives (Zeydan and Turk 2020). In conventional networks, an administrator can convey their intent in a concrete description of network configuration properly stored; on the other hand, in IBN, the intent is expressed abstractly and prescriptively (Leivadeas and Falkner 2023). Specifically, IBN relies on Artificial Intelligence (AI) and ML to prescribe and perform routine tasks, set policies, respond to system events (e.g., anomalies), and corroborate that goals and actions have been achieved.

The heterogeneous and diverse infrastructure of scenarios such as the Cloud-to-Things continuum leads to the design of an abstraction layer on top of the network infrastructure where it is possible to define and tune parameters in an agnostic way. This higher management level is the result of the search for a solution for the management of complex, heterogeneous, and distributed networks, and it is known as IBN. Simultaneously, the recent rise of interest in AI/ML platforms thanks to the advances in Deep Learning (DL) and mainly with applications in Natural Language Processing (NLP) promote paradigms such as IBN.

Zero touch network & Service Management (ZSM) emerged from the European Telecommunications Standards Institute (ETSI) to automatically orchestrate and manage network resources while assuring QoS requirements from users (Gallego-Madrid et al. 2022). This paradigm relies on ML as enabling technology to produce intelligent decision-making to network management systems. The type of ML technique to use will depend on the objectives set and on the nature of the information collected, which makes the monitoring and data gathering tasks critical for the success of an automated ML-based network management solution. The orchestration decisions sometimes will reach different (sometimes contradicting) objectives for a diverse set of applications and services (e.g., maximise availability, minimise energy consumption).

Knowledge-Defined Networking (KDN) is the evolution of other well-known paradigm, Software Defined Networking (SDN). While SDN decouples the control plane and the data plate to optimise the forwarding tasks, KDN introduces a novel plane, the knowledge plane, for generating knowledge on data collected from the network (Souza et al. 2022). The knowledge plane provides advantages such as automation and recommendations for operating, optimising, and problem-solving for computer networks. KDN and its knowledge plane become an enabler for other paradigms, like Self-X networks and ZSM, since the intelligence generated can guide the prediction and detection mechanisms needed to recover from failures, enhancement of security, and for automation in general.

These umbrella approaches, namely SON, IBN, ZSM, and KDN, serve as foundational pillars for implementing resilience-enhancing mechanisms within the

Cloud-to-Things continuum. They provide the framework for deploying specific techniques aimed at bolstering the network's ability to withstand and recover from disruptions. Recent proposals in this field leverage the capabilities offered by these paradigms to address challenges such as fault tolerance, security enhancement, and automation, ultimately contributing to the advancement of a more robust and resilient infrastructure for the evolving Cloud-to-Things ecosystem. Some of these proposals are discussed below.

5 Improving Resilience in the Cloud-to-Things Continuum

Cloud-to-Things systems are transforming numerous sectors by providing diverse services and applications, such as smart grids, autonomous vehicles, and eHealth, which require high availability for critical functionalities. Even brief disturbances can cause serious consequences, like power outages, road accidents, and delayed medical interventions (Berger et al. 2021; Xing 2020). Resilience in Cloud-to-Things systems refers to their ability to adapt to changing conditions, recover quickly from disruptions or failures, and protect the system against unpredictable threats, ensuring functionality and reliability (Berger et al. 2021). The following subsections depict some current efforts to improve resilience for fault tolerance and security.

5.1 Resilience for Fault Tolerance

Cloud-to-Things systems, relying on numerous interconnected components and resource-constrained devices, are vulnerable to failures (Xing 2020; Lynn et al. 2020; Volpert et al. 2023). By leveraging AI to handle large datasets and automate decision-making, the Cloud-to-Things continuum can become Self-healing and proactively address disruptions (Samie et al. 2019; Ashtari et al. 2022).

AI, particularly ML and DL, offers significant potential for enhancing Cloud-to-Things resilience through anomaly detection and automated decision-making within the system (Samie et al. 2019). Traditional ML and DL models often struggle with the resource-constrained nature of edge devices within the Cloud-to-Things model. This necessitates exploring techniques to reduce computational demands, memory requirements, and power consumption. Research efforts are actively exploring techniques like lightweight models and early exits within DL architectures to achieve this goal (Mao et al. 2021; Samie et al. 2019; Wang et al. 2020; Verdecchia et al. 2023; Teerapittayanon et al. 2016; Shi et al. 2020). These approaches aim to reduce the computational demands of AI models, making them more suitable for deployment on resource-constrained edge devices within the Cloud-to-Things continuum.

Cloud-to-Things's dispersed nature provides a unique opportunity to overcome resource constraints and complex decision-making processes. Distributed AI approaches, which break down large models and process them over several devices, are gaining traction (Arzovs et al. 2024; Carnero et al. 2021; Chen et al. 2021; Hosseinalipour et al. 2020). Distributed AI may help further minimise the computational strain on individual devices, resulting in lower total energy consumption within the Cloud-to-Things continuum. This concern is especially relevant in light of the European Green Deal[1] initiatives that emphasise energy efficiency. Additionally, processing data closer to the source considerably reduces latency, which is critical for real-time applications in areas like Virtual Reality (VR), where even minor latency can cause dizziness and nausea for users (Cortés et al. 2020). Furthermore, distributed AI architectures improve fault tolerance by allowing the system to function even when individual devices fail. This inherent redundancy contrasts with centralised architectures, where a single point of failure can cripple the entire system.

Self-healing capabilities are critical for ensuring a smooth operation within these complex ecosystems, enabling it to autonomously identify, diagnose, and potentially rectify issues like device malfunctions or infrastructure disruptions (Alonso et al. 2021). In this context, Zero-touch capacity remains an ongoing challenge (Coronado et al. 2022). Anomaly detection techniques play a crucial role in enabling Self-healing functionalities. Research in this area explores various methods, including statistical methods, ML, and DL approaches, to effectively identify deviations from normal system behaviour that might indicate potential problems (Caleb and Thangaraj 2023; Xing 2020; Nouioua et al. 2021).

5.2 *Resilience for Security*

Recent applications and services rely on the Cloud-to-Things continuum model, where most of the computation and data storage are handled by distributed servers. This model allows for scalable and accessible services, but raises some risks regarding security, privacy, and data sovereignty, among others (Durán et al. 2023).

As more devices join the network in an ever-increasing landscape, complex network security requirements emerge and more data breaches, vulnerabilities, and hacking attacks are reported. Previously existing solutions including Virtual Private Networks (VPNs) and Network Access Controls (NACs) validate that all internal network users are trusted, while outside users/devices/services are considered untrustworthy. Considering that intruders can exploit security vulnerabilities by gaining internal access, it becomes clear that more robust solutions ought to be applied (Buck et al. 2021). Zero-trust architectures are a novel network paradigm that promotes the idea that no network, whether internal or external, is trustworthy

[1] https://commission.europa.eu/strategy-and-policy/priorities-2019-2024/european-green-deal

Buck et al. (2021). In Zero-trust architectures, AI algorithms can provide information security in untrusted networks (Ramezanpour and Jagannath 2022). Predefined context-based attributes (e.g., location, timestamp) can be used to identify, verify, and authenticate users and devices in a trusted environment (Xiao et al. 2022; Abdulqadder and Zhou 2022; Bumiller et al. 2023).

Data generated at the Things layer by devices such as smartphones, sensors, or IoT devices might be sensitive (e.g., personal data or confidential commercially sensitive information). The protection of sensitive data in the Cloud-to-Things continuum (a dynamic, heterogeneous, and decentralised environment) is very challenging (Ayed et al. 2022). Authenticity and non-repudiation also need to be considered as key aspects, including technologies such as blockchain and digital signatures (Alagheband and Mashatan 2022). Unique digital identifiers, such as Physical Unclonable Functions (PUFs), derived from inherent manufacturing processes of integrated circuits help to enhance authentication at minimal overhead costs (Barbareschi et al. 2023).

With the levels of data generated in the Cloud-to-Things continuum continuing to grow, AI-based solutions become crucial. Pervasive AI can be used for processing and analysing data in real time, reducing the amount of data that needs to traverse the network (Puthal et al. 2023), with significant potential to enhance the security and efficiency of the Cloud-to-Things continuum (Baccour et al. 2022). Finally, and as it was discussed in the case of fault tolerance, green AI approaches (Hassan et al. 2022) (aligned with the requirements of the European Green Deal) ought to be considered, given the limited battery capacity of IoT devices (Heidari et al. 2023).

6 Use Cases

This section discusses two use cases that could benefit from the improvement of resilience in the Cloud-to-Things continuum; namely, smart water management and eHealth active monitoring.

6.1 *Resilience in Smart Water Management Using Artificial Intelligence and Digital Twins*

Water is a significant resource for socioeconomic growth and the promotion of healthy environments. Properly controlled water administration mechanisms could shape the development of communities. In urban sectors, conventional water system management tries to maximise the water flow active in the piping infrastructure to fulfil consumer demands while trying to reduce water losses. Thus, sustainable water resource management is essential for ensuring the proper use of this critical natural resource. Data-driven applications have become vital to daily living. The combination of intelligent metering and the IoT has enabled the water management system to transition from a legacy and manual approach to an intelligent and Zero-touch ecosystem taking advantage of the data gathered by the meters and IoT sensors deployed along the water pipelines. AI is foreseen as one of the enablers of this digital transition by transforming information into a light-way process, improving decision-making based on data-driven approaches by combining numeric tools, models, and human intellectual skills.

Understanding in detail the infrastructure of a system that wants to be improved is key to achieving optimal results. However, it is not always possible to have access to the system, given its complexity or importance. The water supply system in urban environments fits the scenario mentioned above since the water supply is essential to the citizens. For these scenarios, the Digital Twins (DTs) concept could be applied to explore the infrastructure at issue without jeopardising the real one. A DT is a representation of a physical process, person, place, system, or device to extract data, obtain insight, and achieve better situational awareness of the physical system's performance (Tao et al. 2022).

A smart water management system for urban environments that utilises AI and DT is what is envisioned for this use case. For this particular ecosystem, it is possible to improve aspects related to water management:

- Model physical aspects in virtual ones (DT) to better understand how to optimise them without jeopardising critical infrastructures;
- Improve the interaction between the physical system and the DT to guarantee accuracy in the simulations executed and, consequently, in decision-making;
- Explore the behaviour under specific loads or circumstances to predict outcomes and foresee the proper actions to take;
- Determine the consumption pattern of final users under different variants to provide informed recommendations.

Figure 3 shows the DT system working along the Cloud-to-Things continuum. From the left to the right, there is the hardware that belongs to the water system, which is monitored by sensors and controlled by actuators once an event is triggered (e.g., closing a faucet or water leaks). In the centre, the middleware layer deals with the storage, processing, and modelling of data, using software components depicted on the far right side. This ecosystem requires a high resilience of its components along the Cloud-to-Things continuum since it controls the management of a critical resource.

Using an AI approach and DTs across a resilient Cloud-to-Things continuum makes it possible to recognise weaknesses in the system without jeopardising the water piping infrastructure and improve it to achieve innovative and sustainable water management.

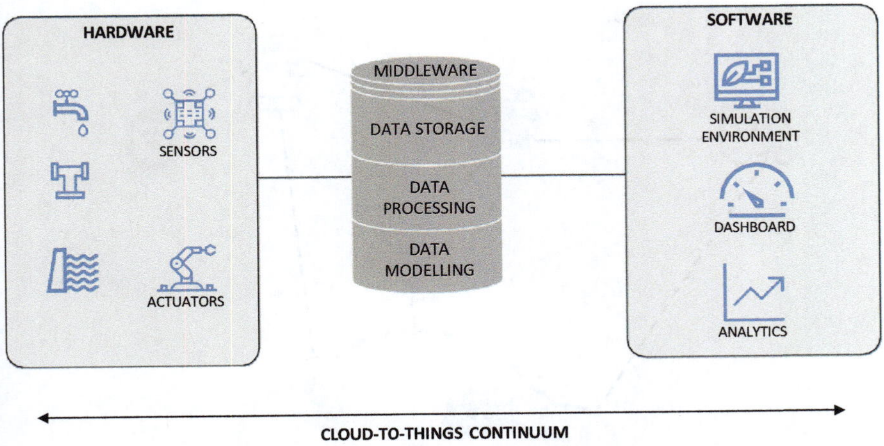

Fig. 3 Smart water management using AI and a DT

6.2 *Resilience in eHealth Scenarios Empowered by the IoT*

Digital Care Angel (DCA) is foreseen as an add-on to Drone Guard Angel (DGA), a service that enables drones (or Unmanned Aerial Vehicle (UAV)) to track and monitor people during their daily activities in urban areas. DGA is now in its final implementation stage in the scope of the ARCADIAN-IoT Horizon 2020 project,[2] focused on cybersecurity, privacy, trust, and recovery in IoT devices and services.

DCA is envisioned as a proactive care service primarily aimed at tracking the health conditions (and secondarily, the overall safety) of a person being followed by the drone, aligned with the smart healthcare paradigm. In this context, a remote user is monitored using an application of the Internet of Medical Things (IoMT), taking advantage of a drone with a high-resolution camera, as well as Image Recognition (IR) supported by AI and AR techniques for processing and visualisation of the collected data. 5G and MEC have a key role in supporting the transmission of the sensed data and its timely processing according to the overall service' and its different modules (i.e., video analytics, AR processor) requirements.

The following is a description of the use case. Consider a user who has requested the DCA service. The drone, which is connected to a 5G infrastructure, would reach the user and start collecting video (with Full High Definition quality) and additional data that might be considered essential (e.g., movement, temperature). In case of unexpected events (e.g., a panic button is activated by the user, the user faints, sudden temperature change), a series of events would be triggered by the service aiming to increase the awareness of the user's exact health status and his/her context (i.e., surroundings). For instance, the streaming resolution of the video is increased

[2] https://www.arcadian-iot.eu

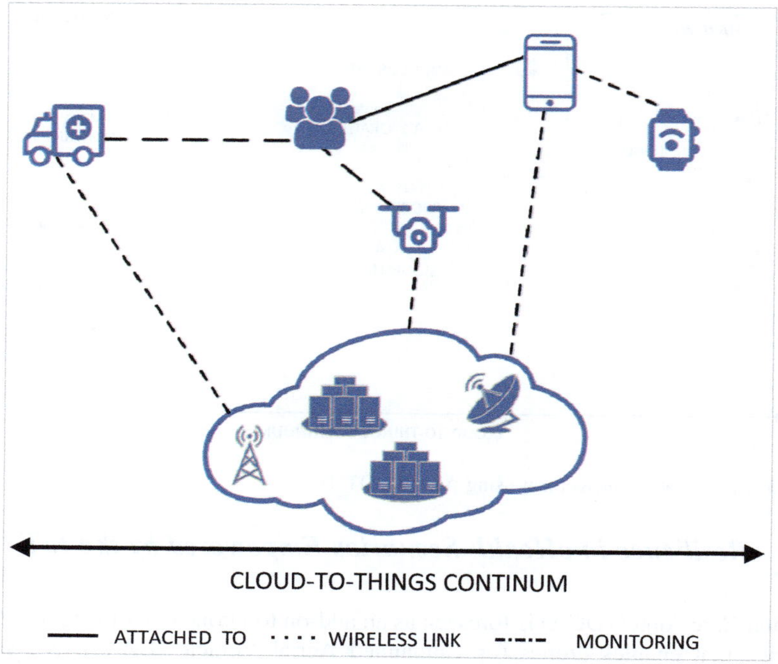

Fig. 4 Digital Care Angel

to get a more precise view. This increase in resolution implies the adaptation of the 5G network slice configuration to meet the new requirements of the UAV's data transmission (i.e., high-resolution surveillance video data rate ranging between 25 and 50 Mbps, End-to-End (E2E) latency of 100 ms). Once the slice is reconfigured, the information is shared with the DCA backend components deployed on the Edge of the network (using the MEC solution), such as AR artefacts overlayed in the video while guaranteeing the new bandwidth and latency requirements, and satisfying the Quality of Experience (QoE) for the medical team—assumed to be located in a central area and their access to the enriched video provides them with a more complete view of the user health status, enables better-informed decisions and reduced response time. DCA service is expected to improve the first triage of the user, increasing the chances of getting a proper and faster diagnosis and treatment, which in some cases might even impact the possibility of having sequelae after the event.

Figure 4 describes the DCA solution and the actors involved. In the figure, fully-dashed lines represent wireless communication (i.e., 5G, Bluetooth, WiFi); partially-dashed lines denote monitoring actions; and straight lines imply that the device is attached to (or carried by) the users. In this scenario, a reliable communication infrastructure is required to fulfil the proper functioning of the application.

The previously described scenario has an inherent complexity that requires the orchestration of resources, not only from the communication network perspective but also from the services that enrich the context using AR and AI-based IR along the Cloud-to-Things continuum, constituting a set of challenges for the DCA service. Another critical aspect for DCA, considering its eHealth action context, is to ensure that its components will be available all the time besides warranting the security of its artefacts. Thus, DCA requires high resilience guarantees from the availability and security perspective from the Cloud to the Things tiers.

7 Discussion and Open Issues

The limitations present in the Cloud-to-Things continuum, such as resource-constrained devices, platform heterogeneity, deployment in adverse environments, and administrative domains, make it susceptible to failures and attacks, compromising its resilience. IoT systems increasingly demand the resilient operability of services as they become part of critical infrastructures. Thus, there is a need for an automated resilience solution for fault tolerance and security that takes into consideration complementary aspects such as energy efficiency. This section presents a discussion about open issues on resilience regarding fault tolerance and security.

7.1 Open Issues in Fault Tolerance

The Cloud-to-Things continuum brings new challenges for fault tolerance. Traditional centralised approaches struggle to keep up with the dynamic nature of the Edge and IoT layers, which is characterised by resource-constrained devices and a constantly changing network topology. Therefore, there is a need for a paradigm shift towards more intelligent and automated solutions.

One promising approach to enhance network resilience is to use SONs that have Self-healing capabilities. These networks can autonomously reconfigure themselves when a node fails, thereby reducing downtime and improving overall reliability. However, incorporating AI and ML algorithms within SONs requires careful consideration of the computational limitations of Edge and IoT devices. To make practical deployment possible, it is essential to use lightweight solutions and offload tasks to nearby nodes with more resources.

A crucial factor to consider is achieving genuine decentralisation of the fault tolerance mechanisms. Relying too heavily on the Cloud leads to more latency and less privacy, factors that later on could impact the Service Level Agreement (SLA) of services. Hence, it is better to have a collaborative Edge network where nodes can identify resources and manage recovery procedures locally without the necessity of a central entity (the Cloud) for the decision-making. This requires additional exploration of local service discovery protocols and promoting cooperation among Edge and IoT entities following a replica set approach. Network Function Virtualisation (NFV) and KDN can play a vital role in managing these decentralised fault tolerance mechanisms. NFV allows for dynamic allocation of virtualised network functions across the Edge, facilitating the creation of redundant paths and real-time adaptation to network changes. KDN provides a centralised view and control over the entire network, enabling the configuration and management of these virtualised network functions empowered by its knowledge plane.

In order to improve decision-making and fault-tolerant mechanisms, it is essential to develop standardised data formats for gathering sensor data, network metrics, and resource utilisation across heterogeneous devices in the Cloud-to-Things. This will facilitate data exchange and the creation of comprehensive datasets for cognitive and intelligent models. Additionally, exploring federated learning techniques where the devices collaboratively train ML models without directly sharing private data while leveraging the collective knowledge of the network is a promising area that needs to be investigated.

Transfer learning, anomaly detection, and consensus protocols are techniques that can enhance fault tolerance in the Cloud-to-Things continuum. Transfer learning can be used to leverage pre-trained models from similar domains and reduce training times on Edge and IoT devices. Anomaly detection algorithms can help identify deviations from normal operating patterns and enable proactive fault mitigation before they cause network failure. Lightweight consensus protocols for Edge and IoT devices can be designed to coordinate recovery actions and ensure a consistent response to failures, preventing conflicting decisions.

Furthermore, the trade-off between fault tolerance and resource efficiency needs to be addressed. While sophisticated fault tolerance mechanisms may enhance resilience, they can also introduce overhead that strains the limited resources of Edge and IoT devices. Thus, striking a balance between fault tolerance and resource efficiency is essential, requiring innovative approaches that optimise performance without sacrificing reliability.

By focusing on these research areas, it is possible to move beyond the current state of the art and achieve a future Cloud-to-Things continuum characterised by exceptional resilience and Self-healing capabilities. This will pave the way for more reliable and trustworthy applications across various domains, ranging from industrial automation to smart cities and connected eHealth.

7.2 Open Issues in Security

The Cloud-to-Things continuum is characterised by a diverse and decentralised context, with multiple layers of technology and data nodes, processes, and objects distributed across them. This leads to an extended attack surface and complex issues of interoperability, scalability, and trust, which are further compounded by trends such as the significant rise in machine-to-machine interactions. These interactions increasingly span across multiple domains and federations, including drones, autonomous vehicles, domestic appliances, and the monitoring and control of automated systems in heterogeneous industrial settings. It is essential to ensure that such federations are designed and implemented with the highest levels of security and data protection by design and by default to truly achieve a *Secure Computing Continuum* in practice.

Europe is striving to create an ecosystem where Cloud, Fog, Edge, and IoT technologies converge towards a computing continuum. This is a key policy line for digital autonomy and the Green Deal. While significant progress has been made in terms of increasing computing capabilities at the Edge and integrating Cloud services, security still remains a major concern. The challenge is to create effective solutions that can improve the security of Cloud-to-Things continuum setups under realistic conditions. This can be achieved through the *Maximum Achievable Security* approach, which involves designing mechanisms and techniques for Cloud-to-Things scenarios in the real world where participating devices have diverse security-enhancing features. The ultimate goal is to create solutions that are compliant with accredited/emerging standards, use mature and widely used enabling technologies, and can be integrated with Data Space ecosystems.

The Secure Computing Continuum aims to extend the concept of Trusted Computing (TC) to the Cloud-to-Things context by providing a flexible framework of composable services, tools, and techniques. In traditional computing environments, TC mechanisms allow for the creation of a Trusted Computing Base (TCB), which is a set of trusted hardware, firmware, and software components that are virtually free of vulnerabilities. This TCB can be used as the initial building block for a secure distributed computing infrastructure.

In a complex, multi-domain, and multi-device environment, TC enables all participating entities to have hardware resources that implement a set of agreed-upon features such as trusted boot, sealed storage, curtained memory, attestation, integrity measurement, and secure I/O. Another important consideration is how to use the Cloud-to-Things decentralised infrastructure for trustworthy data sharing and exchange based on commonly agreed principles. The Data Spaces approach could be useful in this scenario (Otto 2022).

Providing users of the Cloud-to-Things continuum with the ability to apply Maximum Achievable Security at any given point in time is another crucial aspect. This includes the continuous estimation of the security level being guaranteed and automated risk/threat mitigation based on technically sound evaluation techniques that effectively combine measurement and modelling.

A secure Cloud-to-Things can be the base for a fully trusted, automated Public Key Infrastructure (PKI), enabling E2E security with Zero-trust, coupled with seamless access across different subsystems, and secure and seamless ownership transfer. Finally, the goal could be to provide automated, seamless, and transferable PKI-based digital identity support.

To ensure secure authentication in the Cloud-to-Things continuum, it is necessary to further investigate automated and mutually authenticated assurance mechanisms. For this, it is important that the individual entities, such as IoT devices, network Edges, Clouds, and Data Spaces, are in a secure state with no malicious software.

Extend the existing software attestation assurance mechanisms that are typically one-way and customised for individual entities. Additionally, it is necessary to develop attestation and assurance mechanisms that do not require a separate set of trusted third parties and certification bodies by exploring the PKI potential for this.

8 Conclusions

The Cloud computing paradigm was created to support applications and services that are always connected and have wide availability. However, as these services became more complex, it became clear that the Cloud was not sufficient to meet all requirements. In response, other paradigms like Fog and Edge computing emerged. These paradigms aim to reduce the load on the Cloud while providing more privacy and faster response times to end-users. It is important to note that Fog and Edge computing are not supposed to replace the Cloud but to extend its capabilities to other devices closer to the end-users. This approach opens the door to the Cloud-to-Things continuum, which is a complex and highly distributed environment that offers networking, storage, and computing capabilities at different levels of the network infrastructure.

The Cloud-to-Things continuum requires continuous availability and high performance, as well as should not be affected by failures or attacks. Therefore, ensuring its resilience becomes a significant challenge. Resilience can be defined as the capacity to maintain service and network robustness by guaranteeing that resources and components are restored in the event of failures or attacks.

Improving the resilience in the Cloud-to-Things continuum is one of the key challenges for orchestrating these environments. Some paradigms have emerged in recent years, focused on the automation of network and service management operation, including SONs, IBN, ZSM, and KDN, which rely on cognitive approaches to bolster the network's ability to withstand and recover from disruptions. Recent research efforts have been made to improve resilience in the Cloud-to-Things continuum, for both fault tolerance and security. A revision of the State-of-the-Art revealed a particular focus on AI and ML techniques, including DL and pervasive AI.

Some open research issues were identified as more work in this area is required, including networks with Self-healing capabilities, transfer learning for anomaly detection to enhance fault tolerance and Zero-trust architectures. A particular focus should be given to energy-conscious approaches, given the battery limitations of the IoT devices, and to be aligned with initiatives such as the European Green Deal. Some techniques that could be explored include lightweight consensus protocols, breaking down large models with distributed AI, lightweight models, and early exits.

Acknowledgements This work is funded by the FCT—Foundation for Science and Technology, I. P./MCTES through national funds (PIDDAC), within the scope of CISUC R&D Unit—UIDB/00326/2020 or project code UIDP/00326/2020, by the project POWER (grant number POCI-01-

0247-FEDER-070365), co-financed by the European Regional Development Fund (FEDER), through Portugal 2020 (PT2020), and by the Competitiveness and Internationalisation Operational Programme (COMPETE 2020), and by the European Union's Horizon Europe research and innovation program through the project IMAGINE-B5G under grant agreement No. 101096452.

References

Abdali T-AN, Hassan R, Aman AHM, Nguyen QN (2021) Fog computing advancement: concept, architecture, applications, advantages, and open issues. IEEE Access 9(1):75961–75980

Abdulqadder IH, Zhou S (2022) Sliceblock: context-aware authentication handover and secure network slicing using DAG-blockchain in edge-assisted SDN/NFV-6G environment. IEEE Internet Things J 9(18):18079–18097

Ahmadian N, Lim GJ, Cho J, Bora S (2020) A quantitative approach for assessment and improvement of network resilience. Reliab Eng Syst Saf 200(1):106977

Alagheband MR, Mashatan A (2022) Advanced digital signatures for preserving privacy and trust management in hierarchical heterogeneous IoT: taxonomy, capabilities, and objectives. Internet Things 18:100492

Alonso J, Orue-Echevarria L, Osaba E, Lobo JL, Martinez I, de Arcaya JD, Etxaniz I (2021) Optimization and prediction techniques for self-healing and self-learning applications in a trustworthy cloud continuum. Information 12(8):1–17

Amiri Z, Heidari A, Navimipour NJ, Unal M (2023) Resilient and dependability management in distributed environments: a systematic and comprehensive literature review. Clust Comput 26(1):1565–1600

Arzovs A, Krisjanis Nesenbergs JJ, Selavo L (2024) Distributed learning in the IoT–edge-cloud continuum. Mach Learn Knowl Extr 6(1):1–33

Ashtari S, Zhou I, Abolhasan M, Shariati N, Lipman J, Ni W (2022) Knowledge-defined networking: applications, challenges and future work. Array 14(1):100136

Ayed D, Dragan P-A, Félix E, Mann ZA, Salant E, Seidl R, Sidiropoulos A, Taylor S, Vitorino R (2022) Protecting sensitive data in the Cloud-to-Edge continuum: the FogProtect approach. In: 2022 22nd IEEE international symposium on cluster, cloud and internet computing (CCGrid), Taormina, Italy. IEEE, Piscataway, pp 279–288

Baccour E, Mhaisen N, Abdellatif AA, Erbad A, Mohamed A, Hamdi M, Guizani M (2022) Pervasive AI for IoT applications: a survey on resource-efficient distributed artificial intelligence. IEEE Commun Surv Tutorials 24(4):2366–2418

Barbareschi M, Casola V, Lombardi D (2023) Ensuring end-to-end security in computing continuum exploiting physical unclonable functions. In: 2023 IEEE international conference on cloud computing technology and science (CloudCom), Naples, Italy. IEEE, Piscataway, pp 273–278

Benzaid C, Taleb T (2020) AI-driven zero touch network and service management in 5G and beyond: challenges and research directions. IEEE Netw 34(2):186–194

Berger C, Eichhammer P, Reiser HP, Domaschka J, Hauck FJ, Habiger G (2021) A survey on resilience in the IoT: taxonomy, classification, and discussion of resilience mechanisms. ACM Comput Surv 54(7):1–39

Bittencourt L, Immich R, Sakellariou R, Fonseca N, Madeira E, Curado M, Villas L, DaSilva L, Lee C, Rana O (2018) The Internet of Things, fog and cloud continuum: integration and challenges. Internet Things 3-4(1):2542–6605

Buck C, Olenberger C, Schweizer A, Volter F, Eymann T (2021) Never trust, always verify: a multivocal literature review on current knowledge and research gaps of zero-trust. Comput Secur 110:102436

Bumiller A, Challita S, Combemale B, Barais O, Aillery N, Le Lan G (2023) On understanding context modelling for adaptive authentication systems. ACM Trans Auton Adapt Syst 18(1): 1–35

Caleb S, Thangaraj SJJ (2023) Data-driven ML approaches for the concept of self-healing in CWN, including its challenges and possible solutions. In: Proceedings of the 2023 eighth international conference on science technology engineering and mathematics (ICONSTEM), Chennai, India. IEEE, Piscataway

Carnero A, Martín C, Torres DR, Garrido D, Díaz M, Rubio B (2021) Managing and deploying distributed and deep neural models through Kafka-ML in the Cloud-to-Things continuum. IEEE Access 9(1):125478–125495

Chen M, Gündüz D, Saad KHW, Bennis M, Feljan AV, Poor HV (2021) Distributed learning in wireless networks: recent progress and future challenges. IEEE J Sel Areas Commun 39(12): 3579–3605

Chiesa M, Kamisiński A, Rak J, Rétvári G, Schmid S (2021) A survey of fast-recovery mechanisms in packet-switched networks. IEEE Commun Surv Tutorials 23(2):1253–1301

Coronado E, Behravesh R, Subramanya T, Fernández-Fernández A, Siddiqui MS, Costa-Pérez X, Riggio R (2022) Zero touch management: a survey of network automation solutions for 5G and 6G networks. IEEE Commun Surv Tutor 24(4):2535–2578

Cortés C, Pérez P, Gutiérrez J, García N (2020) Influence of video delay on quality, presence, and sickness in viewport adaptive immersive streaming. In: 2020 twelfth international conference on quality of multimedia experience (QoMEX), Athlone, Ireland. IEEE, Piscataway, pp 1–4

Durán A, Fernández P, García JM, Dustdar S (2023) Fides: a proposal for federated accountability in the compute continuum. IEEE Internet Comput 27(5):33–42

Gallego-Madrid J, Sanchez-Iborra R, Ruiz PM, Skarmeta AF (2022) Machine learning-based zero-touch network and service management: a survey. Digit Commun Netw 8(2):105–123

Hassan MB, Saeed RA, Khalifa O, Ali ES, Mokhtar RA, Hashim AA (2022) Green machine learning for green cloud energy efficiency. In: 2022 IEEE 2nd international Maghreb meeting of the conference on sciences and techniques of automatic control and computer engineering (MI-STA), Sabratha, Libya. IEEE, Piscataway, pp 288–294

Heidari A, Navimipour NJ, Jamali MAJ, Akbarpour S (2023) A green, secure, and deep intelligent method for dynamic IoT-edge-cloud offloading scenarios. Sustain Comput Inf Syst 38:100859

Hosseinalipour S, Brinton CG, Aggarwal V, Dai H, Chiang M (2020) From federated to fog learning: distributed machine learning over heterogeneous wireless networks. Commun Mag 50(12):41–47

ITU-T (2009) E.800: definitions of terms related to quality of service. Technical Report E.800, ITU

Leivadeas A, Falkner M (2023) A survey on intent-based networking. IEEE Commun Surv Tutor 25(1):625–655

Luckow A, Rattan K, Jha S (2021) Pilot-edge: distributed resource management along the edge-to-cloud continuum. In: 2021 IEEE international parallel and distributed processing symposium workshops (IPDPSW), Portland, USA. IEEE, Piscataway, pp 874–878

Lynn T, Mooney JG, Lee B, Endo PT (2020) The Cloud-to-Thing continuum. In: Opportunities and challenges in cloud, fog and edge computing, 1st edn. Springer, Cham

Mann ZA (2022) Decentralized application placement in fog computing. IEEE Trans Parallel Distrib Syst 33(12):3262–3273

Mao B, Tang F, Kawamoto Y, Kato N (2021) AI models for green communications towards 6G. IEEE Commun Surv Tutor 24(1):210–247

Moreschini S, Pecorelli F, Li X, Naz S, Hästbacka D, Taibi D (2022) Cloud continuum: the definition. IEEE Access 10:131876–131886

Moysen J, Giupponi L (2018) From 4g to 5g: self-organized network management meets machine learning. Comput Commun 129:248–268

Netes V (2020) Modern network technologies and dependability. In: 2020 international scientific and technical conference modern computer network technologies (MoNeTeC), Moscow, Russia. IEEE, Piscataway, pp 1–10

Nouioua M, Fournier-Viger P, He G, Nouioua F, Min Z (2021) A survey of machine learning for network fault management. Springer, Cham

Otto B (2022) The evolution of data spaces. Springer, Cham, pp 3–15

Papadimitriou D, Mannie E (2006) Analysis of generalized multi-protocol label switching (GMPLS)-based recovery mechanisms (including protection and restoration). RFC 4428

Perez Abreu D (2021) Resilience in the Internet of Things for smart city applications. PhD thesis, University of Coimbra, NCS

Puthal D, Mohanty SP, Yeun CY, Damiani E, Pradhan B (2023) Pervasive AI for secure and scalable IoT-edge-cloud continuum: a big picture. In: 2023 IEEE international conference on high performance computing & communications, data science & systems, smart city & dependability in sensor, cloud & big data systems & application (HPCC/DSS/SmartCity/DependSys), Melbourne, Australia. IEEE, Piscataway, pp 566–573

Ramezanpour K, Jagannath J (2022) Intelligent zero trust architecture for 5G/6G networks: principles, challenges, and the role of machine learning in the context of O-RAN. Comput Netw 217:109358

Sabireen H, Neelanarayanan V (2021) A review on fog computing: architecture, fog with IoT, algorithms and research challenges. ICT Express 7(2):162–176

Samie F, Bauer L, Henkel J (2019) From cloud down to things: an overview of machine learning in Internet of Things. IEEE Internet Things J 24(4):4921–4934

Shi Y, Yang K, Jiang T, Zhang J, Letaief KB (2020) Communication-efficient edge AI: algorithms and systems. Commu Surv Tutor 22(4):2167–2191

Souza R, Santos M, Bonfim M, Dias K, Fernandes S (2022) KDN-based fault-tolerant scheduling for VNFs in data centers. IEEE Trans Netw Serv Manag 19(4):4905–4917

Sunyaev A (2020) Cloud computing. Springer, Cham, pp 195–236

Tao F, Xiao B, Qi Q, Cheng J, Ji P (2022) Digital twin modeling. J Manuf Syst 64:372–389

Teerapittayanon S, McDanel B, Kung HT (2016) BranchyNet: Fast inference via early exiting from deep neural networks. In: 23rd International Conference on Pattern Recognition (ICPR), Cancun, Mexico. IEEE, Piscataway

Verdecchia R, Sallou J, Cruz L (2023) A systematic review of Green AI. Wiley Interdiscip Rev Data Min Knowl Discov 13(4):e1507

Volpert S, Eichhammer P, Held F, Huffert T, Reiser HP, Domaschka J (2023) The view on systems monitoring and its requirements from future Cloud-to-Thing applications and infrastructures. Future Gener Comput Syst 141(1):243–257

Wang F, Zhang M, Wang X, Ma X, Liu J (2020) Deep learning for edge computing applications: a state-of-the-art survey. IEEE Access 8(1):58322–58336

Xiao S, Ye Y, Kanwal N, Newe T, Lee B (2022) SoK: context and risk aware access control for zero trust systems. Secur Commu Netw 2022:7026779–7026720

Xing L (2020) Cascading failures in Internet of Things: review and perspectives on reliability and resilience. IEEE Internet Things J 3-4(1):44–64

Yousefpour A, Fung C, Nguyen T, Kadiyala K, Jalali F, Niakanlahiji A, Kong J, Jue JP (2019) All one needs to know about fog computing and related edge computing paradigms: a complete survey. J Syst Archit 98:289–330

Zeydan E, Turk Y (2020) Recent advances in intent-based networking: A survey. In: 2020 IEEE 91st vehicular technology conference (VTC2020-Spring). IEEE, Piscataway, pp 1–5

Experimental Measurements of Cyber Resilience

Michael J. Weisman, Alexander Kott, Jason E. Ellis, Brian J. Murphy, Travis W. Parker, Sidney Smith, and Joachim Vandekerckhove

1 Introduction

Researchers have discovered vulnerabilities in cyber physical systems and have repeatedly demonstrated ways to confound them. Modern vehicles especially are prone to attack (Hoppe and Dittman 2007; Koscher et al. 2010; Bozdal et al. 2018a, b; Stachowski et al. 2018; Bozdal et al. 2020). Understanding the resilience of these systems has become increasingly important (Alexeev et al. 2017; Henshel et al. 2019; Linkov et al. 2018; Kott and Linkov 2019). One of the key challenges in understanding cyber resilience is measuring resilience (Kott and Linkov 2021; Kott et al. 2021). Developers of systems, such as cars and trucks, need to be able to estimate how each component contributes to the vehicle's resilience. Similarly, consumers of systems need to be able to quantify and compute resilience to understand whether or not the product meets the requirements of its intended use.

In this chapter, we report on a project called *Quantitative Measurement of Cyber Resilience* (QMoCR). The QMoCR program has focused on two research areas:

M. J. Weisman · J. E. Ellis · S. Smith
DEVCOM Army Research Laboratory, Adelphi, MD, USA

A. Kott (✉)
U.S. Army Research Laboratory, Adelphi, MD, USA
e-mail: alexkott@yahoo.com

B. J. Murphy
Pennsylvania State University, University Park, PA, USA

T. W. Parker
ICF International, Columbia, MD, USA

J. Vandekerckhove
University of California, Irvine, CA, USA

© The Author(s), under exclusive license to Springer Nature Switzerland AG 2025

I. Linkov, A. Kott (eds.), *Cyber Resilience: Applied Perspectives*, Risk, Systems and Decisions, https://doi.org/10.1007/978-3-031-90109-6_9

(1) mathematical modeling of cyber resilience and (2) framing an infrastructure for experimentation and measurements.

Our research team developed mathematical models to characterize cyber resistence, defined objective quantitative measures of cyber resilience, and performed repeatable, systematic experiments to identify and quantify characteristics of systems' responses to cyber compromises (Kott et al. 2022; Ellis et al. 2022; Weisman et al. 2023, 2025; Kott et al. 2023). In Kott et al. (2022), we began to develop the tools to model cyber resilience mathematically. In a companion paper Ellis et al. (2022), we described an experimental testbed we have developed, and gave examples of the data we can produce. In Weisman et al. (2023), we continued to develop the mathematical model. We expanded on the piecewise linear model and exhibited its stochastic counterpart.

In the mathematical modeling work, we model the impact on a surrogate vehicle of actions by malware as well as "bonware"—the ensemble of cyber-physical features that defend the vehicle's computer system and allow it to recover from attack. We develop a differential equation that models the effects of these competing forces on the system. We extend this to a stochastic differential equation model that captures the effects of uncertainty and randomness in the activity times of malware and bonware.

In parallel with the mathematical modeling, we have developed an inexpensive experimental environment to test the effects of malware and bonware on the Controller Area Network (CAN) bus of a generic vehicle. Our testbed includes (a) a PASTA platform: a vehicle security testbed developed by Toyota which features a set of connected electrical control units or ECUs), (b) Unity: a popular game development platform and physics engine, and (c) the Active Defense Framework (ADF): a government-developed framework used to quickly produce and test network-based cyber-defense techniques. When running this environment, we capture data that can be used to characterize cyber resilience metrics of the modeled vehicles.

Although we have made significant progress in both of these areas, we have also identified some areas that may benefit from collaboration with research partners. Thus far, we have modeled the impact on a single vehicle type. By considering a network of vehicles, many interesting problems can be formulated. If one vehicle is under attack, what are the impacts on other vehicles that exchange information with this vehicle or that are in physical proximity? What are the probabilities that neighboring vehicles in the network are also under attack? How can our understanding of vehicle networks and robotics inform each other? What other techniques can we draw from to formulate models and compute metrics (for example game theory, neural networks, dueling network architectures, etc.)?

2 Prior Work

In Kott et al. (2022), we reviewed literature related to qualitative and quantitative assessments of a cyber system. We refer the reader to that paper for a more detailed summary. Here, we briefly review the quantitative work to date, and motivate the need for stochastic modeling of cyber resilience.

Most approaches to quantitative measurements of cyber resilience tend to involve the area under the curve (AUC) method (Hosseini et al. 2016; Kott and Linkov 2021). An experimental system engages in a collection of missions where it subject to cyber attack. Data containing the functionality of the system is collected, and a metric based on the area under the functionality curve is computed. Figure 1 illustrates the concept. One or more attacks compromise a system, causing the functionality to diminish. The ratio of the area under the curve and the area under the baseline curve (recorded during a mission where there is no attack) is computed.

This class of measures seems reasonable, however, AUC-based resilience measures are rather simple, and reveal little about the underlying processes. In Kott et al. (2022) we began to explore ways to quantify the resilience effectiveness of the bonware and quantify the effectiveness of malware on a system. We also looked to understand how these values of effectiveness vary over time during an incident.

3 Formal Definition of Concepts

We define goal-relevant resilience as the ability of a system to accomplish its goal—or at least maximize the degree of accomplishment of its goal—in spite of effects of a cyber attack, as a run unfolds over time. To this end, we postulate that for a given run, there exists a function A(t) that represents accomplishment and that is

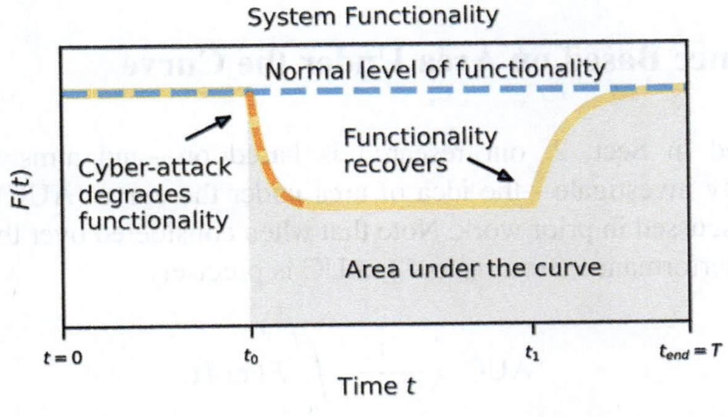

Fig. 1 The functionality $F(t)$ of a system is plotted in yellow and normal functionality is plotted in light blue. In this example, there is a cyber attack at time t_0 which degrades functionality, and the system's resilience allows the functionality to begin recovering at time t_1. The area under the curve (AUC) metric is defined as the normalized area under the curve from time t_0 to time T, the end time of measured system performance

cumulative from the run start time t_0 up until the present time t. We define functionality, $F(t)$, to be the time derivative of goal accomplishment. Thus,

$$F(t) = \frac{dA}{dt}, \quad A(t) = \int_{t_0}^{t} F(\tau)\, d\tau. \tag{1}$$

Note that, in practice, functionality may vary with time, even when the system performs normally and is not experiencing the effects of a cyber attack. To be able to account for this, we will often distinguish between performance under baseline conditions, $F_{\text{baseline}}(t)$ and performance during an attack scenario, $F_{\text{attack}}(t)$. We will require $F_{\text{baseline}}(t) > 0$ everywhere where it is defined.

The purpose of this definition is to take a step towards a more formal treatment of the term "functionality", unlike in prior work where the term is introduced informally and is merely illustrated by domain-specific examples. Also note that we use the expression "goal-relevant" resilience in order to emphasize that the same system may be assigned different goals (e.g., a vehicle might be sent to deliver a maximum weighted cargo at reduced speed no matter the time required or alternatively to get to a destination as quickly as possible), and resilience does depend on a goal the system is pursuing. Needless to say, cyber resilience of a system does depend on cyber attacks imposed on the system. That is, a system may have excellent resilience against one attack, and poor resilience against another attack. Even with the same attack and the same goal pursued under different conditions, the system may exhibit different level of resilience. Therefore, for practical applications, cyber resilience should be measured in the context of defined ensembles of attacks and scenarios of use (see [16] for discussion of practical formulation of such ensembles). Furthermore, a goal may imply multiple objectives. At the same time, the general approach and methodology for measuring resilience does not appear to require changes for applications to different types of systems and scenarios.

4 Resilience Based on Area Under the Curve

As mentioned in Sect. 2, our research is based on—and aims to extend and experimentally investigate—the idea of area under the curve (AUC) that has been frequently discussed in prior work. Note that when considered over the entire period of a system performance (from t_0 to T), AUC is precisely

$$\text{AUC} = \frac{1}{T - t_0} \int_{t_0}^{T} F(\tau)\, d\tau.$$

Here we expand on this concept to make a measure of resilience that calculates the accomplishment—that is, the area under the functionality curve—in a cyber attack scenario relative to the accomplishment in a baseline scenario:

$$R = \frac{\int_{t_0}^{T} F_{\text{attack}}(\tau)\, d\tau}{\int_{t_0}^{T} F_{\text{baseline}}(\tau)\, d\tau} = \frac{A_{\text{attack}}(T)}{A_{\text{baseline}}(T)}. \tag{2}$$

As a measure of resilience, R has a number of advantages. By contrasting behavior in an attack scenario to behavior in a comparable baseline scenario, it is able to account for idiosyncratic differences between vehicles, terrain, or any other features we hold constant between the two scenarios. Imagine two vehicles—one twice bigger and heavier than another, but otherwise with exactly the same design of cyber-relevant systems. The lighter vehicle would have approximately twice greater fuel efficiency (in km per liter) and the impact of a cyber attack on its AUC would be twice greater. This might imply that the cyber resilience of the lighter vehicle is twice worse, which is incorrect. On the other hand, the dimensionless R would be roughly the same for both vehicles, as it should be. R can be interpreted as the *fraction of normal functionality maintained* during a cyber attack. If it is close to 1.0, then the effect of the attack was small; if it is 0.0, then functionality was completely disrupted. However, it is important to note that R is closely related to AUC, as it reflects the ratio of AUC occurring in presence of a cyber attack to the AUC without an attack. In other words, R is a normalized and dimensionless variation of the AUC measure.

5 Continuous Model

We are interested in the functionality of our system, $F(t)$, which we defined in Kott et al. (2022) to be the time derivative of mission accomplishment. We also proposed a baseline functionality (normal functionality), which in general could be time varying, but in our analysis, we take normal functionality, $F_N(t) = F_N$, to be a constant. We also assume that the system, prior to any attack or other deviation from normal operations, is operating normally: $F(t_0) = F_N$. We assume that functionality is differentiable at least once and both malware impact and bonware impact are continuous functions of time: $F \in C^1$ and M, B $\in C^0$. The effectiveness on functionality is the sum of the effectiveness of malware and bonware, and

$$\frac{dF}{dt} + Q(t)F(t) = F_N B(t), \tag{3}$$

where $Q(t) = M(t) + B(t)$. In [13], we reasoned that $B(t)$, $M(t) \geq 0$, $F_N > 0$, and $F_N \geq F(t) \geq 0$ and found the general solution to this first-order linear differential equation with initial condition:

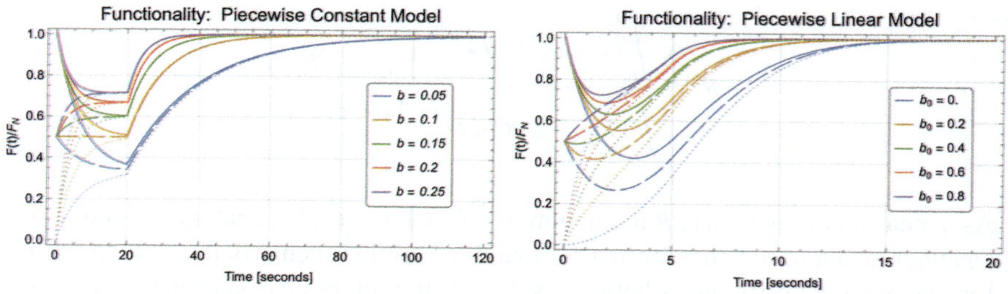

Fig. 2 Left: Normalized functionality, $F(t)/F_N$, is shown for piecewise constant models (Row 3 in Table 1). The malware impact is $M(t) = [u(t) - u(t - 20)]$ and the bonware impact is $B = bu(t - 20)$. Initially, malware attacks at mission time $t = 0$. At time $t = 20$ s, bonware becomes aware of the attack, counters malware, and brings the system back towards normal operations. Right: Normalized functionality, $F(t)/F_N$, is shown for piecewise linear models (Row 5 in Table 1). The malware impact is $M(t) = \max(0, 0.5-0.1t)$ and the bonware impact is $B = b_0 + 0.04t$. Both malware and bonware effectivenesss are initially linear functions of time. When malware effectiveness reaches $M = 0$, then malware effectiveness is zero but bonware effectiveness continues to increase. The function $u(t)$ is the unit step function: $u(t) = 0$ when $t < 0$ and $u(t) = 1$ when $t \geq 0$. The figure on the right is from Kott et al. (2022), used by permission

$$F(t) = e^{-\int_0^t Q(p)\,dp}\left(F(0) + F_N \int_0^t e^{\int_0^\tau Q(p)\,dp} B(\tau)\,d\tau\right).$$

In Kott et al. (2022) we also exhibited solutions for a number of elementary examples. Figure 2 contains plots of piecewise constant and piecewise linear models. The models' differential equations and their solutions are summarized in Table 1.

6 Generating Stochastic Realizations

In Kott et al. (2022) we developed a method to extract the parameters of the stochastic model associated with a continuous model. In the left pane of Fig. 3, we plot a continuous model along with the average over 10,000 runs of its associated stochastic model. After the first two or three initial points, the agreement is excellent. In the right pane of Fig. 3, we show averages of n instantiations of the stochastic model, for n ∈ {10, 100, 1000, 10, 000}. As n increases, we see better agreement with the solution to the continuous differential equation model.

Table 1 Mathematical Models Developed in Kott et al. (2022)

	Differential equation	Solution
Continuous	$\frac{dF}{dt} + Q(t)F(t) = F_{\mathrm{N}}\mathscr{B}(t)$	$F(t) = e^{-\int_0^t Q(p)\ dp}\left(F(0) + F_{\mathrm{N}}\int_0^t e^{\int_0^\tau Q(p)\ dp}\mathscr{B}(\tau)\ d\tau\right)$
Constant	$\frac{dF}{dt} + QF(t) = F_{\mathrm{N}}\mathscr{B}$	$F(t) = \left[F(0) - \frac{F_{\mathrm{N}}\mathscr{B}}{Q}\right]e^{-Qt} + \frac{F_{\mathrm{N}}\mathscr{B}}{Q}$
Piecewise Constant	$\frac{dF}{dt} = \sum_{j=0}^{N-1}(F_{\mathrm{N}} - F(t))\mathscr{B}_j - F(t)\mathcal{M}_j,$ $(t_j \le t < t_{j+1}), \quad (j = 0, \cdots, N-1)$	$F(t) = [F(t_{j-1}) - \frac{F_{\mathrm{N}}\mathscr{B}_j}{Q_j}]e^{-Q_j(t-t_{j-1})} + \frac{F_{\mathrm{N}}\mathscr{B}_j}{Q_j}$
Linear	$\frac{dF}{dt} + (\lambda - \omega t)F(t) = F_{\mathrm{N}}(\alpha - \beta t)$	$\frac{F(t)}{F_{\mathrm{N}}} = \frac{1}{\Omega(t)}\left\{\frac{F(0)}{F_{\mathrm{N}}} - \frac{\beta}{\omega}(1 - \Omega(t)) + (\alpha\omega - \beta\lambda) \times \frac{\sqrt{\frac{\pi}{2}}e^{\Lambda^2}}{\omega^{3/2}}\left[\mathrm{erf}\ (\Lambda) + \mathrm{erf}\left(\frac{\omega t}{\sqrt{2\omega}} - \Lambda\right)\right]\right\}$ where $\Omega(t) = e^{\lambda t - \frac{1}{2}\omega t^2}$ and $\Lambda = \lambda/\sqrt{2\omega}$
Piecewise Linear	$\frac{dF}{dt} = \sum_{j=0}^{N-1}\left[(\lambda_j - \omega_j t)F(t) - F_{\mathrm{N}}(\alpha_j - \beta_j t)\right],$ $(t_j \le t < t_{j+1}), \quad (j = 0, \cdots, N-1)$	$\frac{F(t)}{F_{\mathrm{N}}} = \frac{1}{\Omega(t - t_{j-1})}\left\{\frac{F(t_{j-1})}{F_{\mathrm{N}}} - \frac{\beta}{\omega}(1 - \Omega(t - t_{j-1})) + (\alpha\omega - \beta\lambda) \times \frac{\sqrt{\frac{\pi}{2}}e^{\Lambda^2}}{\omega^{3/2}}\left[\mathrm{erf}\ (\Lambda) + \mathrm{erf}\left(\frac{\omega(t - t_{j-1})}{\sqrt{2\omega}} - \Lambda\right)\right]\right\},$ where $\Omega_j(t) = e^{\lambda_j(t - t_j) - \frac{1}{2}\omega_j(t - t_j)^2}$ and $\Lambda_j = \lambda_j/\sqrt{2\omega_j}$

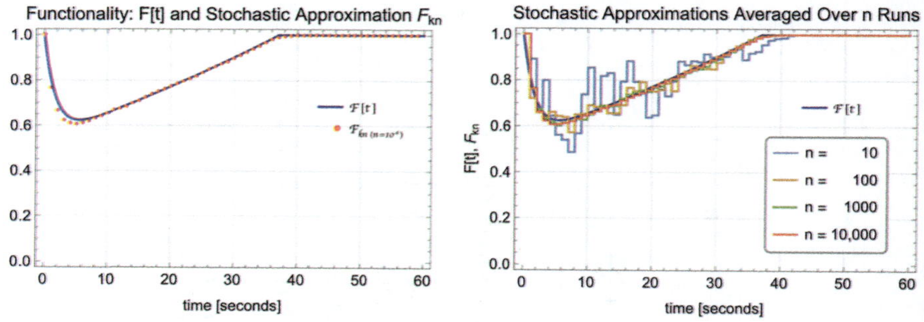

Fig. 3 Left: The functionality for a piecewise linear model is shown in blue. $M(t) = \max(0, 0.25 - 0.00675t)$ and $B(t) = \min(0.5, 0.3 + 0.0045t)$. The average of 10,000 runs for the corresponding stochastic model is shown in orange. Right: The average of $n \in \{10, 100, 1000, 10,000\}$ runs. As the number of runs averaged increases, the aggregate stochastic model approaches the solution of the continuous differential equation

7 Experimental Testbed and Method

In order to validate the QMoCR modeling methodology, data with sufficient fidelity is required. While capturing data straight from physical systems would provide the highest fidelity, gathering such data in a repeatable fashion, with enough statistical variability, and with and without the presence of degradations presents significant hardship. A digital twin of a physical system would be a good alternative; however, these twins must be implemented by the vehicle manufacturers themselves in order to produce realistic data. Another approach, described here, is an experimentation environment that is a blend between simulation and hardware-in-the-loop to reach an equilibrium between expense and fidelity. In this example environment, components include the Toyota Portable Automotive Security Testbed with Adaptability (PASTA) by Toyota Motor Corporation, the Unity game development platform, and the Active Defense Framework (ADF) developed in-house at the DEVCOM Army Research Laboratory. In general, Unity generates messages via the Message Queue Telemetry Transport (MQTT) publish-subscribe network protocol. ADF ingests these messages and translates them to Control Area Network (CAN) format, which are then injected onto the appropriate CAN bus within PASTA. Data flowing in the opposite direction is handled in a similar manner, but in reverse. Figure 4 illustrates the connection between these components.

7.1 Toyota Portable Automotive Security Testbed with Adaptability (PASTA)

PASTA, created by Toyota (Toyama et al. 2018), is a vehicle security testbed designed to allow researchers the ability to develop and evaluate new security

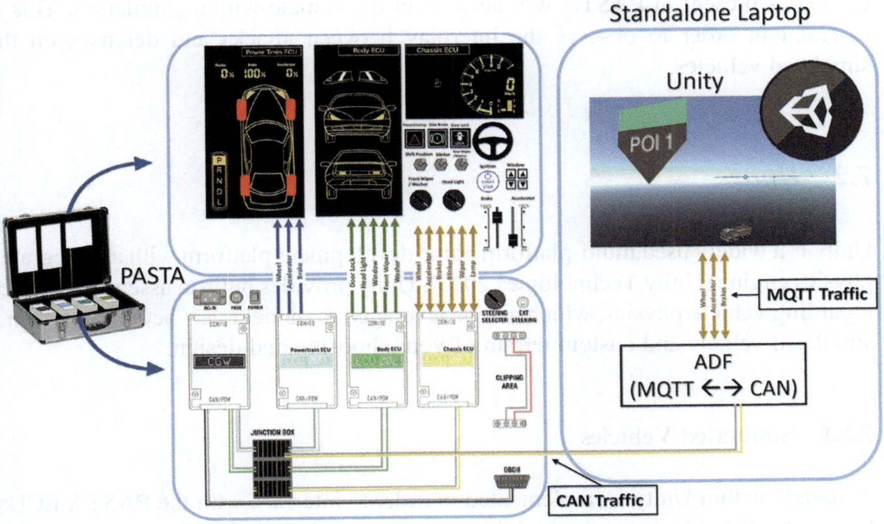

Fig. 4 A high-level overview of the data flow between components. Portions are derived from [20]. Unity handles the interactions between the simulated trucks and the terrain, provides driver inputs to maintain pathing and speed over a pre-determined course across the terrain, and simulates many aspects of the vehicle to provide high-fidelity vehicle performance. PASTA is hardware-in-the-loop that generates actual CAN bus traffic in response to the inputs from Unity, and provides three vehicle ECUs (powertrain, body, and chassis). ADF allows for communication translation between Unity and PASTA, provides simulated ECU plugins for additional sensor information not present in PASTA, and implements simulated cyber attacks and defenses

technology and approaches on realistic open-source electronic control units (ECUs). There are three vehicle ECUs within the testbed (powertrain, body, and chassis) connected via a CAN bus similar to what would be found in a real vehicle. A fourth ECU, the central gateway (CGW), acts as a junction between the three previously mentioned buses. The CGW routes the CAN messages seen on the bus to the appropriate ECU.

PASTA also has the ability to simulate the performance of a hypothetical sedan based on the messages and values being reported by the ECUs. For example, based on the acceleration input received from the chassis ECU, PASTA is able to calculate how the vehicle would respond in terms of increased engine revolutions per minute (RPMs) and change in vehicle ground speed. This updated information is directly fed back into the ECUs, not over the CAN bus. This prevents a mixing of simulation updates and real messages on the CAN bus.

An issue with this simulation capability—in contrast to the ECUs themselves—is that it is closed source and immutable. In essence, the actual performance characteristics of the simulated vehicle to its current state and the terrain it is supposedly traversing cannot be modified. To address this, a configurable physics engine is integrated into the experimentation environment. In this configuration, PASTA becomes hardware-in-the-loop for the physics engine. Attacks or defenses that affect the ECUs present in PASTA will also affect the vehicle within simulation. This is essential in order to observe the interplay between attacks and defenses on the simulated vehicles.

7.2 Unity

Unity is a widely used multi-platform game development platform with an integrated physics engine (Unity Technologies 2021). Unity provides built-in assets and classes regarding vehicle physics, which can be leveraged to model inter- actions between a simulated vehicle and custom terrain of your choosing and design.

7.2.1 Simulated Vehicles

A vehicle within Unity is implemented in order to interface with the PASTA ECUs. In general, the vehicle produces driver responses to the terrain. These responses are sent to the corresponding ECUs within PASTA as if directly generated from within PASTA's own simulation capability. The ECUs update their state based on these responses, which is then sent back to Unity in order to calculate parameters like torque and fuel consumption. With this workflow, attacks or degradations that would affect the ECUs in PASTA will now also be reflected in the simulated vehicle in Unity.

7.2.2 Automated Driver

The automated driver generates steering, acceleration, and braking responses as the vehicle traverses the terrain. Steering responses are based on a waypoint system over a customizable path. A proportional-integral- derivative (PID) controller produces acceleration and braking responses using a target vehicle speed. A pseudo-random number generator with a configurable seed is incorporated to add an effect similar to driver attention drift. This affects the target vehicle speed used by the PID controller, allowing for configurable variability between experiment runs based on the seed chosen.

7.2.3 Vehicle Performance Characteristics

A variety of performance characteristics are generated by considering torque, horsepower, and brake-specific fuel consumption (BSFC) curves. These curves are based off of those corresponding to a real vehicle to enhance fidelity, and can be changed to match other vehicles.

First, engine RPM is calculated by taking into account the current vehicle speed, wheel circumference, and effective gear ratio. This RPM value is then used as an index of sorts to evaluate each of the three previously mentioned curves. Torque is multiplied by the engine throttle received from PASTA and the current effective gear ratio to obtain the total amount of force that can be applied to the wheels. Horsepower and the BSFC value are used together to calculate the amount of fuel that has been used. A gear shifting algorithm was incorporated that favored fuel economy over engine power availability. However, the algorithm will adjust into a more aggressive profile based on the recent history seen from the automated driver accelerator input.

7.2.4 Custom Sensors and Functionality

Additional sensor information that is either not present in PASTA or needs its functionality altered for increased fidelity can be handled by the simulated vehicle. This includes engine coolant temperature, vehicle attitude, and a custom backup engine ECU. Engine coolant temperature is represented within PASTA, but corresponds to its own internal vehicle simulation. For the vehicle in Unity, a basic temperature model is implemented that can be controlled by an external fan controller ECU described further in a later section. The fan controller monitors the coolant temperature reported by the vehicle and dictates the operation of a simulated fan on the vehicle to reduce the coolant temperature. On a real vehicle, the fan reduces the total available horsepower that can be applied toward maintaining vehicle speed. To account for this, when the fan is active, an appropriate reduction in horsepower is realized.

7.2.5 Terrain

The simulated vehicle traverses a custom terrain map that encompasses multiple terrain types: flat main road, float off-road, hilly, sustained incline, and sustained decline. Trigger points on the map are set that change the target speed of the vehicle to one appropriate for that terrain type.

7.3 Active Defense Framework

ADF, a framework developed at the DEVCOM Army Research Laboratory, allows for rapid prototyping of network-based active cyber-defense techniques. ADF supports both traditional enterprise and vehicle network protocols. The framework acts as an intermediary, as depicted in Fig. 1, allowing it to control, inspect, modify, drop, or generate network messages. With this positioning, those utilizing ADF have the ability to implement any network defense, detection, or prevention technique.

In this example experimentation environment, ADF handles message passing between PASTA and Unity by translating CAN messages to and from MQTT, a standard publish-subscribe IP-based messaging protocol. It also provides simulated components in the form of ECU functionality for additional vehicle capabilities, and handles both attack and defense on the CAN bus.

7.3.1 Unity-to-PASTA Message Translation

ADF is connected to the PASTA CAN bus such that it is essentially the new central gate way between PASTA and Unity. This is necessary since Unity is unable to produce or understand CAN traffic on its own. In this configuration, ADF translates CAN messages between the powertrain ECU and Unity in real-time, while also sending vehicle parameters such as current vehicle speed, engine RPMs, and accelerator position to the chassis ECU.

7.3.2 Simulated ECU Functionality

ADF provides the means to add or modify ECU functionality. For instance, PASTA does not have a notion of a hysteresis fan control ECU, but for increased fidelity, this functionality is required. ADF assumes this responsibility by monitoring messages between PASTA and Unity, activating when the engine coolant temperature reported by Unity surpasses an upper threshold, and deactivates when the temperature drops below a lower threshold. Unity monitors the state of this simulate fan controller to This functionality could be extended further with other vehicle functionality such as a windshield defroster or tire pressure sensors.

7.3.3 Attacks and Defenses

ADF provides a mechanism to attack and defend the additional functionalities described previously, as well as the CAN bus itself, due to its position as the new central gateway. This can manifest both offensively and defensively as any of intermediary message functions it provides (control, inspect, modify, drop, or inject), simulating effects such as those from arbitrary malware or defensive tools, or utilizing innovative defensive techniques or approaches as software-in-the-loop. Regardless of the offensive or defensive approach, ADF is used to respond in a user-configurable manner to generate the desired effect.

7.4 *Data Format and Characteristics*

Data is captured from either the perspective of Unity or PASTA, meaning the state of the simulated vehicle or the state maintained by the hardware ECUs in PASTA. From the perspective of the simulated vehicle, additional functionality represented within ADF can be seen and reported, and can be exported in CSV format. From the perspective of the ECUs, only the signals and parameters present in the base PASTA can be reported. However, this will be exported in a CAN-formatted traffic capture with appropriate and realistic message timings. This may be important to work such as training a defensive agent for a real vehicle on the traffic behaviors seen on a CAN bus.

8 Illustrative Experimental Case Study and Discussion

8.1 *Scenario Description*

For the purpose of generating data to use in the validation of the QMoCR methodology, we created a scenario where malware, simulated by ADF, infected the previously described fan controller ECU functionality. The effect of this malware is simple yet noticeable; when active, the fan controller loses the ability to turn off the simulated engine fan, but maintains the ability to turn the fan on. Due to the previously mentioned engine power draw caused by the use of the fan, this malware will cause the vehicle to enter a degraded performance state.

This scenario, over the course of 800 s, captures the interplay between simulated bonware monitoring the vehicle for abnormalities and the malware present on the fan controller. Once alerted to the presence of abnormal behavior regarding the engine coolant fan, the bonware will commence a period of observation prior to acting to reduce the possibility of a false positive. Recovery includes re-flashing the firmware of the fan controller ECU. This re-flashing is simulated by ADF through a short 20 s period of inactivity where the fan controller is unresponsive. After, normal fan operation resumes.

8.2 *Discussion and Analysis*

Figure 5 shows two separate baseline-attack run pairs. All data was captured from the perspective of the vehicle within Unity, since the malware is affecting a component simulated with ADF. The left column represents data from the malware being active starting at 283 s, and the right column represents data with the malware active at 619 s. All data was smoothed with a 72 s moving average window due to the density and variability of data points coming from the physics engine. The graphs in

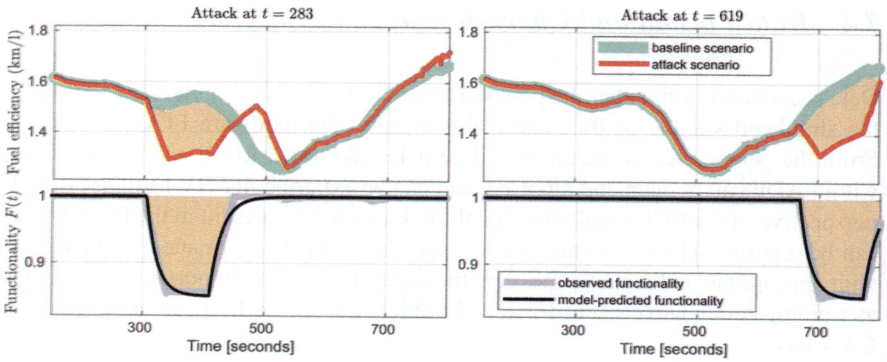

Fig. 5 **Top row**: Fuel efficiency curves under baseline (cyan) and attack (red) scenarios, with the effect of malware shaded orange. **Bottom row**: Observed (blue) and inferred (black) functionality curves for the time series model. A change in the timing of the attack ($t = 283$ vs, $t = 619$) results in a simple translation of the functionality curve

the top row illustrate with a light blue line the observed fuel efficiency of the vehicle during a baseline run with no malware active, while the same is shown with a red line during an attack run with the malware active at the time points previously indicated. The graphs in the bottom row show the functionality over time with a light purple line. This functionality for this scenario is defined as the fuel efficiency of the vehicle in the attack run divided by the fuel efficiency in the baseline or attack run, whichever is lower. This ensures that the functionality cannot excede 1. The bottom graphs also show with a black line the functionality as predicted by the piecewise constant modeling described previously.

The effects of each attack run can be seen starting at 307 s for the first run and 669 s for the second run. The loss in fuel efficiency is shown in orange as the AUC where the baseline and attack runs differ. The loss associated with the attack effect was 10.76% for the first run and 11.42% for the second run. The modeled functionality in the bottom graphs capture the noteworthy aspects of the data produced with the experimental testbed. The most important aspects being the rate of functionality decline when the malware activated and the rate of functionality return when the bonware began the recovery process to thwart the malicious presence.

9 Conclusions and Future Work

In this chapter, we have presented a framework for the analysis of cyber resilience of physical systems. Our framework is applicable to a broad array of systems including, for example, communications, financial, medical, robotics, and transportation systems and employs custom differential equations to model malware attacks and "bonware" defenses. We've also developed a metric to measure resilience that compares behavior of a system under an attack scenario to behavior in a baseline scenario.

We've introduced an illustrative case study which demonstrates that the experimental infrastructure described generates data with sufficient fidelity to validate our model and generate a sensible quantitative measurement of cyber resilience. Analysis of the data by subject matter experts confirm that it is similiar to data that would be generated by real-world systems.

References

Alexeev A, Henshel DS, Levitt K, McDaniel P, Rivera B, Templeton S, Weisman M (2017) Constructing a science of cyber-resilience for military systems. In: NATO IST-153 workshop on cyber resilience, pp 23–25

Bozdal M, Randa M, Samie M, Jennions I (2018a) Hardware trojan enabled denial of service attack on CAN bus. Procedia Manuf 16:47–52

Bozdal M, Samie M, Jennions I (2018b) A survey on CAN bus protocol: attacks, challenges, and potential solutions. In: 2018 International Conference on Computing, Electronics Communications Engineering (iCCECE), pp 201–205

Bozdal M, Samie M, Aslam S, Jennions I (2020) Evaluation of CAN bus security challenges. Sensors 20(8):2364

Ellis J, Parker T, Vandekerckhove J, Murphy B, Smith S, Kott A, Weisman M (2022) Experimental infrastructure for study of measurements of resilience. In: Proceedings of IEEE military communications conference, pp 841–846

Henshel DS, Levitt K, Templeton S, Cains MG, Alexeev A, Blakely B, McDaniel P, Wehner G, Rowell J, Weisman M (2019) The science of cyber resilience: characteristics and initial system taxonomy. In: Fifth world conference on risk

Hoppe T, Dittman J (2007) Sniffing/replay attacks on CAN buses: a simulated attack on the electric window lift classified using an adapted cert taxonomy. In: Proceedings of the 2nd workshop on embedded systems security (WESS), pp 1–6

Hosseini S, Barker K, Ramirez-Marquez JE (2016) A review of definitions and measures of system resilience. Reliab Eng Syst Saf 145:47–61

Koscher K, Czeskis A, Roesner F, Patel S, Kohno T, Checkoway S, McCoy D, Kantor B, Anderson D, Shacham H et al (2010) Experimental security analysis of a modern automobile. In: 2010 IEEE symposium on security and privacy. IEEE, pp 447–462

Kott A, Linkov I (2019) Cyber resilience of systems and networks. Springer, New York

Kott A, Linkov I (2021) To improve cyber resilience, measure it. Computer 54(2):80–85

Kott A, Golan MS, Trump BD, Linkov I (2021) Cyber resilience: by design or by intervention? Computer 54(8):112–117

Kott A, Weisman M, Vandekerckhove J (2022) Mathematical modeling of cyber resilience. In: Proceedings of IEEE military communications conference, pp 835–840

Kott, A., Weisman, M., Vandekerckhove, J., Ellis, J., Parker, T., Murphy, B., and Smith, S. (2023). A methodology for quantitative measurement of cyber resilience (qmocr). ARL-TR-9672

Linkov I, Trump BD, Keisler J (2018) Risk and resilience must be independently managed. Nature 555:7694

Stachowski S, Bielawski R, Weimerskirch A (2018) Cybersecurity research considerations for heavy vehicles, Technical Report Dot Hs 812 636. University of Michigan, Ann Arbor, Transportation Research Institute, Washington DC

Toyama, T., Yoshida, T., Oguma, H., and Matsumoto, T. (2018). PASTA: portable automotive security testbed with adaptability. Black Hat Europe 2018. Dec. 3–6, 2018

Unity Technologies (2021) The leading platform for creating interactive, real-time content. (v2020.3.18f1)

Weisman M, Kott A, Vandekerckhove J (2023) Piecewise linear and stochastic models for the analysis of cyber resilience. In: 57th annual conference on information sciences and systems

Weisman MJ, Kott A, Ellis J, Murphy B, Parker T, Smith S, Vandekerckhove J (2025) Quantitative measurement of cyber resilience: modeling and experimentation. ACM Trans Cyber-Phys Syst 9:1

Governance and Applications of Cyber Resilience

Cyber-Physical Dimensions of Resilience Planning in National Security and Defense

Jesse M. Keenan, Benjamin D. Trump, and Igor Linkov

1 Introduction

As the geopolitical landscape grows increasingly complex and interconnected across state boundaries, intense domestic disruptions from acute shocks and chronic stresses in any given state are capable of leading to global disruptions across a variety of meta-scales (Vina and Liu 2023). This chapter outlines existing and emerging resilience policies and planning regimes in national security and defense that interact across various 'layered' jurisdictional, spatial, and infrastructural scales. This exploration provides a brief survey of how resilience is currently applied in practice by the U.S. Department of Defense (DOD), U.S. national security agencies, and NATO. Building on this diversity of applications, various categorical forms of resilience drawn from the empirical science of resilience are positioned within the state of practice. Notably, this chapter highlights the extent to which social processes of resilience are critical for cyber resilience and managing cyber-physical vulnerabilities. By integrating strategic and operational defense planning with greater scientific and epistemic precision of the various forms of resilience, this chapter argues that a more comprehensive utilization of resilience science, which benefits

---

Note: This chapter is based, in part, on the following work: Keenan, J.M., Trump, B.D., Kytöma, E, Adlakha-Hutcheo, G.,& Linkov, I. (2024). The Role of Science in Resilience Planning for Military-Civilian Domains in the U.S. and NATO. *Defence Studies*. doi: 10.1080/14702436.2024.2365218

J. M. Keenan (✉)
U.S. Army Corps of Engineers' Engineer Research and Development Center, Vicksburg, MS, USA

School of Architecture and Built Environment, Tulane University, New Orleans, LA, USA
e-mail: jkeenan@tulane.edu

B. D. Trump · I. Linkov
Engineering Research and Development Center, U.S. Army Corps of Engineers, Concord, MA, USA

© The Author(s), under exclusive license to Springer Nature Switzerland AG 2025

I. Linkov, A. Kott (eds.), *Cyber Resilience: Applied Perspectives*, Risk, Systems and Decisions, https://doi.org/10.1007/978-3-031-90109-6_10

from indicators, benchmarks, and other metrics, offers a more robust quantification of cyber-physical performance (Colabianchi et al. 2021) within sectors and across military-civilian domains (Wied et al. 2020).

This type of precision is important for not only guiding strategic design, but also stress testing and validating investments in resilience that address a wide range of shocks and stresses—associated with everything from climate change to artificial intelligence (AI) in cyberwarfare—that may impair national security and defense capabilities (Thornton and Miron 2020; Milburn 2023). The meta-landscape of hybrid and cyber-physical warfare extends from the psychological conditioning of labor and military force members to the systems of systems that drive the delivery of critical infrastructure services (Mumford and Carlucci 2023). Applying lessons from various forms of resilience to cyber-physical applications, this chapter reinforces the scientific debate that an optimal orientation to resilience leaves open the door for the transformative adaptation of function and identity when the boundary conditions and thresholds of conventional single-equilibrium resilience are exceeded (Folke et al. 2010; Yu et al. 2020). Finally, this chapter concludes with a normative perspective on how military and civilian resilience planning could support the development of a strategic model for a National Resilience Plan in alignment with existing NATO obligations.

2 National Security, Cross-Domain Deterrence, and Resilience

The primary purpose of national security and defense is to protect and ensure the continuity and integrity of civilian institutions, assets, and infrastructure, as well as the welfare of the general population. Such civilian resources reciprocally provide support for and enable the operations and readiness of the military. Of course, the intensive convergent integration and interdependence of civilian and military domains raises several key questions on the state of democracy, constitutional role of the civilian governance of the military (Brooks 2019) and the role of the state in an open society (Hartmann 2017; Popper 2020).

The magnitude of this relationship has intensified as the world's infrastructure and technology have grown in complexity, particularly as it relates to information technology (Lindsay 2020), artificial intelligence (AI)(Ams 2023), and cyber-physical vulnerabilities (Bolbot et al. 2019). This complexity has given rise to the concept of "cross-domain deterrence" (CDD) wherein military-civilian integration in cyber-physical systems has created a new landscape for both low-scale conflict and the potential for mass destruction that serve as a multi-lateral deterrent to classical forms of warfare (Gartzke and Lindsay 2019). "When an opponent has no incentive to initiate or escalate conflict at any given intervention or escalation threshold in any given domain of warfare—both vertically and horizontally within that domain and laterally into one or more additional domains of warfare—successful cross-domain deterrence can be said to be in effect" (Mallory 2018, 1).

Hybrid warfare (also referred to in literature as 'hybrid threats') are one rapidly emerging vector by which interstate hostility is conducted (Linkov et al. 2019). Hybrid threats embody a complex blend of conventional and unconventional tactics, orchestrated by state or non-state actors to exploit vulnerabilities across a wide spectrum of domains, including military, economic, social, and cyber (Hartmann 2017). These multifaceted threats are characterized by their covert nature, strategic ambiguity, and the deliberate blurring of lines between peace and conflict, aiming to undermine societal cohesion, erode trust in institutions and destabilize targeted nations without triggering a conventional military response. The sophisticated nature of these threats benefits from a resilience-oriented approach, emphasizing the need for agile, adaptive strategies that can respond to the dynamic and often indistinct contours of hybrid warfare.

There are two dimensions to CDD relevant in the context of resilience. First, low-scale indirect cyber warfare presents an alternative to classical forms of military aggression wherein the geopolitical stakes may be lower for forms of aggression that do not present visceral forms of destruction and loss of human life. Such attacks may simply be deniable and hence avoid clear punitive consequences. For instance, the People's Liberation Army of China deems "financial warfare" as something that can be just as destructive through "man-made stock market crashes, attacks on financial instruments, and cyber-attacks on the financial systems" (Reinike 2020, 8). In the context of the global marketplace, attacks associated with financial warfare may be plausibly deniable, as any number of unregulated market-based and/or criminal actors may be at work. This level of anonymity and plausible deniability makes hybrid warfare attractive, as the consequences are lower. It is difficult for the target nation to attribute the attack to a specific actor or nation. To this end, some scholars have sought to frame proactive "economic" resilience measures in the financial system that seek to detect and manage responses to malfeasance in everything from currencies to portfolios (Smalenberger 2015; Constantinescu 2023).

Second, by extension, one could argue that greater the robustness in resilience within cyber-physical systems, the less valuable the target. In such cases, the aggressor is limited in their offensive capacity to classical forms of military assault within a military domain which may be geopolitically and technologically limited. In this sense, resilience itself may be a form of CDD. As Sweijs et al. note, "[r]esilience is directed at dissuading hostile measures by presenting the adversary with the futility of his attacks through recovery" (Sweijs et al. 2021, 16–17). In this light, Hartmann notes that resilience is "a constant process of adaptation [that is] potentially able to undermine the effectiveness of hybrid warfare, and, thus, deter opponents" (Hartmann 2017, 8). Some commentators among the Allies have viewed resilience as key element of deterrence (Roepke and Thankey 2019). They have argued that countries with greater measures of resilience, usually characterized by comprehensive preparedness planning involving government, public, and private sectors, possess fewer vulnerabilities susceptible to exploitation or targeting by adversaries (Id.). Thus, resilience becomes a critical component of CDD through denial, thereby aiming to dissuade adversaries from attacking by demonstrating that such actions would fail to accomplish their objectives while at the same time risking the consequences of the identification of proxy actors and methods.

Resilience performance through military and civilian integration may manifest, not just as a form of CDD, but also as a very practical prerequisite for military mobilization and tactical execution, particularly in the context of communications and control systems. This capability is deemed essential for establishing a credible deterrence posture. In times of direct military conflict, militaries utilize much of the same infrastructure that civilians rely on. The relationship becomes more complicated when one considers the prospects of transnational defense and mobilization wherein military forces operate across multiple cyber-physical infrastructural systems across state-borders (Frizzelle et al. 2022). Hence the challenge is not just cross-jurisdictional interoperability standards, but also the management of vulnerabilities such as foreign direct investment and control of critical infrastructure (Id.).

Resilience in cyber-physical systems is thought to fill a strategic void. As some scholars have noted, there is arguably a "need for much broader political, economic, technological and societal resilience in the face of hostile acts below the threshold of war" (Hall and Sandeman 2022, 5). Indeed, it is the interaction between chronic multivalent stresses that exist below the threshold of war and actual direct military confrontation, capabilities and deterrence that is at the heart of a resilience strategy. In this sense, resilience is both the "first and last line of defense" (Hall and Sandeman 2022, 1). It is the first line of defense insofar as it serves as a CDD and it is the last line of defense in the face of shocks—and shocks who's systemic and asset sensitivity is amplified by regular and prolonged multivalent stresses.

At the 2016 Warsaw Summit of the North Atlantic Treaty Organization (NATO), the member nations (or "Allies") committed to adopting a more integrated and better-coordinated approach to strengthening resilience among NATO members (NATO 2016a). At the Vilnius Summit in 2023, the Allies agreed to each develop a National Resilience Plan (NRP) to meet a new range of civil preparedness requirements (NATO 2023a). Here, the Allies outlined the extent to which the robustness of their respective military capacities requires the designed resilience of social, institutional, and cyber-physical systems that cross over civilian and military domains.

In recent years, the United States and NATO have focused on citizen-level resilience planning that is centered on social processes and communications (Christie and Berzina 2022; NATO 2022a). From one perspective, this focus on the individual and the manager of physical systems is a form of redundancy and risk management that seeks to address the vulnerabilities of cyber-physical systems. As the White House's Federal Cybersecurity Research and Development Strategic Plan noted, there is a need for "[a] greater emphasis…on human-centered approaches to cybersecurity where people's needs, motivations, behaviors, and abilities are at the forefront of determining the design, operation, and security of information technology systems" (NSTC 2023). Future crises that deeply impair the mobilization of the general population emphasize the significance of civil societies in national preparedness and in responses that can react when cyber-physical controls are compromised.

These efforts include facilitating effective communications with civil managers and ensuring access to transparent, timely, and accurate information to combat disinformation among the general public (Reding and Wells 2022).

3 Defining Resilience with U.S. National Security and Defense

There are many different types of resilience, including cyber resilience, which may be applicable to a systems approach to managing cyber-physical vulnerability in national security and defense. The U.S. government has at least eight (8) different formal definitions of resilience (DOD 2019). The U.S. Department of Defense (DOD) does not have a single overarching definition of resilience despite prior proposals for developing such a definition (Herrera 2021). Within DOD, resilience planning appears in three core focus areas, including force readiness, military installations, and cyber security. However, these boundaries are not always so clear, as cyber resilience may play a key element of both force readiness and military installations. Often these focus areas are shaped by congressional legislation and appropriations. For instance, "military installation resilience" is defined as:

> [t]he capability of a military installation to avoid, prepare for, minimize the effect of, adapt to, and recover from extreme weather events, or from anticipated or unanticipated changes in environmental conditions, that do, or have the potential to, adversely affect the military installation or essential transportation, logistical, or other necessary resources outside of the military installation that are necessary in order to maintain, improve, or rapidly reestablish installation mission assurance and mission-essential functions (10 U.S.C § 101(e)(8)).

Engineering resilience and cyber resilience concepts for military installations have also been extended to the Production-based Resilience Program (PRP) and the Industrial Base Resilience (IBR) policies for weapons, advanced technology and other critical supply chains within the defense industrial base (DOD 2020; Nicastro 2023). On a tactical level utilizing the aforementioned statutory authority for military installation resilience, the DOD created the Readiness and Environmental Protection Integration Program (REPI) as a means to understand how engineering resilience, cyber resilience, and ecological resilience efforts may shape force readiness capacities and capabilities through an integration of relevant military and non-military public, private and civic stakeholders (REPI 2023). In the context of climate change, the DOD defines resilience as "[t]he ability to anticipate, prepare for, and adapt to changing conditions and withstand, respond to, and recover rapidly from disruptions" (DOD 2018). Resilience is also thematically applied at a strategic level, as highlighted by the creation of the office of the Deputy Assistant Secretary for Defense for Arctic and Global Resilience. This office is charged with ensuring the DOD's competitive advantage as it relates to sociotechnical and political changes relating to the energy transition, among other aspects of national security relating to environmental change (DOD 2023a).

Different forms of resilience are broadly utilized in everything from climate change planning to force readiness. Across the services of the U.S. military, there is an application of the concepts of resilience originating from the fields of psychology and psychiatry to support everything from mental and physical health to force readiness of personnel (McInerney et al. 2022). For instance, the U.S. Army, within the Office of the Deputy Chief of Staff, has a Directorate of Prevention, Resilience and Readiness that defines resilience as "the ability to persevere, adapt, [and] grow in dynamic or stressful environments" (U.S. Army 2023). In this case, the stress is on the individual. More broadly, the Office of Force Resiliency (OFR) within the Undersecretary of Defense for Personnel and Readiness focuses on a wider variety of issues including, "diversity management and equal opportunity, personnel risk reduction, suicide prevention, [and] sexual assault prevention and response" (OFR 2023).

Across many aspects of the U.S. government, resilience is largely relegated to a qualifiable objective, yet it is viewed by DOD as a critical element of its defense strategy. The 2022 National Defense Strategy (NDS) defines resilience as "the ability to withstand, fight through, and recover quickly from disruption" (DOD 2022, 8). The NDS explicitly reinforces the proposition that resilience planning is an effective mechanism for CDD by noting in a section entitled "Deterrence by Resilience" that "the Department will reduce adversary incentives for early attack by fielding diverse, resilient, and redundant [assets]. We will bolster our ability to fight through disruption by improving defensive capabilities and increasing options for reconstitution." (DOD 2020, 8).

Beyond force readiness and climate change applications driving resilience planning at the DOD, cyber resilience is increasingly the central focus of resilience policy and planning at DOD (Linkov and Kott 2019; DOD 2023b). Early work in cyber resilience defined resilience as an "ability to provide acceptable operations despite disruption: natural or man-made, inadvertent or deliberate" (DOD 2013). Yet, performance standards for resilience at the time, a decade ago, did not exist, with the Defense Science Board warning that "[p]oorly defined and improperly used metrics may prove as harmful as no metrics at all. (DOD 2013, 15). MITRE would later go on to develop resilience metrics that focused on performance before and after a detected or detectable event with a focus on the "time between disruption, detection, response and recovery" (Bodeau et al. 2018).

DOD would later build on the work of the National Institute of Standards and Technology (NIST) at the U.S. Department of Commerce who has led work on cyber security across the U.S. government. While NIST has ten different definitions for resilience, it has two distinct definitions for "cyber resiliency," including "[t]he ability to anticipate, withstand, recover from, and adapt to adverse conditions, stresses, attacks, or compromises on systems that use or are enabled by cyber resources" (NIST 2020). The second definition builds on the first definition by noting that "[c]yber resiliency is intended to enable mission or business objectives that depend on cyber resources to be achieved in a contested cyber environment" (NIST 2021). Cyber resilience activities for both internal networks and external contractors has been extended in defensive posture at units such as the Department of

Defense Cyber Crime Center and more broadly at DOD's Under Secretary of Defense for Research and Engineering.

At the current stage of development, cyber security, with its risk management orientation, and "operational resilience" were programmatically segregated, with operational resilience being defined as "the ability of systems to resist, absorb, and recover from or adapt to an adverse occurrence during operation that may cause harm, destruction, or loss of ability to perform mission-related functions" (DoDI 8500.001(2019)). In this context, cyber resilience has been largely relegated to a principled goal or objective and less of a quantification of designed processes. This finding is reinforced by the National Academies of Sciences, Engineering and Medicine's (NASEM) publications published by the Forum on Cyber Resilience, which do not define resilience and offer little to no direct methodological advancement in the forum's publications (NASEM 2023).

To address this gap in objective performance measures and criteria, the Biden Administration advanced cyber resilience within cyber-physical systems along three distinct tracks of research, planning and policy development. The cornerstone of the administration's policy development is the *2023 National Cybersecurity Strategy*, which covers everything from secure IoT to the quality assurance of private market software products and services (White House 2023). The strategy outlined a policy-focus on vulnerability assessment and intelligence sharing, labor training, and cross-sector partnerships. From their perspective, it is not just about setting standards, one needs to diffuse those standards within a trained workforce and among a class of private sector consumers, particularly owners and operators of critical infrastructure assets and systems.

A more tactical perspective on resilience standards came from the President's Council of Advisors on Science and Technology (PCAST), which produced a research report entitled, *The Strategy for Cyber-Physical Resilience: Fortifying Our Critical Infrastructure for a Digital World* (PCAST 2024). PCAST defines cyber-physical resilience as "the capacity of an integrated system to keep running—even if not at peak performance—should it lose specific functions" (PCAST 2024, 10). They positioned this loss of function to be associated with the "degradation or cessation of one or more aspects of the computational or physical functions due to component failures, human errors, natural disasters, or malicious attacks" (Id.).

In light of these challenges, the report explicitly called on the administration to, "[d]efine sector minimum viable operating capabilities and minimum viable delivery objectives…[e]stablish and measure leading indicators [and commit to] …transparency and stress testing" (Id., 8). The report also called for greater participation in setting international standards. These international efforts present the opportunity for learning among partner countries (CISA 2023), but it also allows for the United States to stay ahead of "China and other nations [that] are driving standards across an array of technologies" (PCAST 2024, 34).

PCAST also called for the Department of Homeland Security's (DHS) Cybersecurity and Infrastructure Security Agency (CISA) to expand its role as a coordinator and to provider common resilience products that cut across agencies responsible for key sectors, otherwise knowns as Sector Risk Management Agencies (SRMAs).

This could be accomplished in two ways. First, it could expand its role by "by pushing sector risk management responsibilities out to an empowered set of SRMAs" who need more staff and budgetary resources. (PCAST 2024, 17). The SRMAs include a wide variety of agencies across major sectors. For example, the Department of Energy has jurisdiction over the energy sector and the Treasury Department has jurisdiction over financial services (CISA 2024). In some cases, SRMAs may need additional Congressional authority in order to address gaps between federal, state, and tribal responsibilities, as well as gaps that exist between the public and private sectors. Right now, these multi-sector priorities are managed by Sector Coordinating Councils (SCCs). But PCAST argued that the SCCs should be chaired by major owners and operators who can drive performance standards based on their own experience, capabilities, and aspirations.

Second, PCAST felt that CISA could serve as a central information and intelligence clearinghouse by "sharing through Information Sharing and Analysis Centers (ISACs), the DHS Joint Cyber Defense Collaborative (JCDC), and the National Security Agency (NSA) Cyber Collaboration Center" (PCAST 2024, 17). This would allow for a more rigorous incident command and to support an incident review process undertaken by DHS's Cyber Safety Review Board. This central flow of information could also support the triaging and prioritization of critical infrastructure systems by the National Risk Management Center (NRMC) and the Office of the National Cyber Director (ONCD).

This consolidated body of work across the administration within cyber-physical resilience culminated in 2024, with the release of the *National Security Memorandum on Critical Infrastructure Security and Resilience* (White House 2024). Among other things, the policy memorandum set-in motion a process for SRMAs to work with various public and private stakeholders to begin to set standards for resilience performance. This wide-ranging initiative is engaged the relevant national laboratories that are driving cutting-edge research, as well as relevant law enforcement agencies who have responsibility for domestic cyber crimes. More importantly, as recommended by PCAST, CISA was tasked with establishing a national coordinator office to centralize the work of the SRMAs. Although, the policy memorandum utilizes a definition of resilience that is substantially similar to the previously cited NIST definition, it is very likely each sector and corresponding SRMA will utilize a different types and measures of resilience that are specific to sector, asset class and system operations. The more fundamental challenge will be consolidating and coordination resilience standards within each sector—not necessarily across sectors.

Beyond the United States, cyber-physical vulnerabilities have been growing in NATO. In recent decades, there has been an increased reliance on private sector actors to provide critical services and infrastructure (Newlove-Eriksson et al. 2018; Radvanovsky and McDougall 2018). In the Cold War era, essential infrastructure like railways, ports, airfields, grids, and airspace were predominantly under state control in Europe, which facilitated their transfer to NATO authority in times of crisis or conflict. By contrast, approximately "90 per cent of NATO's supplies and logistics are moved [today] by private companies and 75 per cent of the host nation support for NATO forces forward deployed on the territory of the eastern Allies comes from private sector contracts" (Shea 2016).

NATO formally defines resilience in two contexts. In the context of cyber-physical applications, resilience is defined as "[t]he ability of a functional unit to continue to perform a required function in the presence of faults or errors" (NATO 2005). In the general terminology of NATO, resilience is officially defined as "[t]he ability of an entity to continue to perform specified functions during and after an attack or an incident" (NATO 2022b). Here, resilience is oriented towards episodes, shocks and known or potentially detectable phenomena. As will be discussed, this might be too limiting of a frame for design systems to adapt to the uncertainties and unknowns of cyberwarfare and cyber-physical vulnerabilities.

While DOD, NIST and U.S. agencies have approached resilience from a variety of bottom-up thematic areas focused on strategic objectives, NATO has approached resilience planning from a largely top-down central planning orientation that starts with a meta-system perspective. For CISA's efforts to be successful as a national coordinator, they likely have something to learn from NATO's efforts to plan systematically across jurisdictions and sectors. Conversely, NATO is poised to benefit from sector-specific activities of the SRMA's under CISA guidance.

4 How Science Can Support 'Layered Resilience'

The science of resilience suggests that there is no one single or universal definition or conceptual framing of resilience. Rather there are a variety of major categories that have emerged over time in core disciplinary areas of science and each of these categorical forms of resilience possesses its own conceptual and analytical autonomy, empirical application and external validity (Davidson et al. 2016; Joseph 2018; Mentges et al. 2023). Key categorical forms of resilience are applied in physical, informational, cognitive, social, and cyber domains (Wells et al. 2022).

Across these domains, many different forms of resilience are programmatically applied across the U.S. government policy, including engineering resilience, cyber resilience, operational resilience, ecological resilience, and community resilience (Keenan 2018a, b; Kang et al. 2023). Even within civil engineering and critical infrastructure systems design and management, these different types of resilience serve different functions, metrics, and assessments (Field et al. 2022). This conceptual plurality within material systems of the built environment and across economic sectors is also widely observed in European policy (Baravikova et al. 2021).

Table 1 provides a non-exhaustive overview of key categorical forms of resilience found in American and European policies and programs. This table provides a heuristic reference to each categories' properties, attributes, and performance criteria. Some forms of resilience, such as cyber resilience, engineering resilience, and disaster resilience are primarily concerned with an elastic and reversionary process that reverts to the pre-perturbation state (Wied et al. 2020). This is often referenced as a single equilibrium process (Zampieri 2021). These forms of

Table 1 Example, Non-Exhaustive List of Categorical Forms of Resilience in Science

Concept	Definition	Performance (Single Equilibrum, Multi-Equilibrum)	Adaptation	Properties and Attributes	Citation
Engineering resilience	"A resilient engineered system is able to successfully complete its planned mission(s) in the face of environmental and adversarial threats, and has capabilities allowing it to flexibly adapt to future missions with evolving threats"	Multi-equilibrium	Yes	Flexibility, reliable performance	Small et al. (2018)
	"Implies the ability of an engineered system to autonomously sense and response to adverse changes in health conditions, to withstand failure events, and to recover from the effects of these unpredicted events"	Multi-equilibrium	No	Anticipate, withstand, recover	Yodo and Wang (2016)
	"..Concentrates on stability near an equilibrium steady state, where resistance to disturbance and speed of return to the equilibrium are used to measure the property"	Single equilibrium	No	Return speed, steady state	Holling (1996)
Cyber resilience	"Cyber resilience refers to the ability of the system to prepare, absorb, recover, and adapt to adverse effects, especially those associated with cyber-attacks."	Multi-equilibrium	Yes	Absorption, recovery	Linkov and Kott (2019)
	"...The ability to anticipate, withstand, recover from, and adapt to adverse conditions, stresses, attacks, or compromises on cyber resources"	Multi-equilibrium	Yes	Anticipate, absorption, recovery	Bodeau et al. (2018)
	"Cyber resilience is the ability of a cyber system to recover from stress that causes a reduction of performance."	Single equilibrium	No	Recovery	Smith (2023)
Disaster resilience	"The ability to prepare and plan for, absorb, recover from or more successfully adapt to actual or potential adverse events"	Multi-equilibrium	Yes	Absorption, recovery	Cutter (2016)

	"Adaptive and integrated disaster resilience ... Is defined as the ability of nations and communities to build resilience in an integrated manner and strengthen mechanisms to build system adaptiveness"	Multi-equilibrium	Yes	Integration, flexibility	Djalante et al. (2013)
	"The ability of a system, community or society to pursue its social, ecological and economic development objectives, while managing its disaster risk over time in a mutually reinforcing way."	Single equilibrium	No	Performance, persistence	Keating et al. (2017)
Ecological resilience	"..Emphasizes conditions far from any equilibrium steady state, where instabilities can flip a system into another regime of behavior— That is, to another stability domain. In this case, the measurement of resilience is the magnitude of disturbance that can be absorbed before the system changes its structure by changing the variables and processes that control that behavior."	Multi-equilibrium	Yes	Absorb, deform, persistence, transform	Holling (1996)
	"The magnitude of disturbance that can be absorbed before a system flips to another state"	Multi-equilibrium	Yes	Absorptionm, transforms	Dakos and Kéfi (2022)
	"The capacity of natural systems subject to instability to absorb disturbances without shifting to a fundamentally different ecosystem domain.	Multi-equilibrium	Yes	Absorption, identity	Grade et al. (2023)
Community resilience	"Community resilience is the ability of communities to withstand, recover, and learn from past disasters, and to learn from past disasters to strengthen future response and recovery efforts. This can include but isnot limited to physical and psychological health of the population, social and economic equity and Well-being of the community, effective risk communication, integration of organizations (governmental and	Multi-equilibrium	Yes	Withstand, recovery, social learning	Keenan (2019)

(continued)

Table 1 (continued)

Concept	Definition	Performance (Single Equilibrum, Multi-Equilibrum)	Adaptation	Properties and Attributes	Citation
	nongovernmental) in planning, response, and recovery, and social connectedness for resource exchange, cohesion, response, and recovery."				
	"Ability of a system, community, or society exposed to hazards to resist, absorb, accommodate to and recover from the effects of a hazard in a timely and efficient manner including through the preservation and restoration of its essential basic structures and functions	Single equilibrium	No	Resistance, absorption, recovery, return time	Ostadtaghizadeh (2015)
	"Community resilience is indicated by evidence of community Well-being following a disaster or crisis. Thus community provides an opportunity for a collective to adaptively cope with the experience of a potentially traumatic event."	Single equilibrium	Yes	Coping	Houston et al. (2015)
Psychological resilience	"Psychological resilience is largely concerned with how people cope with adversity, psychological risk factors, and individual sensitivities to trauma"	Single equilibrium	Yes	Coping	Davidson et al. (2016)
	"Psychological resilience is the ability to adapt positively to life conditions. It is a dynamic process evolving over time that implies a type of adaptive functioning that specifically allows us to face difficulties by recovering an initial balance or bouncing back as an opportunity for growth."	Multi-equilibrium	Yes	Flexibility, recovery, renewal, social learning	Sisto et al. (2019)
	"a process in which an individual's state or functioning bounces back to the previous level following a stressor."	Single equilibrium	No	Recovery	Den Hartigh and Hill (2022)

resilience are focused on designed processes within the context of engineered systems and systems-of-systems (SoSs) wherein fixed levels of performance have been optimized to a robust range for purposes of reliability, safety, efficiency and other limiting factors (Park et al. 2013; Martello et al. 2021).

Conversely, within the context of multi-equilibrium processes, other forms of engineering resilience and cyber resilience may provide an underlying capacity for adaptation wherein the designed resilience performance may result in a superior level of performance across its lifecycle in a post-perturbation state (Woods 2015; Ungar 2021). In either case, scholars have noted the value of utilizing different forms of resilience across systems within SOSs as a means to decomposing and managing case-specific complexities, as Uday and Marais note:

> Although there are advantages to using a single calculated value to define resilience, it is also important to recognize the potential issues associated with doing so [Haimes 2009]. In particular, an overall metric provides little, if any, information regarding specific areas within the SoS that need attention. Also, in the context of SoSs, the uncertainties associated with network operations, evolution, and management are quite large and hence one metric may not be able to capture all the unknowns. To address these concerns, some studies focus on capturing or disentangling the various dimensions of resilience (Uday and Marais 2015, 502).

In a different context of complex adaptive systems, ecological resilience is also a multi-equilibrium phenomenon that is concerned with the degree of external impact that an ecological unit may endure while still maintaining its key identify and performance measures in what might otherwise be shifting responsive structures (Dakos and Kéfi 2022). Other forms of resilience, such as community resilience, are focused on people and social behavior and may be explicitly multi-equilibrium phenomena due to the social learning and the modification of behavior that leads to forms of adaptation (Koliou et al. 2020). Social dimensions of resilience have also found specialized forms across scales—from individuals (i.e., psychological resilience)(Denckla et al. 2020), organizations (i.e., organizational resilience, operational resilience) (Essuman et al. 2020; Galaitsi et al. 2023), and institutions (i.e., economic resilience, financial system resilience) (Keenan et al. 2021; Hynes et al. 2022).

Social aspects of resilience—often referenced as 'community resilience'—specifically addresses the capacity of communities and societies to effectively confront, adapt to, and recover from external perturbations (Koliou et al. 2020), including the complex impacts of hybrid and conventional warfare (Zekulić et al. 2017). This form of resilience is pivotal in safeguarding societal coherence, morale, and the continuity of social functions amidst the psychological, informational, and physical disruptions posed by such conflicts (Hadji-Janev 2018; Lee et al. 2022). The essence of community resilience in this milieu involves the strategic cultivation of societal attributes such as informed awareness, communal solidarity, and adaptive governance structures. These elements are crucial for mitigating the divisive and destabilizing tactics employed in hybrid warfare, which often targets the social fabric through disinformation campaigns, cyber-attacks, and other forms of indirect aggression (Apostol et al. 2022).

For this reason, community resilience and other social aspects are increasingly considered critical for managing cyber-physical vulnerabilities and control systems. Mechanisms for fostering social resilience include community-based education and training programs that enhance disaster readiness, the development of infrastructure, particularly communications infrastructure, capable of withstanding physical and cyber-attacks (Nikolov 2018). Such strategies not only aim to preserve the immediate welfare and security of the population, but they also ensure the long-term viability and cohesion of the social structure, thereby reinforcing the overall resilience performance of the state against the multifarious challenges of modern warfare (Wigell et al. 2021).

Wherein some forms of resilience (i.e., engineering resilience, cyber resilience) are explicitly designed as endogenous processes that may be directly indicated, other forms of resilience may be defined by innate characteristics (i.e., psychological resilience) or normative outcomes (i.e., community resilience). Because some forms of resilience may be defined by objectives (e.g., engineering resilience, cyber resilience) and others may be defined by characteristics (e.g., ecological resilience, psychological resilience), it "becomes imperative to understand and establish interlinkages between the objectives and characteristics," as there may be a misalignment and/or direct conflict of objectives across scales (Kolte et al. 2023, 11). For this reason, it is often necessary to draw analytical distinction between the various forms of resilience across scales in order to 'disentangle' objectives and characteristics and processes and outcomes.

This diverse application of different forms of resilience across sectors has not been widely internalized within national security and defense policies, which has tended to utilize a singular form of single-equilibrium resilience in general and non-specific terms. Unfortunately, early applications of resilience in American national security policy development lacked scientific specificity and therefore were viewed by some scholars as "an exercise in 'occupying' them with ideologically loaded meanings, which can be interpreted as the actualisation of… 'political keywords'" (Selchow 2017, 36). In Europe, the incorporation of resilience into national security raised a more fundamental critique that "resilience [is] a way to 'act on the cheap' removing responsibility from the international community, as local actors become the main protagonists – and in case of failure, the main culprit – of conflict management, also potentially leading to an indirect legitimization of dysfunctional local practices for conflict and security management" (Baldaro and Costantini 2021, 309).

Yet, other scholars have noted that resilience in the context of security policy represents a fundamental shift in a theory of governance. Chandler notes, "[t]he area where resilience-thinking has gone furthest is in the rejection of modernist or liberal approaches to governance, based upon linear or deterministic knowledge assumptions of causality. For liberal universalist approaches, policymaking is constructed in 'top-down' ways, determined by the 'known knowns' of established knowledge and generalisable assumptions of cause and effect" (Chandler 2014, 62). In rejecting a linear and deterministic orientation to external shocks and stresses, an extension of this position would suggest that resilience policy is or should be—at its best—

grounded in a bottom-up orientation that acknowledges unknowns and deep uncertainty in a world that is shaped by increasing interdependencies and rapid rates of global change manifested across scales and sectors.

In 2021, the United States and the Allies endorsed the NATO Warfighting Capstone Concept that introduced 'layered resilience' as one of five development imperatives necessary for current and future security challenges (NATO 2022c; NATO 2023b). In broad terms, layered resilience relies on different sectors across military-civilian domains to each demonstrate their own resilience performance measures and thus enable NATO to sustain, if necessary, long and protracted campaigns (van Mill 2023). This highlights the extent to which cyber resilience in cyber-physical systems is just one dimension that interacts with various others forms of resilience across the military and civilian domains (NATO 2023c). As such, Table 2 provides an illustrative orientation to layered resilience that highlights various scientific applications across select categorical forms of resilience (NATO, 2023d). In many ways, this representation of layered resilience is useful for CISA's oversight of various sectors across the SRMAs.

While some forms of resilience may reinforce reliability and safety dimensions of cyber-physical systems in a single-equilibrium context, other forms acknowledge the potential value and capacity for transformation (i.e., adaptation) when thresholds for single equilibrium recovery are either undesirable, infeasible or otherwise not possible. For example, ongoing advancements at DOD in climate adaptation planning within installations, facilities and force readiness is a prime example of the interaction between engineering resilience (i.e., often but not always single equilibrium) and ecological resilience (i.e., multi-equilibrium) within the MR domain wherein environmental conditions are expected to far exceed current steady state operational parameters (Garfin et al. 2021).

Carrying forward this example, these horizontal connections between different forms of resilience within the MR domain share a convergent scenario where installations must be secured (engineering resilience); people must be secured and protected (disaster resilience); assets must adapt (ecological resilience); forces must take on new operating conditions and health precautions (community resilience); and, force personnel must be prepared to improvise in the face of the stress of change and disasters (psychological resilience). This example represents a singular and unilateral risk vector that may impact military capabilities. When hostile acts compound simultaneously with climate impacts and/or cyber-attacks, for example, then the value of resilience is the capacity to execute planned recovery and "to improvise" beyond the limits of resilience in order to change structures to transform to a new operational state (Pursiainen and Kytömaa 2023, 10). For instance, redundant communications hardware and infrastructure designed for post-disaster coordination may serve as a critical lifeline during a cyberattack, which may also, by extension, increase the cyber resilience of dual-use command, control and communications supporting military capabilities (Afina et al. 2020).

This is not hypothetical, as cyber-attacks have been observed to increase in number following natural disasters and extreme events (Lallie et al. 2021). Years before the Russian military invasion of Ukraine, hackers, widely believed to be

Table 2 Example Applications of Scientific Categories of Resilience Across NATO's Layered Resilience Model

	Cyber Resilience/ Engineering Resilience	Disaster Resilience	Ecological Resilience	Community Resilience	Psychological Resilience
Alliance resilience (AR)	Continuity of Government; Instutional Redudancy and Operations Continuity for Strategic and Logistics Command and Political Communications (Joseph 2018; Hodicky et al. 2020)	Cross-Jurisdictional Mutual Aid Agreements and Processes (DSCA 2021)	Cross-Border Environmental Regulation and Measurement; National Climate Adaptation Planning (Pennington 2020)	Monitoring Emergent Public Health Events, Social Welfare and Vulnerabilty, and Invest in Shared Public Health and Social Support Resources (Chandra et al. 2010; Cutter and Derakhshan 2019)	Monitoring and Countering Malicious Misinformation Campaigns in Health, Crime, Foreign Relations, Elections and Public Trust (Bonnevie et al. 2021; Maronkova 2021)
Military resilience (MR)	Installation and Asset Preparedness and Risk Mitigation (DOD 2018; Herrera 2021)	Event Contigency and Scenario Planning and Training (DSCA 2021)	Measure, Monitor, Mitigate and Adapt Impacts to Ecosystem Services Support Operations (DOD 2018; REPI 2023)	Manage Force Readiness (OFR 2023; U.S. Army 2023)	Manage Force Readiness; Monitor Mental Health and Radicalization (McInerney et al. 2022; OFR 2023; U.S. Army 2023)
Military-civilian resilience (MCR)	Classify and Protect Critical Infrastructure, Production Base, and Cyber Networks; Audit and Test Critial Networks; Cross-Domain Intelligence Gathering and Mitigation; External Utilization of Contractors with Flexible Resilience Capabilities (Cusumano 2018; DOD 2020; NIST 2020; Nicastro 2023; DOD 2023b)	Joint Pre-and Post-Disaster Mobilization; Resource Reserve Management (Puckett 2021)	Deployment of Sustainability, Biodiversity, Climate Mitigation and Climate Adaptation Cross-Sector Protocols and Disclosure Reporting; Shared Data Ecosystem for Water and Air Quality; Environmental Security and Peacebuilding (Schilling et al. 2017; Vogler 2023; Johnson et al. 2023)	Indentify, train, develop and protect key personnel at various stages of deployment and mobilization; support adaptability between military and civilian jobs (Oprins et al. 2018; Stette et al. 2023)	Intelligence and Counter Measures in Cognitive Warfare; Indentify, Develop and Support Key Personnel Through Mobilization, Demobilization and Remobilization; Identify and Monior At-Risk Key Personnel (MacLeish 2020; Claverie and du Cluzel 2022)

Civilian resilience (CR)	Cyber Network Security Performance Standards and Critical Infrastructure Systems; Critical Infrastructure Levels of Service Standards (Liu and Song 2020; NIST 2021)	Business continuity and organizational resilience of dependent producers (Galaitsi et al. 2023)	Environmental impact Management for Industrial Symbiosis (Fraccascia et al. 2017)	Invest in Public Health Institutions and Surveillance; Occupational Health and Exposure; Job Training and Technical Education; Physical Fitness (Sharifi 2016; Patel et al. 2017)	Invest in Mental Health Support; Monitor Malicious Misinformation Campaigns (Manwaring and Holloway 2023)

associated with Russian state-sponsored groups, launched a sophisticated attack in December 2015 that resulted in power outages for several hours in parts of western Ukraine. The attackers used malware to gain access to the control systems of energy distribution companies, manipulating and disrupting the operation of critical infrastructure and highlighting the vulnerabilities inherent in critical infrastructure sectors (Condliffe 2016).

5 Resilience May Support Transformative Adaptation

Adaptation takes over when the limits of resilience have been reached. Resilience narrowly viewed through traditional risk management strategies often focuses on specific, known threats, creating targeted solutions that may become inadequate or obsolete as new challenges emerge (Linkov et al. 2017). It is often impossible to predict the threats that will emerge, their magnitude, or the likelihood or existence of a specific threat scenario (Hynes et al. 2022). Planners attempting to mitigate all known risks will face prohibitive costs due to the sheer number of possible disruptions—most of which will never emerge as actual disruptions (Galaitsi et al. 2022). In contrast, resilience is often threat-agnostic because it processes system performance regardless of the nature or origin of the disruption (Pescaroli et al. 2023). A continuing challenge of investing in resilience is the fact that, in many cases outside of designed single equilibrium forms of engineering resilience, it is not easy to measure, as it depends on unknown timing and the extent of the system degradation. Therefore, probabilistic expected values and net present value optimizations originating from risk management processes are less useful for quantifying resource trade-offs supporting resilience investments than other forms of robust decision-making. In the face of a range of uncertainties, forms of resilience that support the adaptive capacity of a system are the preferred pathway when total and complete system failure increases the likelihood of military defeat.

Adaptation is critical particularly in the context of when frontline defense and resistance to acute shocks fails. Under one common scenario, a defense may simply fail because the acute shock or attack was either unknown or undetectable. For instance, in the NATO description of MR, the deterrence operation to adapt occurs in a linear condition only under circumstances of known and anticipated risks. By the same token, defense may manifest to known or unknown (but detectable) phenomena, but the ultimate outcome under this emergent layered resilience framework is presumed to be the recovery to the pre-event baseline and not necessarily alternative superior state achievable through adaptation through new structures and processes. Both known and, particularly, unknown phenomena may cross thresholds from which pre-event baseline conditions are either impossible or undesirable. Resilience is only useful if military commanders and NATO decision makers can understand and operate from the knowledge that systems can and will eventually fail, and that resilience is largely the capability or pathway to flexibly recover and/or adaptively execute new structures that support core functions and performance measures in the eventual case that such systems fail.

Equally, effective adaptation to conventional and hybrid threats requires an understanding and integration of societal systems into U.S. and NATO doctrine, preparedness, and best practices. Particularly for cyber-physical vulnerabilities and threats, it is incumbent upon policymakers to ensure that societal systems and collective human intelligence is able to adapt to the shifting nature of hybrid threats, particularly in the emerging era of generative artificial intelligence (Terziyan et al. 2021). This necessitates a resilience framework that not only anticipates and withstands attacks but also evolves in response to unknown or anticipated threats, ensuring societal structures can dynamically adjust and maintain critical control functions. Emphasizing societal aspects of resilience policy within U.S. and NATO's strategy involves reinforcing public awareness (Lillemäe et al. 2023), enhancing digital literacy, and fostering a culture of adaptability and innovation (Larsen 2023). By doing so, the U.S. and NATO can create a more robust defense ecosystem, one that leverages the collective resilience of its member states to navigate and overcome the multifaceted challenges posed by both known and unforeseen threats.

This is because the resilience performance is reliant on a set of fixed design parameters and resources that may not scale with the increased intensity (or novelty) of hostile action or the convergence of multiple acute extreme events (e.g., cyber-attacks and climate impacts). At some point, the resilience threshold may be reached and the traditional "V" curve of resilience—where the post-perturbation state equals the pre-perturbation state—is degraded to the extent to that the post-perturbation performance is inferior to the level of pre-perturbation. To this end, an inferior performance of resilience at this stage may also be defined by a longer execution time and at a greater cost. Such an outcome defined by the stress strain on the elastic performance of resilience would suggest that there is a greater capacity for maladaptation as the intensity of hostile acts increases. This may be simply because any attempt at transformation at the edge of a resilience threshold may be based on the same limited assumptions and resources of pre-perturbation designed resilience and/or processes and policies that are "too rigid [so as to] pose any obstacle to improvisation" (Pursiainen and Kytömaa 2023, 10). Indeed, if the resilience function collapses entirely without any subsequent adaptation, then total failure may be the result.

In this context, PCAST positioned cyber-physical resilience (technically, cyber resilience) as elastic single-equilibrium function in response to an all-hazards orientation to risk. They highlighted the near paradoxical situation where "security and resilience elements are tacked on after systems are deployed, often imperfectly and at considerable expense" whereas "technology manufacturers [will need to develop] their systems to be secure and resilient by design [in order] to dramatically reduce the number of flaws that can fail or be exploited by threat actors" (PCAST 2024, 12). This perspective reinforces the value of the social and communications aspects of human managers that are able to adapt systems when resilience thresholds are reached. From PCAST's perspective, these thresholds are often more widespread than people appreciate because cyber resilience is incorporated ex post into systems in manner that potentially expands the total range of potentially vulnerable points of system conflict.

Very often adaptation happens as a function of social and/or machine learning wherein experience is translated into an alternative range of practice. This is in contrast with designed single equilibrium resilience in an engineered system, which, in many cases, is based on an entirely endogenous utilization of intelligence in the initial exercise of the design of the resilience performance. At some point in time, when there is sufficient social learning, machine learning, and/or external intelligence the subject system may begin to reduce the measure of consequence as it adapts to the intensity and/or novelty of hostile acts. Indeed, with greater exposure and experience, the adaptive capacity may significantly increase proportionate to the intelligence. Of course, this assumes on some level that the adverse action has been detected.

There is no fixed measure of certainty associated with maladaptation or adaptation, respectively, in designed systems. Rather, these are measures of capacity for such outcomes. However, with greater degrees of social and machine learning and intelligence, there is a greater capacity for flexibility, and alternative pathways and structures that alter the identity of a system but otherwise maintain desirable processes and/or conditions. This raises the challenging paradox of resilience in convergent cyber-physical critical infrastructure systems. On one level, the cyber domain offers a greater range of novelty for malicious acts, while, at the same time, there is a great capacity for the utilization of machine learning and artificial intelligence to create new structures and pathways that represent transformative adaptation to such novelties (Colabianchi et al. 2021). Yet, paradoxically, machine learning algorithms—and perhaps the agents of social learning—themselves must deploy a measure of resilience (Olowononi et al. 2020). This raises the prospects that alternating states of resilience and adaptation may reach an absolute frontier from which total and absolute failure is the only outcome.

6 Conclusions for Guiding National Resilience Plans (NRPs)

This chapter has highlighted the extent to which resilience represents a strategic policy goal that provides both a front and last line of defense. It is a front-line defense by virtue of its presumed ability to serve as a CDD, and it serves as a last line of defense when dependent systems and SOSs are stressed by convergent and novel shocks and stresses, particularly in the context of cyber-physical vulnerabilities. As a strategic policy goal, resilience represents a fundamental shift in a theory of national security that acknowledges the limitations of deterministic threat mitigation and conventional models of risk management. For the U.S. and NATO, it also represents a critical stop-gap deterrence building measure when the rates and degrees of change are moving faster than many institutions and political processes have historically been able to keep up with.

The current resilience strategy is based on an integrated approach that encourages collaboration among military, civilian, and private sectors, enhancing the collective ability to detect, respond to, and recover from threats. This is particularly key because many cyber-physical vulnerabilities are resident in the private sector, which has been historically under-resourced and under-regulated in terms of cyber resilience investments and performance standards, respectively. By adopting a comprehensive resilience framework, the U.S. and NATO can enhance their adaptive capacities, ensuring they remain robust in the face of evolving challenges stemming from hybrid and cyber threats, among other unknown or uncertain vectors of conflict and disruption. This approach enables the identification and management of vulnerabilities across cyber-physical systems.

The DOD, NIST, and various federal agencies have a great deal of experience in applying resilience across multiple sectors, as well physical and cyber-physical domains. What the U.S. government lacks is a fundamental and overarching strategic deployment of resilience within national security and defense governance bodies. CISA's recent efforts as a national coordinator are a significant step forward. Having such a strategic plan would be arguably critical for horizontal and vertical integration across sectors and military and civilian sectors and physical and cyber-physical domains. By contrast, NATO has a strong strategic orientation to resilience with its emerging 'layered resilience' framework. It simply lacks the specificity of scientific application of resilience within and across the domains. In this, NATO likely has a great deal to learn from CISA's and SRMA's sector-specific performance standards. By working together, the U.S. and its NATO Allies can merge bottom-up experience with top-down planning toward a more comprehensive deployment of resilience for strategic goals.

The United States and NATO Allies have committed to developing a NRP that connects domestic and international processes, capabilities, and performance standards. In developing a comprehensive NRP, it is crucial to recognize the complex interdependencies that exist across jurisdictions and shared cyber-physical infrastructure. From transportation to energy and healthcare to defense, each sector faces unique challenges, yet they share many structural underlying similarities and interconnected vulnerabilities. To craft an integrated NRP that capitalizes on these commonalities, one should align efforts in significant areas where collaborations and information can be exchanged and sustained. Given the dynamic nature of change, ongoing networks for people and information are going to need to be built alongside the NRP process. This chapter has highlighted the extent to which human agency is critical for cyber resilience and managing cyber-physical vulnerabilities.

The following steps may be taken to advance greater cooperative development of resilience standards by and between the U.S. federal government NATO Allies. First, the respective parties should engage the private owners and operators of every major sector to support sector-specific standards for resilience, with the acknowledgement that each sector is likely to vary a great deal This is similar to the outreach that CISA and SRMA's are engaged in already. However, by engaging other countries, there is an opportunity to understand not only endogenous processes but also exogenous factors, particularly in the context of cross-sector dependencies. By extension, the respective standard setting stakeholders should acknowledge the epistemological diversity of resilience. For instance, a pure application of cyber resilience definitions that are more or less a function of single-equilibrium engineering resilience performance setting may overlook social learning and other forms of adaptive behaviors that are key for when thresholds are crossed.

Second, the respective SRMAs, DOD and NATO allies should synchronize and level-set standards by sector. This will be important for defining the appropriate metrics and proxy indicators that should be measured. It will also facilitate cross-jurisdiction learning and highlight risk and uncertainties in cross-sector resilience performance that have otherwise been unobserved. This exercise in level-setting may also include stress testing across sectors and using AI to simulate novel attacks that stress the capacity for signal intelligence and response. Finally, once standards are synchronized across jurisdictions, existing cyber intelligence centers and post-event review boards should work collaboratively to regularly test and update these centers. As hybrid threats evolve and vulnerabilities shift, resilience standards will need to adapt pursuant to a rigorous process of testing and evaluation. This evaluation will be critical for engaging and supporting private and public sector stakeholders who will be able to make informed decisions about balancing costs with preemptive resilience designs and investments. Together, these steps will form the basis of NRPs development processes within and across NATO allies.

Resilience planning requires acknowledging that SOSs, systems, and non-systems are interconnected across sectors and domains in ways that are often non-linear, unintuitive, and counter to prevailing institutional wisdom. This means that a conventional orientation to single-equilibrium forms of resilience (e.g., engineering and cyber resilience) is likely to be maladaptive given the convergent vulnerabilities across scales, sectors, and cyber-physical domains. As such, engineering and cyber resilience should be conceptualized as dynamic and adaptive processes that offer a pathway for a sustainable response, recovery, and maybe even transformation. To this end, success might very well depend on a full recognition of the epistemological diversity found in empirical science for the different types of resilience. As the United States and NATO Allies prepare their NRPs, they have the opportunity to incorporate adaptive processes that lean towards bolstering resilience in the deployment of resources that seek to guard against a very unpredictable future.

Disclosure The authors report there are no competing interests to declare.

Funding Details This chapter is based upon work supported by the U.S. Department of Defense Operational Resilience International Cooperation Program (DORIC) and U.S. Army Engineer Research and Development Center FLEX Project on Compounding Threats. The views and opinions expressed in this chapter are those of the individual authors and not those of the U.S. Army, or other sponsor organizations.

References

Afina Y, Inverarity C, Unal B (2020) Ensuring cyber resilience in NATO's command, control and communications systems. Research Paper, International Security Programme. The Royal Institute of International Affairs, Chatham House, London, UK. https://www.chathamhouse.org/sites/default/files/2020-07-17-cyber-resilience-nato-command-control-communication-afina-inverarity-unal.pdf

Ams S (2023) Blurred lines: the convergence of military and civilian uses of AI & data use and its impact on liberal democracy. Int Politics 60(4):879–896. https://doi.org/10.1057/s41311-021-00351-y

Apostol AC, Cristache N, Năstase M (2022) Societal resilience, a key factor in combating hybrid threats. Int Conf Knowl-Based Organ 28(2):107–115. https://doi.org/10.2478/kbo-2022-0057

Baldaro E, Costantini I (2021) Fragility and resilience in the European Union's security strategy: comparing policy paradigms. Ital Polit Sci Rev/Rivista Italiana di Scienza Politica 51(3): 305–320. https://doi.org/10.1017/ipo.2020.22

Baravikova A, Coppola A, Terenzi A (2021) Operationalizing urban resilience: insights from the science-policy interface in the European Union. Eur Plan Stud 29(2):241–258. https://doi.org/10.1080/09654313.2020.1729346

Bodeau DJ, Graubart RD, McQuaid RM, Woodill J (2018) Cyber resiliency metrics, measures of effectiveness and score. MITRE Corporation, Bedford. Retrieved from https://apps.dtic.mil/sti/citations/trecms/AD1108054

Bolbot V, Theotokatos G, Bujorianu LM, Boulougouris E, Vassalos D (2019) Vulnerabilities and safety assurance methods in cyber-physical systems: a comprehensive review. Reliab Eng Syst Saf 182:179–193. https://doi.org/10.1016/j.ress.2018.09.004

Bonnevie E, Sittig J, Smyser J (2021) The case for tracking misinformation the way we track disease. Big Data Soc 8(1):20539517211013867. https://doi.org/10.1177/20539517211013867

Brooks RA (2019) Integrating the civil–military relations subfield. Annu Rev Polit Sci 22:379–398. https://doi.org/10.1146/annurev-polisci-060518-025407

Chandler D (2014) Beyond neoliberalism: resilience, the new art of governing complexity. Resilience 2(1):47–63. https://doi.org/10.1080/21693293.2013.878544

Chandra A, Acosta J, Meredith LS, Sanches K, Stern S, Uscher-Pines L, Yeung D (2010) Understanding community resilience in the context of National Health Security. RAND Corporation, Santa Monica. https://www.rand.org/content/dam/rand/pubs/working_papers/2010/RAND_WR737.pdf

Christie EH, Berzina K (2022) *NATO and societal resilience: all hands-on deck in the age of war*, Policy brief. The German Marshall Fund of the United States, Washington, D.C.. https://www.gmfus.org/sites/default/files/2022-07/NATO%20and%20Societal%20Resilience%20All%20Hands%20on%20Deck%20in%20an%20Age%20of%20War.pdf

Claverie B, du Cluzel F (2022) The cognitive warfare concept. In: Innovation hub working paper. NATO Allied Command Transformation, Norfolk. https://innovationhub-act.org/wp-content/uploads/2023/12/CW-article-Claverie-du-Cluzel-final_0.pdf

Colabianchi S, Costantino F, Di Gravio G, Nonino F, Patriarca R (2021) Discussing resilience in the context of cyber physical systems. Comput Ind Eng 160:107534. https://doi.org/10.1016/j.cie.2021.107534

Condliffe J (2016) Ukraine's power grid gets hacked again, a worrying sign for infrastructure attacks. MIT Technol Rev. December 22nd, 2016. Retrieved from https://www.technologyreview.com/2016/12/22/5969/ukraines-power-grid-gets-hacked-again-a-worrying-sign-for-infrastructure-attacks/

Constantinescu M (2023) Measuring economic resilience for the CEE and Black Sea countries in the framework of comprehensive defence. Secur Def Q 44(4):55–83. https://doi.org/10.35467/sdq/175379

Cusumano E (2018) Resilience for hire? NATO contractor support in Afghanistan examined. In: Cusumano E, Corbe M (eds) A civil-military response to hybrid threats. Palgrave Macmillan, Cham. https://doi.org/10.1007/978-3-319-60798-6_6

Cutter SL (2016) The landscape of disaster resilience indicators in the USA. Nat Hazards 80(2): 741–758. https://doi.org/10.1007/s11069-015-1993-2

Cutter SL, Derakhshan S (2019) Implementing disaster policy: exploring scale and measurement schemes for disaster resilience. J Homeland Secur Emerg Manage 16(3):20180029. https://doi.org/10.1515/jhsem-2018-0029

Cybersecurity and Infrastructure Security Agency (CISA) (2023) U.S. and international partners publish secure-by-design and -default principles and approaches. U.S. Department of Homeland Security, Washington, D.C.. 2023, April 13th. https://www.cisa.gov/news-events/news/us-and-international-partners-publish-secure-design-and-default-principles-and-approaches

Cybersecurity and Infrastructure Security Agency (CISA) (2024) Sector risk management agencies. U.S. Department of Homeland Security, Washington, D.C.. https://www.cisa.gov/topics/critical-infrastructure-security-and-resilience/critical-infrastructure-sectors/sector-risk-management-agencies

Dakos V, Kéfi S (2022) Ecological resilience: what to measure and how. Environ Res Lett 17(4): 043003. https://doi.org/10.1088/1748-9326/ac5767

Davidson JL, Jacobson C, Lyth A, Dedekorkut-Howes A, Baldwin CL, Ellison JC, Smith TF (2016) Interrogating resilience: toward a typology to improve its operationalization. Ecol Soc 21(2) https://www.jstor.org/stable/26270410

Defense Security Cooperation Agency (DSCA) (2021) Building resilience for National Defense. Institute for Security Governance, Monterey, CA. Retrieved from https://nps.edu/documents/113018911/125185574/ISG+Resilience.pdf/bbbe0227-207d-873a-3864-c4d19d7bdc14?t=1622747034982

Den Hartigh RJ, Hill Y (2022) Conceptualizing and measuring psychological resilience: what can we learn from physics? New Ideas Psychol 66:100934. https://doi.org/10.1016/j.newideapsych.2022.100934

Denckla CA, Cicchetti D, Kubzansky LD, Seedat S, Teicher MH, Williams DR, Koenen KC (2020) Psychological resilience: an update on definitions, a critical appraisal, and research recommendations. Eur J Psychotraumatol 11(1):1822064. https://doi.org/10.1080/20008198.2020.1822064

Djalante R, Holley C, Thomalla F, Carnegie M (2013) Pathways for adaptive and integrated disaster resilience. Nat Hazards 69(3):2105–2135. https://doi.org/10.1007/s11069-013-0797-5

Essuman D, Boso N, Annan J (2020) Operational resilience, disruption, and efficiency: conceptual and empirical analyses. Int J Prod Econ 229:107762. https://doi.org/10.1016/j.ijpe.2020.107762

Field C, Sutley E, Naderpajouh N, van de Lindt JW, Butry D, Keenan JM, Koliou M (2022) Incorporating socioeconomic metrics in civil engineering projects: the resilience perspective. Nat Hazard Rev 23(1):04021064. https://doi.org/10.1061/(ASCE)NH.1527-6996.000053

Folke C, Carpenter SR, Walker B, Scheffer M, Chapin T, Rockström J (2010) Resilience thinking: integrating resilience, adaptability and transformability. Ecol Soc 15(4) https://www.jstor.org/stable/26268226

Fraccascia L, Giannoccaro I, Albino V (2017) Rethinking resilience in industrial symbiosis: conceptualization and measurements. Ecol Econ 137:148–162
Frizzelle B, Garey J, Kulalic I (2022) NATO's national resilience mandate: challenges and opportunities. Def Stud 22(3):525–532. https://doi.org/10.1080/14702436.2022.2082954
Galaitsi S, Kurth M, Fries S, Linkov I (2022) Mainstreaming resilience analytics: 10 years after the Fukushima disaster. Integr Environ Assess Manag 18(6):1551–1554. https://doi.org/10.1002/ieam.4623
Galaitsi SE, Pinigina E, Keisler JM, Pescaroli G, Keenan JM, Linkov I (2023) Business continuity management, operational resilience, and organizational resilience: commonalities, distinctions, and synthesis. Int J Disaster Risk Sci 14(5):713–721
Garfin G, Falk DA, O'Connor CD, Jacobs K, Sagarin RD, Haverland AC, Zuniga-Teran AA (2021) A new mission: mainstreaming climate adaptation in the US Department of Defense. Clim Serv 22:100230. https://doi.org/10.1016/j.cliser.2021.100230
Gartzke E, Lindsay JR (eds) (2019) Cross-domain deterrence: strategy in an era of complexity. Oxford University Press, New York
Grade AM, Crimmins AR, Basile S, Essig MR, Goldsmith L, Lustig AR, Maycock TK, McCarrick A, Scheetz A (2023) Appendix 5. Glossary. In: Crimmins AR, Avery CW, Easterling DR, Kunkel KE, Stewart BC, Maycock TK (eds) Fifth National Climate Assessment. U.S. Global Change Research Program, Washington, D.C. https://doi.org/10.7930/NCA5.2023.A5
Hadji-Janev M (2018) Toward community resilience against disinformation & propaganda with a focus. J Commun 68(6):1007–1028
Haimes YY (2009) On the definition of resilience in systems. Risk Analysis: An International Journal 29(4):498–501. https://doi.org/10.1111/j.1539-6924.2009.01216.x
Hall J, Sandeman H (2022) NATO's resilience: the First and last line of defence, Strategic update. The London School of Economics and Political Science, London. https://www.lse.ac.uk/ideas/Assets/Documents/updates/2022-SU-NATO-HallSandeman.pdf
Hartmann U (2017) The evolution of the hybrid threat, and resilience as a countermeasure, Research paper 139. NATO Defense College, Research Division, Rome. Retrieved from https://www.ndc.nato.int/news/news.php?icode=1083
Herrera JG (2021) Military installation resilience: what does it mean? Congressional Research Service, Washington, D.C.. Retrieved from https://apps.dtic.mil/sti/pdfs/AD1147490.pdf
Hodicky J, Özkan G, Özdemir H, Stodola P, Drozd J, Buck W (2020) Dynamic modeling for resilience measurement: NATO resilience decision support model. Appl Sci 10(8):2639. https://doi.org/10.3390/app10082639
Holling CS (1996) Engineering resilience versus ecological resilience. In: Schulze E (ed) Engineering within ecological constraints. National Academy of Engineering, Washington, DC
Houston JB, Spialek ML, Cox J, Greenwood MM, First J (2015) The centrality of communication and media in fostering community resilience: a framework for assessment and intervention. Am Behav Sci 59(2):270–283. https://doi.org/10.1177/0002764214548563
Hynes W, Trump BD, Kirman A, Haldane A, Linkov I (2022) Systemic resilience in economics. Nat Phys 18(4):381–384. https://doi.org/10.1038/s41567-022-01581-4
Johnson MF, Ide T, Cruz JG (2023) Conceptualizing resilience within environmental peacebuilding. Curr Opin Environ Sustain 65:101362. https://doi.org/10.1016/j.cosust.2023.101362
Joseph J (2018) Varieties of resilience: studies in governmentality. Cambridge University Press, Cambridge.
Kang KE, Bowman AOM, Hannibal B, Woodruff S, Portney K (2023) Ecological, engineering and community resilience policy adoption in large US cities. Urban Aff Rev 59(6):1973–2004. https://doi.org/10.1177/10780874221150793

Keating A, Campbell K, Mechler R, Magnuszewski P, Mochizuki J, Liu W, Szoenyi M, McQuistan C (2017) Disaster resilience: what it is and how it can engender a meaningful change in development policy. Dev Policy Rev 35(1):65–91. https://doi.org/10.1111/dpr.12201

Keenan JM (2018a) From climate change to national security: analysis of the Obama administration's federal resilience mandates and measures. Nat Hazard Rev 19(1):04017022. https://doi.org/10.1061/(ASCE)NH.1527-6996.0000273

Keenan JM (2018b) Types and forms of resilience in local planning in the US: who does what? Environ Sci Pol 88:116–123. https://doi.org/10.1016/j.envsci.2018.06.015

Keenan JM (2019) Climate adaptation finance and Investment in California. Routledge, London

Keenan JM, Trump BD, Hynes W, Linkov I (2021) Exploring the convergence of resilience processes and sustainable outcomes in post-covid, post-Glasgow economies. Sustain For 13(23):13415. https://doi.org/10.3390/su132313415

Koliou M, van de Lindt JW, McAllister TP, Ellingwood BR, Dillard M, Cutler H (2020) State of the research in community resilience: progress and challenges. Sustain Resilient Infrastruct 5(3): 131–151. https://doi.org/10.1080/23789689.2017.1418547

Kolte R, Goswami S, Kumar A, Pipralia S (2023) Challenges in practical implementation of the concept of urban resilience in cities. Int J Disaster Risk Reduct 99:104142. https://doi.org/10.1016/j.ijdrr.2023.104142

Lallie HS, Shepherd LA, Nurse JR, Erola A, Epiphaniou G, Maple C, Bellekens X (2021) Cyber security in the age of COVID-19: a timeline and analysis of cyber-crime and cyber-attacks during the pandemic. Comput Secur 105:102248. https://doi.org/10.1016/j.cose.2021.102248

Larsen JA (2023) The United States and NATO: is there a western strategic culture? In: Routledge handbook of strategic culture. Routledge, New York, pp 91–105

Lee JH, Santero N, Bhattacharya A, May E, Spiro ES (2022) Community-based strategies for combating misinformation: learning from a popular culture fandom. Harvard Kennedy School Misinformation Review. Retrieved from https://misinforeview.hks.harvard.edu/article/community-based-strategies-for-combating-misinformation-learning-from-a-popular-culture-fandom/

Lillemäe E, Talves K, Wagner W (2023) Public perception of military AI in the context of techno-optimistic society. AI & Soc 1-15. https://doi.org/10.1007/s00146-023-01785-z

Lindsay JR (2020) Information technology and military power. Cornell University Press, Ithaca

Linkov I, Kott A (2019) Fundamental concepts of cyber resilience: introduction and overview. In: Kott A, Linkov I (eds) Cyber resilience of systems and networks. Springer, Cham, pp 1–25. https://doi.org/10.1007/978-3-319-77492-3_1

Linkov I, Trump BD, Fox-Lent C (2017) Resilience: approaches to risk analysis and governance. In: Florin MV, Linkov I (eds) Resource guide on resilience. EPFL International Risk Governance Center (IRGC), Lausanne

Linkov I, Baiardi F, Florin MV, Greer S, Lambert JH, Pollock M, Trump BD (2019) Applying resilience to hybrid threats. IEEE Secur Privacy 17(5):78–83. https://doi.org/10.1109/MSEC.2019.2922866

Liu W, Song Z (2020) Review of studies on the resilience of urban critical infrastructure networks. Reliab Eng Syst Saf 193:106617. https://doi.org/10.1016/j.ress.2019.106617

MacLeish K (2020) Churn: mobilization–demobilization and the fungibility of American military life. Secur Dialogue 51(2–3):194–210. https://doi.org/10.1177/0967010619889469

Mallory K (2018) New challenges in cross-domain deterrence, Perspective. RAND Corporation, Santa Monica. Retrieved from https://www.jstor.org/stable/pdf/resrep17639.pdf

Manwaring R, Holloway J (2023) Resilience to cyber-enabled foreign interference: citizen understanding and threat perceptions. Def Stud 23(2):334–357. https://doi.org/10.1080/14702436.2022.2138349

Maronkova B (2021) NATO amidst hybrid warfare threats: effective strategic communications as a tool against disinformation and propaganda. In: Jayakumar S, Ang B, Anwar ND (eds) Disinformation and fake news. Palgrave Macmillan, Singapore. https://doi.org/10.1007/978-981-15-5876-4_9

Martello MV, Whittle AJ, Keenan JM, Salvucci FP (2021) Evaluation of climate change resilience for Boston's rail rapid transit network. Transp Res Part D: Transp Environ 97:102908. https://doi.org/10.1016/j.trd.2021.102908

McInerney SA, Waldrep E, Benight CC (2022) Resilience enhancing programs in the US military: an exploration of theory and applied practice. Mil Psychol 36:1–12. https://doi.org/10.1080/08995605.2022.2086418

Mentges A, Halekotte L, Schneider M, Demmer T, Lichte D (2023) A resilience glossary shaped by context: reviewing resilience-related terms for critical infrastructures. Int J Disast Risk Reduct 96:103893. https://doi.org/10.1016/j.ijdrr.2023.103893

Milburn R (2023) Carbon warriors: enhancing NATO's response to climate change. RUSI J 168(4): 30–39. https://doi.org/10.1080/03071847.2023.2235152

Mumford A, Carlucci P (2023) Hybrid warfare: the continuation of ambiguity by other means. Eur J Int Secur 8(2):192–206. https://doi.org/10.1017/eis.2022.19

National Academies of Sciences, Engineering and Medicine's (NASEM) (2023) Forum on cyber resilience. Retrieved from https://www.nationalacademies.org/our-work/forum-on-cyber-resilience

National Institute of Standards and Technology (NIST) (2020) Protecting controlled unclassified information in nonfederal systems and organizations, NIST Special Publication 800-171, Revision 2. National Institute of Standards and Technology, U.S. Department of Commerce, Gaithersburg, MD. https://doi.org/10.6028/NIST.SP.800-171r2

National Institute of Standards and Technology (NIST) (2021) Developing cyber-resilient systems: a systems security engineering approach, NIST Special Publication 800-160, Revision 1. National Institute of Standards and Technology, U.S. Department of Commerce, Gaithersburg. https://doi.org/10.6028/NIST.SP.800-160v2r1

National Science and Technology Council (NSTC) (2023) Federal cybersecurity research and development strategic plan. Executive Office of the President, Washington, D.C.

NATO (2005) Science & technology: information technology: resilience. The Official NATO Terminology Database, NATO Term. Retrieved from https://nso.nato.int/natoterm/Web.mvc

NATO (2016a) Commitment to enhance resilience. Press Release #116. July 8th, 2016. Retrieved from https://www.nato.int/cps/en/natohq/official_texts_133180.htm

NATO (2022a) Call for proposals: increasing societal resilience: innovative ways to counter Disinformation and hostile activities. NATO Headquarters, Brussels. Retrieved from https://www.nato.int/structur/pdd/2022/220411-ResilienceContentGuidelines.pdf

NATO (2022b) General terminology: resilience. NATO Term: The Official NATO Terminology Database. Retrieved from https://nso.nato.int/natoterm/Web.mvc

NATO (2022c) NATO standard AJP-01 allied joint doctrine. Edited F version 1 with UK National Elements. NATO Standardization Office, Brussels

NATO (2023a) Vilnius Summit Communiqué. Press Release #001. July 11th, 2023. Retrieved from https://www.nato.int/cps/en/natohq/official_texts_217320.htm

NATO (2023b) NWCC: NATO warfighting capstone concept. Allied Command Transformation, North Atlantic Treaty Organization, Norfolk. Retrieved from https://www.act.nato.int/wp-content/uploads/2023/06/NWCC-Glossy-18-MAY.pdf

NATO (2023c) What is military resilience (MR)?. Layered resilience team working paper. NATO HQ Supreme Allied Commander Transformation, Norfolk

NATO (2023d) Layered resilience interdependencies between the civil and military layers, Layered resilience team working paper. NATO HQ Supreme Allied Commander Transformation, Norfolk, VA

Newlove-Eriksson L, Giacomello G, Eriksson J (2018) The invisible hand? Critical information infrastructures, commercialisation and national security. Int Spect 53(2):124–140. https://doi.org/10.1080/03932729.2018.1458445

Nicastro LA (2023) The U.S. defense industrial base: background and issues for congress, R47751. Congressional Research Service, Washington, D.C.. October 12th, 2023. https://crsreports.congress.gov/product/pdf/R/R47751

Nikolov O (2018) Building societal resilience against hybrid threat. Inf Secur 39(1):91–109. https://doi.org/10.11610/isij.3908

Office of Force Resiliency (OFR), Office of the Under Secretary of Defense for Personnel & Readiness 2023. Office of force resiliency: overview. Retrieved from https://prhome.defense.gov/Home/Organization/OFR/

Olowononi FO, Rawat DB, Liu C (2020) Resilient machine learning for networked cyber physical systems: a survey for machine learning security to securing machine learning for CPS. IEEE Commun Surv Tutor 23(1):524–552. https://doi.org/10.1109/COMST.2020.3036778

Oprins EA, Bosch KVD, Venrooij W (2018) Measuring adaptability demands of jobs and the adaptability of military and civilians. Mil Psychol 30(6):576–589. https://doi.org/10.1080/08995605.2018.1521689

Ostadtaghizadeh A, Ardalan A, Paton D, Jabbari H, Khankeh HR (2015) Community disaster resilience: a systematic review on assessment models and tools. PLoS Curr Disasters 7. https://www.ncbi.nlm.nih.gov/pmc/articles/PMC4395373/

Park J, Seager TP, Rao PSC, Convertino M, Linkov I (2013) Integrating risk and resilience approaches to catastrophe management in engineering systems. Risk Anal 33(3):356–367. https://doi.org/10.1111/j.1539-6924.2012.01885.x

Patel SS, Rogers MB, Amlôt R, Rubin GJ (2017) What do we mean by 'community resilience'? A systematic literature review of how it is defined in the literature. PLoS Curr 9. https://doi.org/10.1371/currents.dis.db775aff25efc5ac4f0660ad9c9f7db2

Pennington C (2020) Strengthening our resilience to environmental change is a National Security Imperative. MITRE Center for Data-Driven Policy, McLean. Retrieved from https://apps.dtic.mil/sti/trecms/pdf/AD1137491.pdf

Pescaroli G, Guida K, Reynolds J, Pulwarty RS, Linkov I, Alexander DE (2023) Managing systemic risk in emergency management, organizational resilience and climate change adaptation. Disaster Prev Manage 32(1):234–251. https://doi.org/10.1108/DPM-08-2022-0179

Popper KR (2020) The open society and its enemies. Princeton University Press, Princeton

President's Council of Advisors on Science and Technology (PCAST) (2024) Strategy for cyber-physical resilience: fortifying our critical infrastructure for a digital world. Executive Office of the President, Washington, DC

Puckett LM (2021) Civil-military coordination in disaster preparedness and response. Nat Hazard Rev 22(2):04021005. https://doi.org/10.1061/(ASCE)NH.1527-6996.0000446

Pursiainen C, Kytömaa E (2023) From European critical infrastructure protection to the resilience of European critical entities: what does it mean? Sustain Resilient Infrastruct 8(sup1):85–101. https://doi.org/10.1080/23789689.2022.2128562

Radvanovsky RS, McDougall A (2018) Critical infrastructure: homeland security and emergency preparedness. CRC Press, Boca Raton

Readiness and Environmental Protection Integration Program (REPI) (2023) Using 10 U.S.C. §2684a and the Sikes Act (16 U.S.C. §670c-1) to maintain or improve military installation resilience. U.S. Department of Defense, Washington, D.C.. Retrieved from https://www.repi.mil/Portals/44/Documents/Resilience/ResilienceAuthorities_OCT2020.pdf

Reding DF, Wells B (2022) Cognitive warfare: NATO, COVID-19 and the impact of emerging and disruptive technologies. In: COVID-19 Disinformation: a multi-national, whole of society perspective. Springer, Cham, pp 25–45. https://doi.org/10.1007/978-3-030-94825-2_2

Reinike WJ (2020) The U.S. financial system as a network: insight and implications for hybrid warfare, Master's Thesis. Naval Postgraduate School, Monterey. https://apps.dtic.mil/sti/citations/AD1114705

Roepke, W. and Thankey, H. 2019. "Resilience: the first line of defense" NATO Review. Retrieved from https://www.nato.int/docu/review/articles/2019/02/27/resilience-the-first-line-of-defence/index.html

Schilling J, Nash SL, Ide T, Scheffran J, Froese R, Von Prondzinski P (2017) Resilience and environmental security: towards joint application in peacebuilding. Global Change Peace Secur 29(2):107–127. https://doi.org/10.1080/14781158.2017.1305347

Selchow S (2017) Resilience and resilient in Obama's National Security Strategy 2010: enter two 'political keywords'. Politics 37(1):36–51. https://doi.org/10.1177/0263395715614847

Sharifi A (2016) A critical review of selected tools for assessing community resilience. Ecol Indic 69:629–647. https://doi.org/10.1016/j.ecolind.2016.05.023

Shea J (2016) Resilience: a core element of collective defense. NATO Review. Retrieved from https://www.nato.int/docu/review/articles/2016/03/30/resilience-a-core-element-of-collective-defence/index.html

Sisto A, Vicinanza F, Campanozzi LL, Ricci G, Tartaglini D, Tambone V (2019) Towards a transversal definition of psychological resilience: a literature review. Medicina 55(11):745. https://doi.org/10.3390/medicina55110745

Smalenberger DM (2015) On pecuniary resiliency, early warning, and market imitation under unrestricted warfare. Doctoral dissertation. Wright-Patterson Air Force Base: Air Force Institute of Technology. https://scholar.afit.edu/etd/227/

Small C, Parnell G, Pohl E, Goerger S, Cottam B, Specking E, Wade Z (2018) Engineering resilience for complex systems. In: Madni AM, Boehm B, Ghanem RG, Erwin D, Wheaton MJ (eds) Disciplinary convergence in systems engineering research. Springer, pp 3–15. https://doi.org/10.1007/978-3-319-62217-0_1

Smith S (2023) Towards a scientific definition of cyber resilience. Proc Int Conf Cyber Warfare Secur 18(1):379–386. https://doi.org/10.34190/iccws.18.1.960

Stette M, Porath K, Muehlich S (2023) Civil-military cooperation for the countering of threats: protection of civilians during the development of a threat. In: Balomenos KP, Fytopoulos A, Pardalos PM (eds) Handbook for management of threats, Springer optimization and its applications. Springer, Cham. https://doi.org/10.1007/978-3-031-39542-0_23

Sweijs T, Zilincik S, Bekkers F, Meessen R (2021) A framework for cross-domain strategies against hybrid threats. Hague Centre for Strategic Studies, Den Haag. Retrieved from https://euhybnet.eu/wp-content/uploads/2021/01/Framework-for-Cross-Domain-Strategies-against-Hybrid-Threats.pdf

Terziyan V, Gavriushenko M, Girka A, Gontarenko A, Kaikova O (2021) Cloning and training collective intelligence with generative adversarial networks. IET Collab Intell Manuf 3(1): 64–74. https://doi.org/10.1049/cim2.12008

Thornton R, Miron M (2020) Towards the 'third revolution in military affairs' the Russian military's use of AI-enabled cyber warfare. RUSI J 165(3):12–21. https://doi.org/10.1080/03071847.2020.1765514

U.S. Army, Office of the Deputy Chief of Staff, Directorate of Prevent, Resilience and Readiness 2023. About the directorate of prevent, resilience and readiness. Retrieved from https://www.armyresilience.army.mil/ard/about-ARD.html

U.S. Department of Defense (DOD) (2013) Task force report: resilient military systems and the advanced cyber threat. Office of the Under Secretary of Defense for Acquisition, Technology and Logistics, Washington, D.C.. Retrieved https://apps.dtic.mil/sti/pdfs/ADA569975.pdf

U.S. Department of Defense (DOD) (2018) DOD directive 4715.21: climate change adaptation and resilience. Office of the Under Secretary of Defense for Acquisition and Sustainment, Washington, D.C.. Retrieved from https://www.esd.whs.mil/Portals/54/Documents/DD/issuances/dodd/471521p.pdf

U.S. Department of Defense (DOD) (2019) United state government compendium of interagency and associated terms. Joint Chiefs of Staff, U.S. Department of Defense, Washington, D.C.. Retrieved from https://www.jcs.mil/Portals/36/Documents/Doctrine/dictionary/repository/usg_compendium.pdf?ver=2019-11-04-174229-423

U.S. Department of Defense (DOD) (2022) National Defense Strategy of The United States of America. U.S. Department of Defense, Washington, D.C.. Retrieve from https://media.defense.gov/2022/Oct/27/2003103845/-1/-1/1/2022-NATIONAL-DEFENSE-STRATEGY-NPR-MDR.PDF

U.S. Department of Defense (DOD) 2023a. Deputy assistant secretary of defense for arctic and global resilience. Retrieved from https://policy.defense.gov/OUSDP-Offices/ASD-for-Homeland-Defense-and-Hemispheric-Affairs/Arctic-and-Global-Resilience/
U.S. Department of Defense (DOD) (2023b) 2023 cyber strategy of the department of defense. Department of Defense, Washington, D.C.. Retrieved from https://media.defense.gov/2023/Sep/12/2003299076/-1/-1/1/2023_DOD_Cyber_Strategy_Summary.PDF
Uday P, Marais K (2015) Designing resilient systems-of-systems: a survey of metrics, methods, and challenges. Syst Eng 18(5):491–510. https://doi.org/10.1002/sys.21325
Ungar M (ed) (2021) Multisystemic resilience: adaptation and transformation in contexts of change. Oxford University Press, New York
van Mill J (2023) The future of NATO's resilience concepts and wargaming, The three swords. Concepts, Capability Integration and Experimentation Branch, NATO Joint Warfare Centre, Stavanger. Retrieved from https://www.jwc.nato.int/application/files/9616/9804/8578/RESILIENCE.pdf
Vina A, Liu J (2023) Effects of global shocks on the evolution of an interconnected world. Ambio 52(1):95–106. https://doi.org/10.1007/s13280-022-01778-0
Vogler A (2023) Tracking climate securitization: framings of climate security by civil and defense ministries. Int Stud Rev 25(2):viad010. https://doi.org/10.1093/isr/viad010
Wells EM, Boden M, Tseytlin I, Linkov I (2022) Modeling critical infrastructure resilience under compounding threats: a systematic literature review. Prog Disaster Sci 100244:100244. https://doi.org/10.1016/j.pdisas.2022.100244
White House (2023) 2023 national cybersecurity strategy. Executive Office of the President, Washington, DC
White House (2024) National security memorandum on critical infrastructure security and resilience. Executive Office of the President, Washington, D.C.
Wied M, Oehmen J, Welo T (2020) Conceptualizing resilience in engineering systems: an analysis of the literature. Syst Eng 23(1):3–13. https://doi.org/10.1002/sys.21491
Wigell M, Mikkola H, Juntunen T (2021) Best practices in the whole-of-society approach in countering hybrid threats, Study for European parliament. Directorate General for External Policies of the European Union, Brussels. https://doi.org/10.2861/702047
Woods DD (2015) Four concepts for resilience and the implications for the future of resilience engineering. Reliab Eng Syst Saf 141:5–9. https://doi.org/10.1016/j.ress.2015.03.018
Yodo N, Wang P (2016) Engineering resilience quantification and system design implications: a literature survey. J Mech Des 138(11):111408. https://doi.org/10.1115/1.4034223
Yu DJ, Schoon ML, Hawes JK, Lee S, Park J, Rao PSC, Ukkusuri SV (2020) Toward general principles for resilience engineering. Risk Anal 40(8):1509–1537. https://doi.org/10.1111/risa.13494
Zampieri M (2021) Reconciling the ecological and engineering definitions of resilience. Ecosphere 12(2):e03375. https://doi.org/10.1002/ecs2.3375
Zekulić V, Godwin C, Cole J (2017) Reinvigorating civil–military relationships in building national resilience. RUSI J 162(4):30–38. https://doi.org/10.1080/03071847.2017.1380376

Regional Critical Infrastructure: A Cyber-Physical Resilience Assessment Methodology

Nathaniel Evans and Logan Sturm

1 Introduction

Regions rely upon critical infrastructure systems and services as well as the people and cyber systems that operate, monitor, and supply them. The critical infrastructure and services provide water, electricity, natural gas, communications, transportation, and healthcare as well as emergency response services such as police, fire, and ambulance. Cyber systems operate or support most of these systems and services. Cyber systems in turn depend upon critical infrastructure, especially electricity, communications, and water.

This interdependent web of systems is made more complex by the diversity of entities that build, own, and operate these systems. The entities range from local, state, tribal, and federal governments to small and large private sector firms. The interconnected, interdependent infrastructure allows individual incidents to spread and damage other systems and perhaps lead to systemic, cascading failure. This possibility is illustrated by the Northeast blackout of 2003.

In 2003, an overloaded electricity transmission line in the northeastern region of the United States (US) sagged because of high summer heat, and hit an overgrown tree, causing a fault in the transmission line. Normally, grid monitoring software would detect the fault and alert system operators who would reroute the electricity. However, a software error prevented the alarm to the operators. Without the alarm, the operators failed to respond appropriately, and other lines failed, ultimately causing a cascade of failures throughout southeastern Canada and northeastern United States. High summer heat caused a physical incident that started the problem, but a cyber incident prevented operators from responding appropriately.

N. Evans · L. Sturm (✉)
Oak Ridge National Laboratory, Oak Ridge, TN, USA
e-mail: sturmld@ornl.gov

© The Author(s), under exclusive license to Springer Nature Switzerland AG 2025

I. Linkov, A. Kott (eds.), *Cyber Resilience: Applied Perspectives*, Risk, Systems and Decisions, https://doi.org/10.1007/978-3-031-90109-6_11

The 2003 blackout affected millions of people and many types of infrastructure ranging from power and water to transportation and communication, including the internet infrastructure required by cyber systems (Cowie et al. 2003).

Another major power outage was the 2021 winter blackout in Texas. Abnormally cold weather caused record demand for electricity while simultaneously reducing the generation capability. This resulted in the Electric Reliability Council of Texas (ERCOT) shutting off power to more than 4.5 million households to avoid a potential total grid collapse. Economic losses from the storm were estimated to be $130 billion for the state of Texas and at least 111 people lost their lives from causes such as hypothermia and carbon monoxide poisoning.

One of the key factors contributing to the blackouts was the underperformance of natural gas power generation, with approximately 40% of gas power production being unavailable. Natural gas power plants rely on pressurized supply lines from natural gas producers, many of which require significant amounts of electricity for their operations. Many natural gas suppliers used interruptible power contracts and were not classified as critical infrastructure, resulting in them being unintentionally shut down during load shedding, reducing the gas production available and further reducing generation capacity in a negative feedback loop. These underlying flaws in Texas' power grid, combined with its regional infrastructure being unable to draw significant power from the larger Eastern/Western Interconnects in the USA led to the widespread blackouts in a state with significant energy reserves. Busby et al. (2021).

Many types of hazards pose a threat to critical infrastructure. Categories of threats include natural, such as earthquakes, or environmental, such as tornadoes or excessive heat; human-caused, such as sabotage or terrorism; and accidental such as software faults or equipment failure. Threats can occur suddenly: consider tornadoes and earthquakes, terrorist and denial-of-service attacks, and bridge collapse or computer failure. Threats can also occur slowly as happens during flooding or increasing summer temperatures, computer intrusions known as advanced persistent threats, or the slow degradation of infrastructure.

Most hazards arise locally, within the regions, but cyber threats can be delivered remotely from outside the region. For example, Russian operatives have launched cyber attacks against control systems to cause electricity outages in the Ukraine (ICS-CERT 2018a) and Georgia (Armed Forces Communications and Electronics Association (AFCEA) 2012).

In May 2021 one of the most significant cyber-attacks in US history occurred when the Colonial Pipeline that delivers oil from Houston, TX up the East Coast was hit with a cyber-attack. This attack shut down the pipeline for several days which caused localized shortages and panic buying as well as impacting air traffic due to fuel shortages at several airports in the region.

The shutdown occurred when the business information technology (BIT) systems of Colonial Pipeline were infected by Darkside ransomware. An inability to process billing information and concerns that the ransomware could spread from the BIT systems to the operational technology systems (OT) caused the Colonial Pipeline to shut down its pumping infrastructure as well until the attack could be resolved (UNECE 2023).

The Colonial Pipeline attack demonstrated how vulnerabilities in seemingly unconnected systems can indirectly target critical regional infrastructure. While the Texas Winter Blackout was not caused by a cyber-attack, the cascading failure seen in natural gas production and power generation could also be exploited in malicious actor scenario, leading to major disruptions in the regional infrastructure and beyond.

To assess regional resilience, analysts benefit from identifying and assessing the complex interrelationship between cyber and physical systems. The analysis needs to understand the types of threats to and the vulnerabilities of the critical infrastructure. Understanding the cyber dependency of critical systems and the dependency of cyber systems upon the critical infrastructure of the region is essential for understanding the impact of incidents on the resilience of the region.

This chapter presents a cyber resiliency assessment (CRA) methodology to examine a region by bringing together interested stakeholders as well as infrastructure, simulation, and resilience experts to identify and analyze that region's dependency upon critical infrastructure. The CRA aims to improve regional resilience by finding areas of concern and offering resilience enhancement options. This chapter provides an overview of the CRA process, some of the analytical and modeling techniques that are used for cyber assessments, and a case study that illustrates some of the possible results of a CRA.

2 CRA Overview

The goal of a cyber resiliency assessment is to increase cyber infrastructure resilience by generating greater understanding and action among public and private sector partners to improve the resilience of a region's cyber critical infrastructure. CRAs assess specific infrastructure systems within a designated geographic area and address a range of hazards that could have regionally and nationally significant consequences.

CRAs focus on improving cyber infrastructure resilience by conducting several cyber-specific data collections. One type of data collection concerns the physical dependencies of cyber infrastructure, such as data centers and communication networks, with special attention paid to identifying whether their failures could cascade and cause national or global consequences. The case study presented later in this chapter shows how a local physical disaster in the eastern region of the United States, near the city of Ashburn in the state of Virginia, could produce national or global outages affecting many organizations.

Another type of investigation focuses on the supply chain that provides critical infrastructure with cyber equipment, services, and possibly personnel. The computerized control systems that operate critical infrastructure are often supplied by only one or a very few vendors. If damaged, can they be replaced, either from spares or the vendor? If not available, are functionally equivalent replacements available? Do these require customization or a redesign of the control system? If the control systems contain security vulnerabilities, can they be mitigated? Does the vendor provide security patches? Can those patches be applied in a timely manner?

When identifying cyber resilience gaps and mitigation strategies, the CRA also assesses whether new cyber-specific policies and procedures, especially for cyber incident response, can be formulated or applied. A goal of the investigation is ensuring cooperation among regional organizations on reporting and response to cyber incidents.

2.1 Methodology

A CRA can incorporate various components, including voluntary facility vulnerability assessments, targeted studies and modeling, first responder capability assessments, subject-matter expert workshops, and other exchanges of information.

CRA studies can produce a wide variety of products, but usually include a written report that summarizes the research and analysis results and provides key findings, including key regional resilience gaps and options for addressing those shortfalls. Select CRA participants receive the output products along with any supporting documentation and information. Facility owners and operators, regional organizations, and government agencies can use the results to help guide strategic investments in equipment, planning, training, and resources to enhance the resilience and protection of facilities, surrounding communities, and entire regions.

CRAs are conducted in a multi-step process that includes the following:

- Assess critical infrastructure on a regional level, focusing on threats, vulnerabilities, and consequences from an all-hazards perspective
- Identify dependencies, interdependencies, and cascading effects by developing awareness of how this system can be disrupted through facility site assessments, cyber assessments, one-on-one interviews, and facilitated meetings with a wide range of stakeholders
- Coordinate efforts to enhance resilience by developing “options for resilience enhancements” to be implemented by the state and local stakeholders

The CRA incorporates vulnerability assessments, capabilities assessments, and infrastructure protection planning efforts to assemble an analysis of a region’s critical infrastructure and relevant preparedness capabilities. Specific efforts associated with each CRA project may include:

- On-site physical security and regional assessments
- Cyber resilience reviews
- Cyber assessments
- Geospatial products
- Workshops

- Facilitated discussions with public and private stakeholders
- Modeling and analysis
- Risk mitigation and awareness training
- Tabletop exercises

3 Cyber Analysis at the Regional Level

One of the initial hurdles for a CRA is understanding how to look at cyber through a regional perspective. Many examples of network and data infrastructure exist independently of the Internet; however, the world increasingly sees cyber as only Internet-connected devices and networks. Internet connected, by definition, means globally accessible. The CRA considers network-connected infrastructure differently for different types of assessments as seen below.

3.1 *Defining Cyber in Regional Terms*

Characterization efforts in a regional assessment typically include identifying cyber assets that are unique to a region or to an infrastructure system in that region. It is important to understand the difference between these two concepts. A content provider such as Google or Facebook may have a data center in a specific region that would cause significant impact to their operations in the event of an outage but that impact would not be localized to the region (even though its cause is in the region). In contrast, a public safety answering point may depend on data and network services that are hosted in another state. In this case, a cyber consequence may be region specific, even if the cause of an outage may not be related to the region. In the case of a local government entity, it may be that they depend on data and network services that are primarily local to the region (for instance, if they run their own data centers); in this case, these concepts overlap.

Dependency and interdependency studies allow for the greatest amount of flexibility in how regionalism is approached. A dependency cyber resiliency assessment can look at what entities (public and private sector data owners) have data and network dependencies within a region and how an outage to those resources would impact those entities. It examines how data locality exhibits itself in current failure models, both from the perspective of cyber-attacks and physical threats to the underlying infrastructure.

Tabletop exercises and scenarios can be a useful tool to explore the interrelated dependencies of critical infrastructure. Since the CRA takes an all-hazards approach to risks and vulnerabilities, scenarios can include those where the threats are local, and the consequence are national (or global) as easily as scenarios where the threats are global and the consequences are local or regional.

3.2 Unique Challenges of Cyber Dependencies Between Regions and Cyber Systems, Networks, and Infrastructure

The Internet and its underlying protocols were designed with resilience to failure as a primary goal. The properties that make this infrastructure resilient also make it difficult to gather data and conduct data-driven analyses of potential failures. Even if an Internet company claims that it has conducted failure analyses to show that a regional outage would have no national or global consequences for Internet routing, any such study would be limited to the company's own data and, thus, incomplete. Moreover, Internet infrastructure comprises many different private and public sector entities. Many companies own different parts of the infrastructure, and there are many paths for Internet traffic to travel. These properties make the study of Internet infrastructure and dependencies through the CRA methodology particularly compelling, since any study done by private sector entities is at this time incomplete.

The characterization of both traditional physical infrastructure or information technology (IT) resources and network infrastructure are typically well-understood problem areas. However, characterizing how physical systems interact with networked systems is often difficult. Industrial control systems (ICS) and supervisory control and data acquisition (SCADA) systems are vital components of traditional infrastructure systems in energy, water, transportation, and other critical infrastructure sectors. The 25–30-year lifecycles of these systems conflict with the average 3–5-year lifecycle in the IT domain. Many such systems were designed before the advent of the Internet yet are being connected to the Internet without adequate security protections. These haphazardly connected devices represent a misunderstood but significant vulnerability. The problem arises in part because expertise and management of the two types of systems do not overlap. ICS and SCADA systems, and their requirements, tend to be outside of the area of expertise of the majority of IT security professionals. Conversely, specific critical infrastructure domain experts may know the ICS/SCADA operations and requirements well but do not adequately understand the vulnerabilities that result from connecting these systems to a network. Unfortunately, few cybersecurity personnel bridge these two worlds, and thus, the vulnerabilities of these infrastructure systems tend to be poorly understood until an assessment like the CRA is conducted.

Current marketing trends lead customers to believe that putting data in the cloud makes it more available, safer, and more secure. Though this may be the case in some situations, it is highly dependent upon configuration and deployment options. Cloud services complicate the study of regional resilience, especially when critical infrastructure owner/operators depend on those services for normal operations of their systems. Data center and cloud services can be deployed in a very resilient fashion, but due to cost or operational constraints are just as often deployed in ways that are less fault tolerant. Customers and providers must establish resilience requirements as part of any service deployment; however, the lack of a shared vocabulary means that attempts to ensure that resilience needs are met are often inadequate. Data centers tend to be very transparent with their customers, but as that relationship becomes more obscured (through cloud services and third-party contracts), this transparency is often lost. To complicate matters, customers and providers of network and data center services often place responsibility on the other entity for a lack of information required to build resilient infrastructure. Stakeholders that regard uptime (the amount of time a host, service, or data center is running without any outages) as critical, such as government and lifeline sector entities, pay for private managed networks with the assumption that a private network isolates them from the Internet's physical vulnerabilities. Private networks and "dark" fiber (dark fiber is unused fiber that is privately owned and/or left for capacity expansion of existing networks) often run through the same conduits as the public Internet backbone. This colocation means that private managed networks share the same physical vulnerability profile as the Internet.

Emergency services personnel require more information to be able to evaluate the criticality of data and network infrastructure. Likewise, data centers and network operators need better methods of sharing information and communicating with emergency services personnel and with each other during an emergency. This needed exchange can be accomplished through an ongoing process of developing relationships, platforms, and communication channels.

4 Cyber Resiliency Assessment Tools

A CRA employs various tools to aid in the assessment process. This section describes the various assessment and modeling tools that are used. Each assessment uses a different subset of tools depending on its specific needs and stakeholder requirements.

Three types of tools are described below: cybersecurity assessment tools, stakeholder improvement tools, and modeling and simulation tools.

4.1 Cybersecurity Assessment Tools

To help owners and operators of critical infrastructure assess and enhance their cybersecurity, CRAs employ a variety of cybersecurity assessment tools. These tools ask organizations a series of questions to help them evaluate various aspects of their cybersecurity posture, addressing both protective and resilience measures. The questions are generally derived from widely accepted standards, guidelines, and best practices but differ in order to meet the objectives of the assessment tools. The results of each tool give the organization an overall evaluation of its cybersecurity posture and can also be used to compare the organization's performance to other organizations to identify its comparative strengths and weaknesses.

4.1.1 Cyber Security Evaluation Tool (CSET)

CSET® (ICS-CERT 2018b) serves as a self-assessment tool to allow an organization to evaluate the cybersecurity of its industrial control systems (ICSs) or other information technology systems. CSET uses a questionnaire, customized for the organization, based upon standards and guidelines chosen by the organization, the architecture of its systems and networks, and the worst-case consequence of a cyber-attack. The results summarize the organization's compliance with their chosen standards, identify security gaps ranked by vulnerability, and provide a prioritized list of recommendations, solutions, common practices, or compensating controls.

Assessment Process

The self-assessment is performed, using a stand-alone software application, by staff with knowledge of the critical assets; risk assessments; system, software, and network architecture; security implementation; data exchanges; and policies and procedures.

The automated process asks the organization to choose one or more standards and guidelines, to import or create a system and network diagram, and to answer a series of worst-case consequence scenarios. The results of the assessment are automatically generated based on these inputs.

The organization customizes its evaluation by selecting from a set of standards and guidelines that are published by the DHS, National Institute of Standards and Technology (NIST), Department of State (DOS), International Organization for Standardization (ISO), North American Electric Reliability Corporation (NERC), and Nuclear Regulatory Commission (NRC). Derived from this selection, CSET creates a customized set of questions and user-selectable answers.

Applying the organization's answers to a series of worst-case consequence scenarios, CSET establishes an organization's security assurance level (SAL) to determine the level of cybersecurity required by the organization. CSET uses the SAL to identify security gaps and prioritize recommendations.

Results

CSET compares the organization's answers with the requirements of the user-selected security standards to identify which answers fail to meet the required security assurances. These security gaps are ranked by severity, listed by question, and summarized by subject area.

From an analysis of the security gaps and the organization's security assurance level (SA), CSET provides a prioritized list of recommendations, solutions, common practices, or compensating controls. The CSET results summarize the compliance of the organization with respect to the selected standards, components (based upon system and network architecture), and overall performance.

4.1.2 Cyber Resilience Review (CRR)

The CRR (US-CERT 2018) is the assessment and analysis of an organization's cyber resilience—its ability to handle disruptive cyber events and continue to manage and operate their critical services. The CRR uses a standardized questionnaire, based on the CERT (Computer Emergency Readiness Team) Resilience Management Model, to measure operational resilience by examining the organization's cybersecurity assets and practices. The questions are designed to measure how the organization manages processes, services, and assets, not whether it implements technical controls or complies with standards.

The CRR addresses ten domains and four asset types. The ten domains are asset management, configuration and change management, risk management, controls management, vulnerability management, incident management, service continuity management, external dependencies management, training awareness, and situational awareness. The four asset types are people, information, technology, and facilities.

Assessment Process

The review is performed either as a facilitated interview by a team or as a self-assessment with a stand-alone questionnaire. The on-site facilitated interview is a 1-day meeting with key cybersecurity personnel, including business leaders who understand the business continuity requirements, chief information officers, chief information security officers, and other information technology security, operations, and management personnel. The results are delivered within 30 calendar days after the visit.

Results

The results of a CRR include a performance score indicating an organization's level of maturity, overall and by domain, illustrated in heat maps and graphs that summarize its strengths and weaknesses. The report provides options for consideration (recommendations) that contain guidance and activities to help an organization improve.

The results also compare an organization's overall performance to other organizations who have taken the CRR. The organization can compare its overall results as well as results on each of the ten domains.

4.1.3 Cyber Survey Tool (CST)

The Cyber Survey Tool (CST), previously known as the Cyber Infrastructure Survey Tool (C-IST) serves as a facilitated assessment tool to measure the cybersecurity protection and resilience practices that are employed by owner-operators to manage and secure their critical cyber services (CCSs). The methodology allows organizations to retake the assessment and to assess their progress and benchmark themselves against other organizations by comparing their scores and practices.

A CCS is a cyber-dependent service upon which an organization depends to ensure delivery of critical infrastructure services. The loss of a CCS could damage or interrupt the organization's ability to provide critical infrastructure services. The loss or damage of critical infrastructure services "would have a debilitating impact on security, national economic security, national public health or safety, or any combination of those matters (United States DHS 2013)."

The CST was developed by using questions from other, existing security assessment tools: Chemical Facility Anti-Terrorism Standards (CFATS) (Chemical Facility Anti-Terrorism Standards 2018), Infrastructure Survey Tool (IST) (Infrastructure Survey Tool 2018), Cybersecurity Evaluation Tool (CSET) (Cyber Security Evaluation Tool (CSET®) 2018), and the Cyber Resilience Review (CRR) (Cyber Resilience Review 2018).

The CST categorizes questions into five key cyber-related functional areas: (1) cybersecurity management, (2) cybersecurity forces, (3) cybersecurity controls, (4) incident response, and (5) dependencies. Each of the five key areas is expanded into a multi-level hierarchy containing increasingly more specific questions.

To evaluate the assessments, the CST methodology scores each question and its answers by using weights that have been elicited from subject-matter experts in cybersecurity from (1) national laboratories, (2) Cybersecurity Advisors (CSAs), (3) industry, and (4) state, local, tribal, or territorial governments (SLTTs). The subject-matter experts assigned a relative importance-based weight for each level of the assessment hierarchy along with each question and its possible answers. Each assessment question contains a fixed set of possible response options (a forced choice format) to improve the reproducibility of the assessment.

The CST also includes a scoring method that considers cyber threats such as distributed denial of service, natural disaster, remote access, and destructive malware. The subject-matter experts re-weighted each level of the assessment hierarchy for each threat. This permits CST to evaluate an organization not only for protection against and resilience to non-specific threats but also when faced with specific threats, known as a threat "overlay."

Assessment Process

The CST assessment is conducted by a cybersecurity analyst as a 2-h interview with an organization's key cybersecurity personnel.

Results

The results show an organization's cyber protection and resilience score along with comparisons (high, median, low) to other organizations. The score is termed its Cyber Protection and Resilience Index (CPRI).

The organization can benchmark its scores with all other critical infrastructure, with others in its sector (or other sectors), and with organizations that operate similar CCSs. The organization can benchmark its scores at multiple levels of detail, ranging from overall scores to one of the five key areas (cybersecurity management, cybersecurity forces, cybersecurity controls, incident response, and dependencies) to further levels that contain increasingly more specific questions.

In addition to viewing and benchmarking its scores without considering the type of threat, the organization can select one of four threat overlays to assess its protective and resilience measures against that particular threat (distributed denial of service, natural disaster, remote access, and destructive malware).

4.1.4 Risk and Vulnerability Assessments (RVA)

The Cybersecurity and Infrastructure Security Agency (CISA) conducts Risk and Vulnerability Assessments (RVAs) for the federal civilian executive branch, high priority private and public sector critical infrastructure operators, and select state, local, tribal and territorial (SLTT) stakeholders. Additionally, the United States Coast Guard conducts RVAs on maritime critical infrastructure operated by SLTT and private-sector organizations. The RVA program scans a given operating system (OS) for vulnerabilities and, upon detection, tests whether these vulnerabilities can be exploited to compromise the OS. Participating agencies are also tested for compromised wireless devices, proper configuration of applications and databases, and a sound network operations center (NOC) and security operations center (SOC). Furthermore, individual agents are tested for their resistance to "social engineering" tactics that are employed by phishing schemes.

4.1.5 Cybersecurity Capability Maturity Model (C2M2)

C2M2 is a tool that helps organizations evaluate their cybersecurity capabilities and optimize security investments. It uses a set of industry-vetted cybersecurity practices focused on both information technology (IT) and operations technology (OT) assets and environments. It is used heavily in the U.S. Energy Infrastructure and is a self-evaluation model that can be completed offline via a PDF or online via a web portal.

The model contains more than 350 cybersecurity practices, which are grouped by objective into 10 logical domains. Each practice is assigned a maturity indicator level (MIL) that indicates the progression of practices within a domain. A domain contains a structured set of cybersecurity practices focused on a specific subject area.

And practices within each domain are organized into objectives that can be achieved by implementing the practices in the domain.

Similar to the CRR above, to measure progression, the C2M2 uses a scale of maturity indicator levels, each representing maturity attributes.

4.1.6 Open-Source Intelligence Tool (OSINT)

OSINT can provide a useful way to approach the overall state of a system by taking a look at how outsiders may perceive and interact with it. By leveraging freely available tools within the public space, a stakeholder can gain insight into possible misconfigurations or vulnerabilities that are otherwise obscured by insider knowledge and assumptions. Moreover, the constant change within IT systems creates a gap between documentation and system understanding when compared with actual system functionality. System vulnerabilities can readily manifest themselves within this gap and go undetected. This tool simulates the information gathering, or reconnaissance, step of a cyber-attack which correlates the first step in different attack frameworks such as the cyber kill chain.

Assessment Process

The OSINT takes a series, or range, of IP addresses as input and uses a combination of crawling, scraping, and openly available tool output to generate a holistic report of the specified sites. Examples of this include an enumeration of all sites within the IP address and all related IP addresses; job postings for the company(ies) contained within the IP range; parsed information from third-party port scanners; social engineering information such as names, phone numbers, email addresses, etc.; software and versioning information; and enumeration of non-spidered sites (such as robots.txt).

Results

Upon completion, the OSINT presents the operator with a dashboard containing several different views to convey the information gathered. The dashboard enables the operator to either see the results at a high, aggregate level based on category (social engineering, vulnerability, etc.) or drill down and view details at finer granularity.

4.1.7 Industrial Assessment Center (IAC) Control Systems Cybersecurity Assessment Tool

USDOE Industrial Assessment Centers are university-based teams of engineering students and faculty who conduct free site visits to small and medium manufacturers to conduct technical assessments and provide reports on opportunities for improving competitiveness. As part of this program there is an Industrial Control Systems Cybersecurity Assessment Tool that manufacturers can use for risk assessment.

Assessment Process

The assessment process is a short 20 question assessment used to "characterize industrial control systems and plant operators" and has a companion user guide to provide additional content for the tool, other guidance can be found in NIST's Manufacturing Extension Partnership (MEP).

Results

The tool provides "a high-level assessment of risk (high, medium, or low)" and generates a custom list of action items associated with the risks identified. After this assessment the manufacture can consider using additional more detailed tools for addressing any cybersecurity concerns that they may have (IAC University Models (n.d.)).

4.2 *Modeling and Simulation*

The National Laboratories have developed resilience and restoration tools in order to prepare for, mitigate, respond to, and recover from crises that affect critical systems. Modeling various potential disruptions to critical systems helps planners anticipate the types of scenarios that may occur. Empowered with this knowledge, the case can be made for resilience improvements such as additional planning, redundancy, and interagency agreements.

4.2.1 North American Energy Resilience Model (NAERM)

The Nation's energy resilience depends on national-scale energy planning and situational awareness capabilities for rigorous and quantitative assessment, prediction, and improvement. An ambitious initiative led by U.S. Department of Energy (DOE)—the North American Energy Resilience Model (NAERM)—provides this unique and ground-breaking capability to ensure reliable and resilient energy delivery across multiple sectors, spanning multiple organizations and authorities, while considering a range of large-scale, emerging threats.

NAERM is a comprehensive resilience modeling system for the North American energy sector infrastructure. The United States is increasingly experiencing threats, natural and man-made. The need to proactively plan and prepare for events that threaten this Nation's wellbeing and security necessitates the ambitious span of scope. NAERM will predict the impact of threats, evaluate and identify effective mitigation strategies, and support black start planning, benefiting the U.S. by advancing energy and economic security.

NAERM advances existing capabilities to model, simulate, and assess the behavior of electric power systems, as well as its associated dependencies on natural gas, telecommunications, and other critical infrastructures. Integration of significant and unique expertise at the national laboratories, plus data integration and collaboration from all stakeholders, will support threat characterization for the energy sector across varying geographic areas and supporting sectors. NAERM will engage with industry experts to ensure threat and consequence models are realistic and representative of actual system responses.

4.3 Cyber Resilience Matrix

The resilience matrix as it applies to cybersecurity was developed by Linkov et al. (2013). The assessment tool consists of individual "cells" (see Table 1), each of which addresses a system's ability to handle cyber incidents that pertain to preparation, absorption, recovery, and adaptation to cyber threats within the physical, information, cognitive, and social domains.

The matrix suggests that the best evaluation of a system's cyber resilience is to incorporate both quantitative and qualitative metrics that are identified for that specific system by subject-matter experts and stakeholders. Table 1 provides an example of such metrics:

4.4 Cyber Maturity Improvement Framework

Enterprises performing cybersecurity assessments and baselining may question the efficacy of addressing all cybersecurity concerns simultaneously without a useful ordering. Oftentimes security recommendations of various levels are recommended without a focus on securing previous "levels." By ignoring basic cybersecurity facets and focusing on more sophisticated aspects, an enterprise may provide a more superficial solution to a symptom rather than mitigate the root cause, which may be much more fundamental. The illustration below depicts many cybersecurity tools and assessments that occur at different levels of the Department of Energy's.

Table 1 Example of cyber metrics per resilience matrix

	Plan and prepare for	Absorb	Recover from	Adapt to
Physical	Implement controls/sensors for critical assets Implement controls/sensors for critical services Assessment of network structure and interconnection to system components and to the environment Redundancy of critical physical infrastructure Redundancy of data physically or logically separated from the network	Signal the compromise of assets or services Use redundant assets to continue service Dedicate cyber resources to defend against attack	Investigate and repair malfunctioning controls or sensors Assess service/asset damage Assess distance to functional recovery Safely dispose of irreparable assets	Review asset and service configuration in response to recent event Phase out obsolete assets and introduce new assets
Information	Categorize assets and services based on sensitivity or resilience requirements Documentation of certifications, qualifications, and pedigree of critical hardware and/or software providers Prepare plans for storage and containment of classified or sensitive information Identify external system dependencies (i.e., internet providers, electricity, water) Identify internal system dependencies	Observe sensors for critical services and assets Effectively and efficiently transmit relevant data to responsible stakeholders/decision-makers	Log events and sensors during event Review and compare systems before and after the event	Document incident's impact and cause Document time between problem and discovery/discovery and recovery Anticipate future system states post-recovery Document point of entry (attack)
Cognitive	Anticipate and plan for system states and events Understand performance trade-offs of	Use a decision-making protocol or aid to determine when event can be considered "contained"	Review critical points of physical and information failure in order to make informed decisions	Review management response and decision-making processes Determine motive of event (attack)

(continued)

Table 1 (continued)

	Plan and prepare for	Absorb	Recover from	Adapt to
	organizational goals Scenario-based cyber war gaming	The ability to evaluate performance impact to determine if mission can continue Focus effort on identified critical assets and services Utilize applicable plans for system state when available	Establish decision-making protocols or aids to select recovery options	
Social	Identify and coordinate with external entities that may influence or be influenced by internal cyber-attacks (establish point of contact) Educate/train employees about resilience and organization's resilience plan Delegate all assets and services to particular employees Prepare/establish resilience communications Establish a cyber-aware culture	Locate and contact identified experts and resilience responsible personnel	Follow resilience communications plan Determine liability for the organization	Evaluate employees response to event in order to determine preparedness and communications effectiveness Assign employees to critical areas that were previously overlooked Stay informed about latest threats and state-of-the-art protection methods/ share with organization

Linkov et al. (2013)

Cybersecurity Capability Maturity Model (C2M2). By applying this approach at a graduated testing and assessment methodology, enterprises can ensure that they are looking at security aspects that are at their current working level and are building a strong, fundamental cybersecurity base (Fig. 1) (CESER 2022).

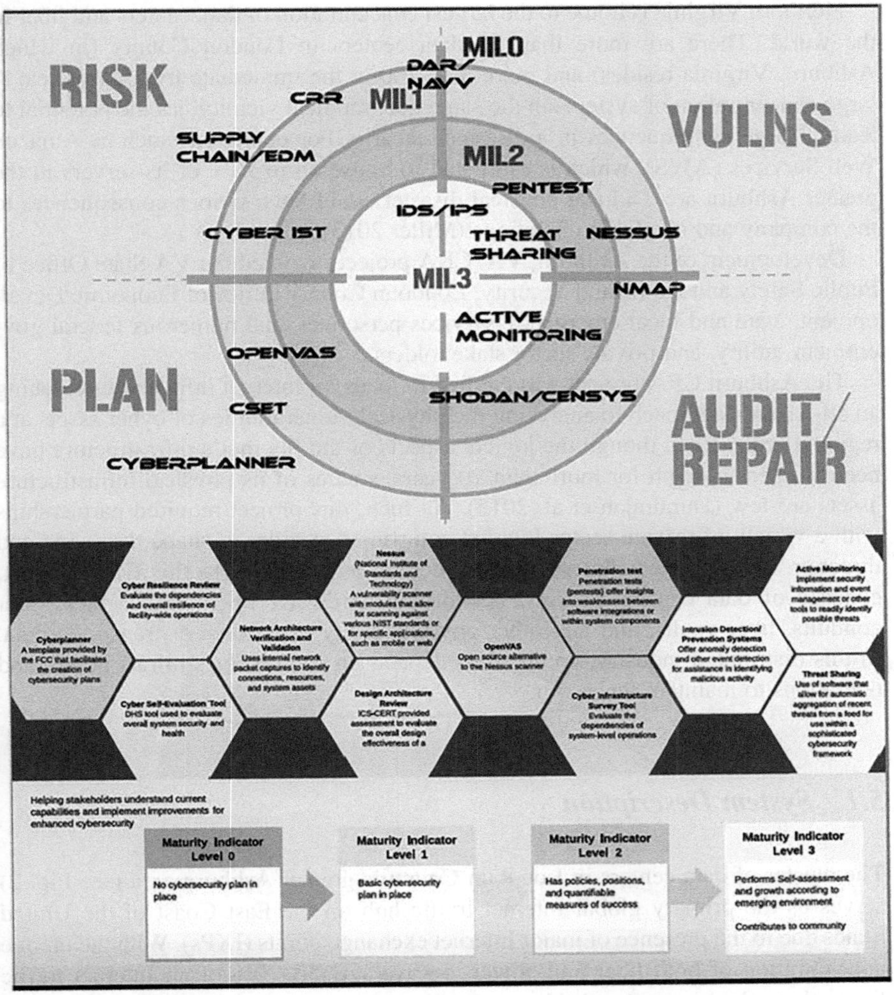

Fig. 1 The cyber maturity improvement framework

5 Case Study: Ashburn, Virginia, CRA

A CRA was conducted in 2015 to characterize and assess the resilience of a concentration of data centers in the northern region of the US state of Virginia (VA). The methodology in the referenced study is a specific instantiation of a CRA under the Department of Homeland Security's RRAP program (Thompson and Evans 2016). Below is a brief summary of the systems that were studied and the key findings from the assessment.

Northern Virginia is home to the largest concentration of data centers and fiber in the world. There are more than 50 data centers in Loudon County (in which Ashburn, Virginia resides) and more than 100 in the immediate tri-county area. A large concentration of systems in the same geographical vicinity has the potential to lead to large consequences in a disaster scenario. For operations such as Amazon Web Services (AWS), which is estimated to house up to 70% of its servers in the greater Ashburn area, a local physical disaster could have serious consequences to the company and its global operations (Miller 2013).

Development of the Ashburn, VA, CRA project involved the VA State Office of Public Safety and Homeland Security, Loudoun County Office of Economic Development, state and local emergency services personnel, and numerous federal government, utility, and private sector stakeholders.

The Ashburn CRA project was the first to focus on Internet infrastructure, taking an all-hazards approach to analyzing the physical vulnerabilities of cyber assets at a regional scale. Even though the logical aspects of the Internet's infrastructure have been studied in depth for more than 20 years, studies of its physical infrastructure assets are few (Durairajan et al. 2015). As such, this project required partnerships with a variety of private-sector Internet companies in order to shape the scope and direction of the project. Internet infrastructure, as it applies to this CRA project, consists of data centers, servers, networking hardware, fiber optic cables, fiber conduits, fiber vaults, and any other physical entity that stores or transports data. In this distributed industry, competitors depend on each other's infrastructure and operations to maintain their own.

5.1 System Description

The cluster of data centers in Loudoun County's greater Ashburn area (see Fig. 2) serves as the primary global Internet traffic hub on the East Coast of the United States due to the presence of major Internet exchange points (IXPs). With the unique concentration of both fiber and power, on average, 50–70% of all Internet traffic flows through the greater Ashburn-area data centers. These facilities contain IT infrastructure and support governmental agencies and private companies that, in turn, supply day-to-day services to utilities and the public.

5.1.1 Data Centers

Throughout the Ashburn, VA, CRA report, the term "data center" was used generically to refer to a wide variety of facilities that have subtle but important differences. In actuality, a single data center can rarely be categorized as only one of the following logical types; however, for the purposes of illustration and explanation, such distinctions remain important. The following figure depicts how these facilities are typically connected to one another (Fig. 3).

Fig. 2 Data centers concentrate near the major internet exchange points in Ashburn, VA

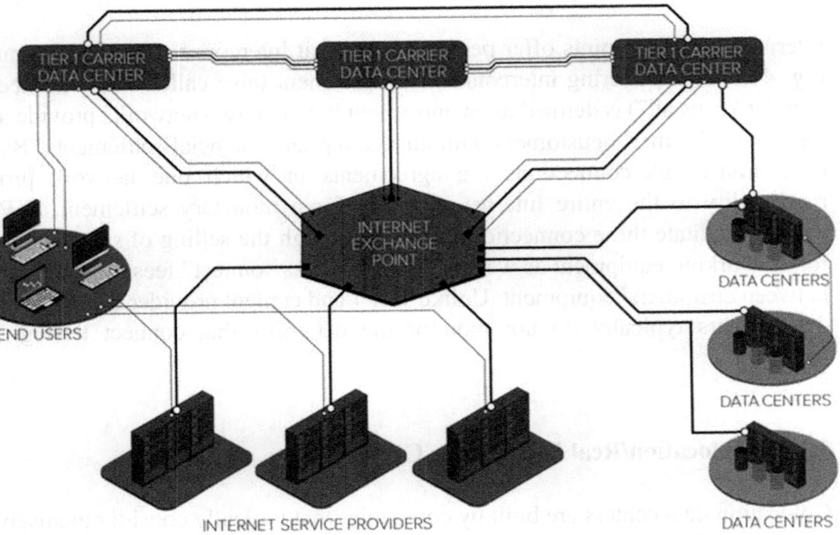

Fig. 3 The internet connectivity of data centers

5.1.2 Tier 1 Network Provider Data Centers

Tier 1 network providers are those that have access to every other network on the Internet without paying for transit. Tier 1 network providers have their own data centers where content providers can rent space to deliver increased performance to a variety of customers (known colloquially as "hosting content closer to the eye-balls"). Because Tier 1 providers typically offer other services (e.g., cell phone service, home Internet service), they may house other services in their data centers, in addition to peering with content providers and other networks.

5.1.3 Content Provider Data Centers

Large content providers and technology companies (e.g., Google, Microsoft, and Facebook) often build their own data centers. While these data centers frequently bring in transit from several major network providers, they usually house only one company's data and infrastructure. Although the data centers may be built by a real estate company, they are typically operated by the content provider itself.

5.1.4 Internet Exchange Points (IXPs, Also Known as Network Access Points [NAPs])

Internet exchange points offer peering and transit interconnections as illustrated in Fig. 4 below. A peering interconnection agreement (also called "bill and keep" or "sender keeps all") is defined as an agreement in which two networks provide access only to each other's customers without making any financial settlements. By contrast, transit interconnections are agreements in which one network provides reachability to the entire Internet in return for a monetary settlement. IXP data centers facilitate these connections, usually through the selling of space and power for networking equipment as well as through "cross connect" fees for running cable between customers' equipment. Unlike Tier 1 and content provider data centers, IXP data centers typically do not monitor the networks that connect through their facilities.

5.1.5 Colocation/Real Estate Data Centers

Colocation data centers are built by companies that typically consider themselves to be in the real estate market. Although these companies may sell power, cooling, and network in addition to physical space, they typically offer no network services (e.g., security, monitoring). They generally view their relationship with the customer as a traditional landlord/tenant relationship. Customers typically install and maintain their own equipment and contract their own network services. Most Tier 1 network data centers and IXPs also offer colocation services.

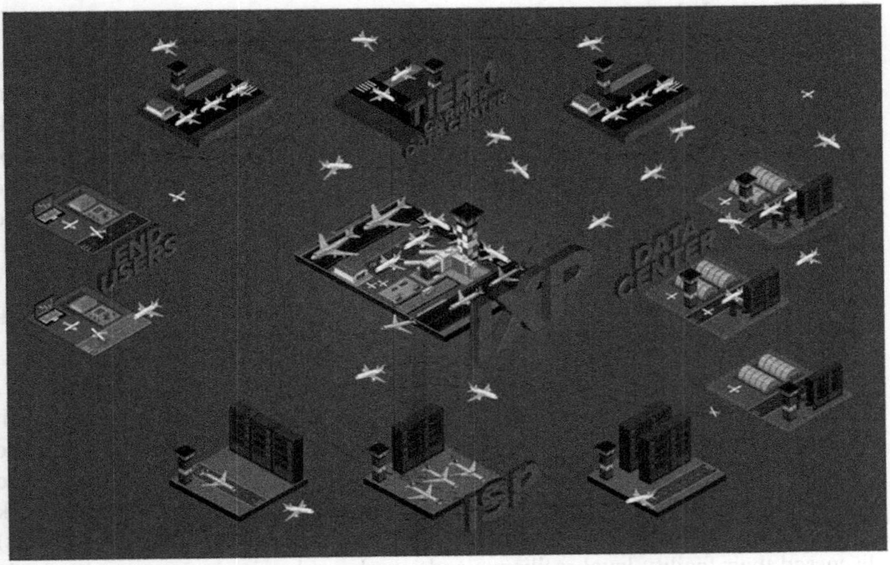

Fig. 4 The connectivity of internet exchange points

5.1.6 Internet Service Provider Data Centers

Although many customers connect directly through Tier 1 providers to the Internet, Tier 2 and Tier 3 Internet service providers (ISPs) often have data centers of their own as well. A Tier 2 network is an ISP that engages in the practice of peering with other networks and also purchases Internet protocol transit to reach some portion of the Internet. Tier 2 providers are the most common ISPs, as it is much easier to purchase transit from a Tier 1 network than to peer with it and attempt to become a Tier 1 carrier. Tier 3 networks purchase Internet protocol transit solely from other networks to reach the Internet. These data centers are smaller and often only house services for their direct customers.

5.1.7 Meet-Me Rooms

Most data centers that offer colocation services also have a "meet-me room," which is not strictly a type of data center but is akin to an IXP within a data center. It is usually a separate room or data center "cage" in which customers can house equipment that is used exclusively for connecting to other customers also located in that facility. Peering and transit agreements can be connected through meet-me rooms as well.

5.2 Methodology

The methodology for data collection in the Ashburn CRA was largely centered around stakeholder interviews and focused facilitated discussion. Data-driven studies around much of the key findings are not currently possible due to the lack of publicly available information on routes, capacity, and throughput. It should be noted that one of the express goals of a CRA project is to improve and facilitate communication between critical infrastructure owner/operators, local state and federal governments, and other stakeholders to a region's resilience. In the spirit of this goal, many of the Ashburn activities and findings were geared toward building consensus for future data gathering and analysis activities.

Stakeholder Interviews
An important source of data for the Ashburn, VA, CRA project was interviews with public and private sector organizations representing essential Federal response agencies, private sector owners and operators of critical infrastructure, and state and local government officials. Participants from the data center and network communities (including telecommunications companies and major carriers) discussed their facility level resilience and interdependencies between each other's infrastructure, as well as their dependence on the energy, water, and transportation sectors for normal operations and the emergency services sector during disaster response and recovery. Energy and water sector participants discussed both their own resilience planning and their dependence on data and network infrastructure for their operations. Emergency services personnel discussed their operational needs for network and technology resources, their prioritization plans during disaster recovery, and their perceived views of their connections with data and network providers in their jurisdictions. All entities presented opinions and ideas for improvements to enhance regional resilience and disaster recovery.

Facilitated Discussion
The Ashburn, VA, CRA project included a facilitated discussion to (1) elicit information on preparedness efforts and restoration needs and (2) elucidate the vulnerabilities to critical dependencies and interdependencies that exist between major stakeholders connected to data center and network infrastructure. The discussions used three major scenarios to drive the dialog among participants: an airplane crash, a severe weather event, and an EMP event. Project participants used the scenarios to examine the criticality, connections, and capabilities of the infrastructure systems and operations of their own facilities and the supporting infrastructure systems and operations within the region.

Other Data Gathering Activities
The Ashburn, VA, CRA project team also gathered data by attending and presenting at two summits held by the North American Network Operators' Group (NANOG) in San Antonio, Texas, and San Francisco, California. NANOG's official charter describes its purpose as existing "to promote dialog between people concerning the creation, maintenance, and operation of Internet Protocol networks"(NANOG

(North American Network Operators' Group) 2010). NANOG provides a forum where people from the network research community, the network operator community, and the network vendor community can identify and solve problems that arise in operating and growing the Internet. The Ashburn project team's involvement in NANOG helped refine and confirm hypotheses and inform key findings.

Throughout the data gathering process, the Ashburn project team also participated in meetings of the Northern Virginia Technology Council and the National Cybersecurity and Communications Integration Center (NCCIC). DHS used these meetings both to inform stakeholders of CRA activities and to solicit feedback on the project's direction and potentially interested parties.

5.3 Key Findings

The Ashburn, VA, CRA identified vulnerabilities that may affect the Internet community's ability to prepare for and recover from the impacts of a variety of natural and man-made threats to its infrastructure assets. The associated resilience enhancement options may be considered for implementation by Ashburn, VA, CRA stakeholders and partners to address the resilience gaps that are described in the key findings.

Though many of the points encompassed by the key findings of the Ashburn CRA have been well known by network operators for some time, they are important points of discussion for senior decision-makers in the private sector and in government.

Internet resilience is contingent on a limited number of centralized Internet exchange points (IXPs).

A study should be conducted that simulates the outage of an IXP. Such a simulation would include modeling the traffic and Transmission Control Protocol congestion during an outage of IXP facilities in the greater Ashburn area.

Transparency in both network and data center infrastructure would enhance resilience planning.

The state, in concert with private sector stakeholders, should coordinate a workshop on the development of cloud and data center service taxonomies and assessments. Such taxonomies and/or assessments should allow for equal comparison of resilience features across providers and empower customers by fostering open, honest competition.

Local law enforcement personnel would benefit from training and the exchange of information concerning How to recognize suspicious activity around IT infrastructure assets.

Law enforcement personnel should engage industry stakeholders to facilitate training and education on fiber routes and suspicious activity. This training should also address how and when to approach maintenance personnel and how to confirm that they are authorized to work in a given area.

Data center and content providers may not have a pathway to contribute to resilience efforts and/or communicate criticality during an emergency.

A workshop should be conducted for the data center community so that all parties can communicate their needs for points of contact, access to emergency operation center (EOC) resources, and communication pathways during an emergency.

Transportation infrastructure and private trucking companies are necessary to supply diesel fuel to data center facilities in the event of an extended electric interruption.

An in-depth study of the region's fuel supply chain should be conducted specifically to address concerns relating to fuel supplies and the ability to deliver fuel during an extended power outage to data center facilities. Such a study should include best- and worst-case scenarios for road conditions after a disaster. Data center and network providers should be present at the state's annual exercise within the region. Their participation would improve decision-makers' understanding concerning the critical roles of data centers in response efforts.

Data centers and network providers should consider electromagnetic pulse (EMP) and radio frequency (RF) generator effects in developing resilience and protective measures plans.

Additional workshops should be conducted to support Ashburn-area data centers in their efforts to improve resilience to EMP/RF effects. A process should be established to keep data centers and local law enforcement up to date on new EMP/RF threats. Discussion about whether and to what extent EMP/RF protection should be deployed should be a standard risk management topic for data centers.

Communication and education efforts between data center providers and fire department personnel are necessary to support resilience planning.

Data centers and fire department personnel should work together to arrange training and education sessions to help ensure that fire department personnel are aware of distinctive data center needs and environments. Data centers should consider installing radio equipment that operates on frequencies that are used by fire and police department radios to assist with operations during emergency situations.

6 Conclusion: Cyber Resiliency Assessment Benefits

The promise of the cyber resiliency assessment lies in an assessor's ability to play neutral broker between infrastructure owners and advocate for the public good in its mission of infrastructure protection. Currently, industry is driving the conversation with regard to network and data infrastructure, transparency, and dependencies. The CRA program is in an excellent position to study the problems of regional resilience and data locality in terms of how usage of new technology and the connection of old technology to new networks could both benefit and harm the public.

No single entity in the private sector is in a position to study network and data dependencies at this scale without risking a compromise of their intellectual property, though it is clear from examples like the Ashburn CRA that individual providers have as much to gain from their competitors' resilience as the public at large. Cyber resiliency assessment projects are an important contribution to the field of infrastructure resilience as they bridge the gap between cyber systems and physical infrastructure. The following is a list of recommendations for a direction in which to take regional cyber resilience research and practice:

1. Focus the assessment on an important regional asset, such as a critical lifeline infrastructure or valuable economic asset. Lifelines include energy, communications, water, and transportation infrastructure. Assets range from water supplies to seaports.
2. Identify the infrastructure that provides or supports the focus of the assessment.
   (a) Identify the owners and operators of the infrastructure.
3. Identify the cyber dependencies of that infrastructure.
   (a) Identify the owners and operators of the cyber infrastructure if different from the infrastructure owner.
4. Identify the cyber dependencies of the lifeline sectors—the energy, communications, water, and transportation systems that provide essential products and services required throughout the region.
   (a) Identify the consequences to lifeline sectors from the loss their cyber dependencies.
   (b) Identify the lifeline infrastructure's dependencies of the essential cyber system.
5. Estimate the consequence of losing essential infrastructure.
   (a) Estimates should consider geographic extent of impact, the number of people affected, and the infrastructure affected.
6. Identify the interdependencies of the infrastructure.
   (a) Identify the cascading consequences of infrastructure damage or loss.
7. Identify mitigation strategies.
   (a) Evaluate emergency response plans, recovery priorities, and resources required for response and recovery.
8. Assess regional cooperation among infrastructure owners, government, and incident response organizations.
   (a) Promote the importance of partnership between the owners and operators, both public and private, of the critical and cyber infrastructure.
   (b) Assess information sharing on infrastructure dependencies.

References

Armed Foryces Communications and Electronics Association (AFCEA) (2012) The Russo-Georgian War 2008: the role of the cyber attacks in the conflict. Available at http://www.afcea.org/committees/cyber/documents/TheRusso-GeorgianWar2008.pdf

Busby J, Baker K et al (2021) Cascading risks: understanding the 2021 winter blackout in Texas. Available at https://www.sciencedirect.com/science/article/pii/S2214629621001997

CESER (2022) Cybersecurity Capability Model. Available at: https://www.energy.gov/ceser/cybersecurity-capability-maturity-model-c2m2

Chemical Facility Anti-Terrorism Standards (2018) Available at http://www.dhs.gov/chemical-facility-anti-terrorism-standards

Cowie JH, Ogielski AT, Premore BJ, Smith EA, Underwood T (2003) Impact of the 2003 blackouts on internet communications. Available at https://pdfs.semanticscholar.org/f837/0739e127e5a65a7f95a4b99390ab173a3de0.pdf

Cyber Resilience Review (2018). Available at https://www.us-cert.gov/ccubedvp/self-service-crr

Cyber Security Evaluation Tool (CSET®) (2018) Available at https://ics-cert.us-cert.gov/Assessments

Durairajan R, Barford P, Sommers J, Willinger W (2015) InterTubes: a study of the US long-haul fiber-optic infrastructure. Available at http://conferences.sigcomm.org/sigcomm/2015/pdf/papers/p565.pdf

IAC University Models. Available at: https://iac.university/cybersecurity#additonalResources

ICS-CERT (2018a) Cyber-attack against Ukrainian critical infrastructure. Available at https://ics-cert.us-cert.gov/alerts/IR-ALERT-H-16-056-01

ICS-CERT (2018b) Undated, Cyber security evaluation tool (CSET®). Available at https://ics-cert.us-cert.gov/Assessments

Infrastructure Survey Tool (2018) Available at http://www.dhs.gov/xlibrary/assets/NIPP_Plan.pdf

Linkov I et al (2013) Resilience metrics for cyber systems. Environ Syst Decis 33(4):471

Miller R (2013) Amazon adding cloud capacity in northern virginia. Data Center Knowledge, January 15. Available at http://www.datacenterknowledge.com/archives/2013/01/15/amazon-to-add-capacity-to-us-east-region/

NANOG (North American Network Operators' Group) (2010) Current charter, October 6. Available at https://www.nanog.org/history/charter

Thompson M, Evans N (2016) Internet resilience in Ashburn, Virginia. Resilience Week. Available at http://ieeexplore.ieee.org/document/7573323/

UNECE (United Nations Economic Commission for Europe Group of Experts on Energy Efficiency Task Force on Digitalization in Energy) (2023) Case study "cyber resilience of critical energy infrastructure". Available at https://unece.org/sustainable-energy/publications/digitalization-energy-case-study-cyber-resilience-critical-energy

United States Department of Homeland Security (2013) Partnering for critical infrastructure security and resilience. Available at https://www.dhs.gov/xlibrary/assets/nipp_consolidated_snapshot.pdf

US-CERT (2018) Undated[a], Cyber resilience review (CRR). Available at https://www.us-cert.gov/ccubedvp/self-service-crr

Supply Chains of Computer and Electronics Hardware with Resilience to Counterfeiting and Other Disruptions

Zachary A. Collier, Madison L. Hassler, James H. Lambert, Daniel DiMase, Catherine A. Lambert, and Igor Linkov

1 Introduction

This chapter offers perspectives for enhancing the resilience of the supply chain for the ubiquitous electronic hardware embedded within modern cyber-physical systems. It begins by introducing the problem and discusses the importance of cyber resilience in the modern, globally interconnected world that is subject to numerous sources of disruption such as operating environments, material shortages, etc. The chapter then provides an overview of the electronics supply chain, focusing on the threat of counterfeiting and how it affects the reliability and security of critical infrastructure. Next, the work explores the concept of resilience, distinguishing it from traditional risk-based approaches, and identifies a set of metrics that can be used to assess and measure supply chain resilience. These metrics can be used by practitioners to identify vulnerabilities, quantify risks, and implement effective mitigation strategies to address disruptions in the supply chain. The chapter explores the broader concept of cyber-physical systems (CPS) and discusses the 10 key areas of concern for securing these complex systems. It presents a pinwheel diagram that

Z. A. Collier
Department of Management, Radford University, Radford, VA, USA

M. L. Hassler · J. H. Lambert (✉)
Department of Systems & Information Engineering, University of Virginia, Charlottesville, VA, USA
e-mail: lambert@virginia.edu

D. DiMase
Aerocyonics, Inc, East Greenwich, RI, USA

C. A. Lambert
Contractor to Keisler Consulting, Boston, MA, USA

I. Linkov
Engineer Research and Development Center, US Army Corps of Engineers, Concord, MA, USA

© The Author(s), under exclusive license to Springer Nature Switzerland AG 2025

I. Linkov, A. Kott (eds.), *Cyber Resilience: Applied Perspectives*, Risk, Systems and Decisions, https://doi.org/10.1007/978-3-031-90109-6_12

illustrates the relationship between these areas and how they can contribute to or hinder overall resilience. A hypothetical example is provided of enhancing resilience of the supply chain of insulin pumps. The chapter concludes with a discussion of future work, including the need for developing new risk and resilience analysis tools and methods, integrating qualitative methods, and recognizing the need for interdisciplinary collaboration across various domains. By integrating these different sections, the chapter offers a set of perspectives and approaches for understanding and enhancing the cyber resilience of supply chains, ultimately contributing to the development of more secure and robust systems.

Stressors such as hurricanes, floods, and droughts also directly and indirectly affect supply chain operations, causing disruptions that increase costs and delay deliveries. Natural disasters, for example, can significantly hinder the efficiency and consistency of raw material allocation, posing risks to business operations, costs, earnings, and customer satisfaction (Kanike 2023). Understanding and mitigating these impacts is essential for managing overall supply chain risks. Supply chains are integral to global commerce but are increasingly vulnerable to stressors including extreme weather, which exacerbate shortages of energy resources, increases emissions and disrupt operations (Er et al. 2020). For example, the semiconductor supply chain, vital for technological advancement, faces significant risks from weather-induced disruptions. Historical records track the occurrences of extreme weather events, which can severely impact supply chain efficiency, especially in manufacturing and transportation sectors (Dasaklis and Pappis 2013). Understanding these vulnerabilities and developing robust risk modeling methods are crucial for ensuring supply chain resilience.

Several raw materials are consumed in the semiconductor manufacturing process, and shortfalls and delays in their availability can pose significant risks to the supply chain as well. With various stressors threatening the supply of these materials, supply chain managers need ways to quantify the impacts of material shortages on manufacturing and other nodes within the supply chain. Effective risk management strategies related to procurement, quality management, and inventory management, are important for mitigating the impacts of material shortages. The COVID-19 pandemic highlighted vulnerabilities in the global supply chain and increased the demand for technology, further straining supply chains responsible for producing these devices and their raw materials (Daly and Singh 2020). Major risks include natural disasters, raw material unavailability, regulatory changes, geopolitical changes, technology breakdowns, labor shortages, and transportation disruptions. These disruptions can lead to higher prices, manufacturing delays, decreased efficiency, and reduced customer satisfaction (Kim and Kim 2018).

As industrial advancements in systems network integration, high performance computing, and software grow, so does cyber vulnerability (DiMase et al. 2015). Advancements in microelectronics manufacturing technology that do not incorporate cyber-physical security measures also increase the risk to cyber systems. The rapid growth of the industry has given cyber attackers an easy target due to unintended vulnerabilities introduced with the integration of complex hardware, software, and firmware, and the potential use of maliciously tampered electronic parts. As systems are assembled from off-the-shelf components with unknown supply chain histories, it becomes important to understand and mitigate the associated technical risks (successful function), operational risks (achievement of business objectives), and programmatic risks (cost and schedule delays) (Horowitz and Lambert 2006).

Traditional approaches to securing the supply chain have been risk-based. Risk is classically defined by Kaplan and Garrick (1981) as a triplet of what can go wrong, what is the likelihood, and what are the consequences. However, traditional risk assessment and management are not well suited to certain contexts. For instance, risk assessment requires one to identify hazards to the system, which is not possible when hazards cannot be foreseen (Park et al. 2013), as is the case in rapidly evolving and adversarial technological fields such as cyber security. Similarly, there is a lack of data on likelihood of occurrence and consequences of many hazards, making risk assessment and management difficult (Fiksel et al. 2015). The field of risk has historically been difficult in the environment of low-probability, high-consequence events (Park et al. 2013; Fiksel et al. 2015). Additionally, one cannot screen every conceivable component within the global supply chain. These shortfalls require taking a different approach.

Instead, a resilience-based perspective is needed. The National Academy of Science defines resilience as "*the ability to prepare and plan for, absorb, recover from, or more successfully adapt to actual or potential adverse events*" (National Research Council 2012). A recent National Research Council (NRC) report describes a resilient system as "*one whose performance degrades gradually rather than catastrophically when its other defensive mechanisms are insufficient to stem an attack. A resilient system will still continue to perform some of its intended functions, although perhaps more slowly or for fewer people or with fewer applications. Features of resilient systems include redundancies and the absence of single points of failure*" (Clark et al. 2014).

This concept of resilience is distinct from that of risk, although the two are related. Linkov et al. (2014) characterize risk as a static snapshot in time of the maximum degradation of performance within the dynamic resilience process. Resilience, on the other hand, is a description of system behavior related to the ability to quickly recover from disruptions to an acceptable level of performance, and adapt to changing environments.

Common themes of supply chain resilience include the ability to return to a desirable state of operation after disruption, speed with which it takes to recover, flexibility, adaptability, contingency planning, and taking a forward-looking, anticipatory view to scan the horizon for potential threats before they disrupt operations. Collier et al. (2015) related trust to supply chain resilience. Trust is based on the trustworthiness of the trustee and the trust propensity of the trustor. When trust exists between links in the supply chain, actors respond faster to disruptions, reducing recovery time and therefore enhancing resilience. Several authors have proposed similar definitions of supply chain resilience, with supporting factors, capabilities, or properties which support resilience. Some of these definitions are described in Table 1.

Table 1 Selected definitions of supply chain resilience

Resilience definition	Resilience enablers	Source
The ability of a system to return to its original state or move to a new, more desirable state after being disturbed	Supply chain re-engineering, Supply chain collaboration, Supply chain risk management culture, Agility	Christopher and Peck (2004)
The ability of a company to bounce back from a large disruption—this includes, for instance, the speed with which it returns to normal performance levels	Redundancy, Flexibility, Corporate culture	Sheffi (2005)
The ability of a supply chain system to reduce the probabilities of a disruption, to reduce the consequences of those disruptions once they occur, and to reduce the time to recover normal performance	Supply chain density, Supply chain complexity, Node criticality	Falasca et al. (2008)
Resilience refers to the ability of the supply chain to cope with unexpected disturbances. It is concerned with the system ability to return to its original state or to a new one, more desirable, after experiencing a disturbance, and avoiding the occurrence of failure modes	Capacity buffers, Risk sharing, Responsiveness, Strategic inventory, Small batch sizes, Flexible transportation, Demand visibility	Carvalho and Cruz-Machado (2011)
The capacity of an enterprise to survive, adapt and grow in the face of turbulent change	Flexibility in sourcing, Flexibility in manufacturing, Flexibility in order fulfillment, Production capacity, Efficiency, Visibility, Adaptability, Anticipation, Recovery, Dispersion, Collaboration, Organization, Market position, Security, Financial strength, Product stewardship	Fiksel et al. (2015)

Adapted from Collier et al. (2015)

Lambert et al. (2013) have described that an approach to resilience of supply chains for embedded hardware should include the following:

- Identifying factors that influence risk and resilience for the hardware supply chain
- Developing risk metrics with attention to phases of the systems life cycle
- Developing a resilience model of the supply chain
- Analyzing scenarios and extreme events for the hardware supply chain, addressing unlikely events by combining inter-disciplinary perspectives
- Applying qualitative and quantitative metrics through the systems life cycle, extended to include multiple stakeholders and time horizons
- Developing recommendations for industry, military, and governance based on integrated examples and case studies

To achieve the goal of supply chain resilience, following the points identified by Lambert et al. (2013), research is needed to develop risk and resilience analysis methods and decision making approaches along with relevant economic, technological, and social indicators for hardware networks and supply chains. Additionally, relevant decision making parameters related to the broader hardware supply chain including identification of specific resilience-enhancing initiatives and their relevant benefits, costs, and time constraints is needed. Research is needed to develop a lexicon of common terms, vulnerabilities, and metrics of security for cyber-physical systems. This chapter aims to lay the groundwork for understanding factors relevant to resilience of the electronic hardware supply chain, and provide guidance on the identification of performance metrics to assess key areas of concern within the system.

The organization of this chapter is as follows. First, we provide a brief background on cyber resilience of supply chains. Next, a discussion of resilience metrics is presented for the supply chain, followed by a discussion of resilience across the entire cyber-physical ecosystem. Finally, future work describing how to move from risk-based approaches to a more holistic view of the supply chain through the lens of resilience is discussed.

2 Background

The concept of cyber resilience within supply chains is pivotal to ensuring the robustness and security of interconnected systems, particularly in the face of increasing cyber threats and the complexities of modern technology. Bonato et al. (2022) discuss the prioritization of resilience initiatives for weather-related disasters in the Metropolitan City of Venice, which is critical in understanding how to prioritize and implement resilience measures in the face of various threats. This framework can be applied to the cyber domain, emphasizing the need for strategic prioritization in resilience planning. Collier et al. (2021) introduce a decision model that quantifies buyer-supplier trust in advanced technology enterprises. This model is significant for cyber resilience as trust is a fundamental component in securing supply chains against cyber threats, where the reliability and integrity of suppliers directly impact the overall security posture. Al-Mutairi et al. (2022) present scenario-based preferences modeling to investigate container port initiatives' resilience. This work is particularly relevant as ports are critical nodes in supply chains, and their resilience directly impacts the security and reliability of supply chains globally, including their cyber aspects. Beteto et al. (2022) focus on anomaly and cyber fraud detection in pipelines and supply chains for liquid fuels. This research is crucial as it highlights specific methods for detecting and mitigating cyber threats, which are essential for maintaining the integrity and security of supply chains. Moghadasi et al. (2022) explore the trust and security of electric vehicle-to-grid systems and hardware supply chains. Their findings underscore the importance of securing hardware components in supply chains, which is a critical aspect of overall cyber resilience. Wheeler et al. (2022) investigate information management for a lifeline infrastructure for the mobility of people and goods. Effective information management is a cornerstone of cyber resilience, ensuring that critical data is protected and resilient against cyber disruptions. Andrews et al. (2023) examine enterprise risk management for automation in correctional facilities with pandemic and other stressors. Although focused on correctional facilities, the principles of enterprise risk management are directly applicable to managing cyber risks in supply chains. Collier et al. (2022) discuss business process modeling for semiconductor production risk analysis using IDEF0. This methodology provides a structured approach to identifying and mitigating risks in semiconductor supply chains, which are vital to many modern technologies and their cyber resilience. Sambo et al. (2023) provide a framework for multi-risk scenarios across system receptors, applied to Venice. This comprehensive approach to risk analysis can be extended to include cyber risks, demonstrating the need for a holistic view of resilience. Collier et al. (2023a) address economic metrics and prioritization of investments in hardware security. This work is directly relevant to cyber resilience as it provides a quantitative basis for prioritizing security investments in supply chains based on costs and benefits. Loose et al. (2023a) present systems modeling and resilience of schedules at maritime container ports. Their work emphasizes the need for robust systems modeling to enhance the resilience of critical infrastructure against cyber threats. Moghadasi and Lambert (2023) explore evaluating system resilience by the degree of order disruption. This concept is crucial for understanding how disruptions, including cyber attacks, can impact supply chains and how resilience can be quantitatively assessed. Loose et al. (2023b) explore reinforcement learning and automatic control for resilience at maritime container ports. The application of advanced technologies like reinforcement learning is essential for enhancing the cyber resilience of critical supply chain nodes. Davis et al. (2023) conduct a geospatially explicit technoeconomic assessment of sustainable aviation fuel production. While focused on aviation fuels, their methodology can be adapted to assess and improve the resilience of supply chains involved in fuel production and distribution. Collier et al. (2023b) explore stress testing for the resilience of semiconductor supply chains. Stress testing is a vital tool for assessing how supply chains can withstand various disruptions, including cyber threats. Baker et al. (2023) use mixed-integer programming with enterprise risk analysis for vehicle electrification at maritime container ports. This approach integrates risk analysis into decision-making processes, enhancing the resilience of supply chains. Moghadasi et al. (2024a, b) analyze the risk of artificial intelligence in medicine with a multilayer concept of system order. Their findings on system order and resilience can inform strategies for enhancing cyber resilience in supply chains involving AI technologies. Johnson et al. (2023) integrate risk sources and risk controls into SysML requirements diagrams with applications to sustainable aviation fuels. This integration is crucial for visualizing and managing risks, including cyber risks, in complex supply chains. Marcellin et al. (2024) conduct systems analysis for energy assets in Iraq influenced by water scarcity. Their systems approach can be adapted to analyze and improve the resilience of energy supply chains. Bazemore et al. (2024) discuss the electrification of utility tractors at maritime container ports. Electrification introduces new cyber vulnerabilities, and their research provides insights into mitigating these risks. Moghadasi et al. (2024a, b) present systems modeling of trust in AI-enabled medical diagnosis.

Trust modeling is essential for ensuring the security and resilience of supply chains involving AI technologies.

This literature collectively underscores the multifaceted nature of cyber resilience in supply chains, highlighting the importance of strategic prioritization, trust, anomaly detection, advanced modeling, and comprehensive risk management. These elements are critical in developing robust and secure supply chains capable of withstanding various cyber threats.

3 Focus on the Electronics Supply Chain

3.1 The Threat of Counterfeiting

The nation's critical infrastructure is enabled through a vast number of potentially vulnerable embedded electronic hardware components. The trend toward increased connectivity and the "internet of things" (IoT) necessitates the security of the electronics that support our critical infrastructure, as well as the supply chains through which these electronics are transported. A major threat to the supply chain is posed by counterfeit electronics—electrical, electronic, and electromechanical (EEE) parts, including those that are relabeled, refurbished, or repackaged to misrepresent the component's authenticity (Sood et al. 2011). There are seven types of counterfeit electronic parts in the framework created to assess test coverage for counterfeiting (Guin et al. 2014). These counterfeit parts types include recycled, remarked, overproduced, out of specification/defective, cloned, forged documentation/part substitution, and tampered. In a 2010 US Department of Commerce report, over 8000 incidents of counterfeit electronics were recorded in 1 year from 387 companies surveyed—a 141% increase in the number of reported incidents over a four-year period (US Department of Commerce 2010). Research is needed to improve avoidance and detection of overproduced, cloned, and tampered counterfeit types. The incorporation of tampered counterfeit electronic parts in assemblies introduces vulnerabilities in the cyber-physical system that could be used to compromise system function or gain access to critical and sensitive system information. Approaching cyber-physical systems security through the lens of resilience will enable the application of both integrated and targeted security measures and policies that ensure the continued functionality of critical safety and security of our cyber-physical systems.

All links of the supply chain face risks from these counterfeit parts. Those firms that produce electronics face lost revenue, increased costs to identify and repair systems compromised by counterfeits, damaged branding, and potential legal liabilities (Pecht and Tiku 2006; DiMase and Zulueta 2009). It is estimated that counterfeits cost the US semiconductor industry around \$7.5 billion in replacement and repair costs, and the downstream costs to other affected companies could be as high as \$200 billion (Wood 2016). Additionally, there are risks to consumers in degraded reliability, and potential health and safety impacts (Pecht and Tiku 2006).

Government and military consumers in particular face compromised mission effectiveness via degraded hardware functionality (US Government Accountability Office 2010). Reliability is especially of concern to the US military, where precision instrumentation and highly mission-critical defense systems can be compromised by counterfeits, most of which are collected and recycled from electronic waste (e-waste) and resold as new (Coalition for American Electronics Recycling 2016). While it is evident that the supply chain has been infected with a significant amount of recycled and remarked counterfeit electronic parts, the amount of tampered counterfeit electronic parts that has permeated into the supply chain is not widely known. The consequences of a tampered counterfeit electronic part could be devastating to cyber-physical systems where these parts are installed. In one example, a German Patriot missile system reportedly was hacked by a foreign source and carried out unexpected commands suspected to be enabled through a tampered microprocessor (Gaffey 2015).

Another example of a vulnerable critical infrastructure system is the electric grid, where a failure may result in compromised national defense and security missions (Karvetski and Lambert 2012; Karvetski et al. 2011), and has prompted the shift toward energy islanding when uninterrupted electric power is a requirement (Hamilton et al. 2013, 2016). Related to energy, Martinez et al. (2013) and Martinez and Lambert (2012) describe several failure scenarios in liquid natural gas systems stemming from failures in pumps, compressors, and other equipment. Even seemingly non-critical systems such as pumps can be at risk from counterfeit parts, leading to cascading failures within the larger system.

Due to ever-increasing threats to the national cyber-enabled infrastructure, there have been calls for improved resilience in our cyber capabilities. Former President Barack Obama in 2013 and Former President Donald Trump in 2017 have both called for increased security and resilience in the cyber domain (Executive Order 13636 2013; Executive Order 13800 2017). The CHIPS Act, passed in 2022 by President Joe Biden, provides over $52 billion in incentives to stimulate domestic chip manufacturing and strengthen the US electronics ecosystem.

3.2 *Supply Chain Actors, Practices, and Policies*

The electronics manufacturing industry has been changing with increased costs of building fabrication facilities and the quickened pace of technological innovation, which has given rise to contract manufacturing (Mason et al. 2002). A horizontal vs. vertical supply chain requires controls to address trust, security, and resilience. This outsourcing leads to systems becoming dependent not only on their own cyber security but on the security of the entire global supply chain, as there is increased necessity to purchase parts from potentially untrusted sources (Pecht and Tiku 2006; Rojo et al. 2010; Villasenor 2013). Buying from untrusted sources vastly increases the likelihood of counterfeit parts being introduced into the supply chain. Collier et al. (2014a) indicates that rather than simply analyzing risk in the supply

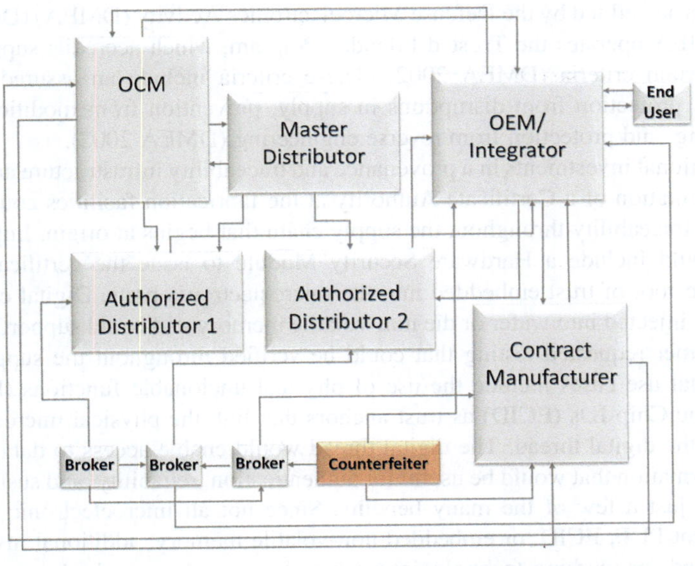

Fig. 1 Conceptual model of the electronics supply chain. (Adapted from DiMase et al. 2016)

chain it is important to consider a resilience based approach that does not solely focus on prevention, but also seeks to allay the consequences of inevitable failures within the supply chain.

DiMase et al. (2016) describes the complexity of the hardware supply chain (Fig. 1). Original equipment manufacturers (OEMs) can purchase components not only from original component manufacturers (OCMs) but also contract manufacturers, master distributors, authorized distributors, brokers and independent distributors. These brokers and independent distributors open an avenue for counterfeit parts to enter the supply chain. At any link in the supply chain components can be redistributed inter-divisionally or resold back into the market. The traceability of these components to the OCM becomes ever more complicated as they are transferred and/or resold again and again. It becomes increasingly difficult to maintain the traceability even from the most reputable authorized/franchised distributors, thus increasing the likelihood of counterfeit material being introduced into a supply chain.

An effective method to trace products through the supply chain is key to assessing the risks of counterfeit parts, but it is cost prohibitive to require the name and location of all supply chain intermediaries. To avoid the costliness of complete traceability for all parts, policies which prioritize acquiring material from authorized sources such as the original manufacturer or its authorized/franchised distributor are proposed. For instance, the Department of Defense (DoD) requires that contractors and subcontractors "acquire electronic parts from trusted suppliers in order to further address the avoidance of counterfeit electronic parts" (DoD 2016). DoD Instruction 5200.44 specifies requirements to procure integrated circuit-related products from suppliers accredited by the Defense Microelectronics Activity (DMEA) (DoD 2012). The DMEA operates the Trusted Foundry Program, which accredits suppliers that meet certain criteria (DMEA 2002). These criteria include an assured chain of custody, protection from disruptions in supply, prevention from modification and tampering, and protection from reverse engineering (DMEA 2002).

Additional investments in a provenance and traceability infrastructure are needed. Implementation of a Certificate Authority at the fabrication facilities could deliver scalable traceability throughout the supply chain that begins at origin. Implementation should include a Hardware Security Module to issue the certificate, and a hardware root of trust embedded into the microelectronic parts. Digital certificates could be injected into wafer or die non-volatile memory and could support use cases of customer requested testing that could be verified throughout the supply chain. Additional use cases include the use of physical unclonable functions (PUF) and Electronic Chip IDs (ECID) as trust anchors that link the physical microelectronic parts to the digital thread. The digital thread would enable access to data analytics and information that would be useful for authentication, reliability, and sustainability to name just a few of the many benefits. Since not all microelectronic parts can implement PUF, ECID, or embedded non-volatile memory, additional investments are needed for marking technologies such as laser marking and e-beam tools that could be integrated into the foundries to establish a trust anchor that can be read in the field. Investments in technology that can perform quick scan reading of the marked devices are also needed. Investments in an infrastructure for provenance and traceability at the microelectronics fabricators would enable better visibility of the end-to-end supply chain and allow for a more robust resilience-based framework for managing supply chain risks.

While it is ideal to purchase parts from trusted suppliers, it is not always possible, therefore it is sometimes necessary to purchase from untrusted sources. In these cases, a risk-based approach must be taken. To address this need, the Society of Automotive Engineers (SAE) G-19 committee is developing a suite of industry standards related to the management of the electronics supply chain. SAE recently published SAE AS6171, which "ensures consistency across the supply chain for test techniques and requirements based on assessed risk associated with the application, component, supplier, and other relevant risk factors" (SAE 2017). By standardizing practices to assess the risk and test suspect counterfeit hardware components, AS6171 can help reduce counterfeit proliferation from a risk-based perspective through industry standard work that intends to improve consistency of test laboratories performing counterfeit electronic parts testing and screening.

The AS6171 includes a risk assessment methodology that assigns a risk score to a batch of electronic hardware components based on relevant criteria. A semi-quantitative model is used to generate a risk score that assesses the risk the user of the component incurs in the application being assessed (Collier et al. 2014b). This risk assessment method has the goal of identifying the recommended level of testing to be performed on components that originate from untrusted or unauthorized sources. Higher risk components require more rigorous level of testing. Following a risk-based approach, there are three main factors contributing to overall risk:

1. *Product risk*—related to the criticality of the product in which the component will be used, and the consequences associated with product failure.
2. *Component risk*—related to the risk given the failure of the component within the system, and is associated with the effect of the component failure on the operation of the product.
3. *Supplier risk*—related to the risks incurred from the selected component distributor, such as a non-authorized supplier.

The previous examples illustrate *risk-based,* rather than a *resilience-based,* approaches. However, given the unprecedented pace of technological evolution and a lack of relevant and reliable historical data in the realm of counterfeit protection, the typical risk analysis triplet of identifying hazards, likelihood of hazard occurrence, and the magnitudes of impact is nearly impossible to quantify reliably (Kaplan and Garrick 1981; Collier et al. 2014a).

The Society of Automotive Engineers (SAE) G-32 committee is developing a series of standards specific to cyber physical systems security, hardware assurance, and software assurance. The suite of standards establishes an assurance framework of cyber physical systems through systems engineering that includes evaluation of the software, firmware, and hardware components integrated into the system, along with the risks due to supply chain vulnerabilities. The suite of standards provides a risk-based framework to address confidentiality, integrity, availability and nonrepudiation of the system, software, hardware, and supply chain. Additional risk evaluations are required based on a triggering event that could include a known security breach or periodic reassessments that could include newly discovered weaknesses and vulnerabilities of the system, hardware, software, or supply chain responsible for the fabrication of these devices. The systems engineering approach for cyber physical systems security requires mitigations of unacceptable risks throughout the lifecycle of the system and its component parts. The *JA7496 Cyber Physical Systems Security Engineering Plan* standard utilizes the NIST framework defined in *SP 800–160 Vol. 1 Engineering Trustworthy Secure Systems* and interweaves domains of consideration defined in the document to achieve security. Future plans for the committee include incorporating the NIST *SP 800–160 Vol. 2, Developing Cyber-Resilient Systems* into the standards' framework. Work from the committee includes the development of assurance framework that identifies threats, weaknesses, and vulnerabilities and their corresponding countermeasures utilizing assurance claims. Given a detailed list of claims and their corresponding efficacy and cost, then metrics could be established that could enable a more robust supply chain resilience-based framework. Additional research and development is needed to build a detailed list of threats, weaknesses, and vulnerabilities across the supply chain, and their corresponding mitigations that includes efficacy and cost.

4 From Risk to Resilience in Global Supply Chains

4.1 Identifying Metrics of Supply Chain Resilience

Increasingly, hardware supply chains have been evolving toward supply webs, moving away from linear procurement and toward systems that have been characterized as "dynamic, hyper-connected, and collaborative" (Kelly and Marchese 2015). Fiksel et al. (2015) and Sheffi (2005) propose supply chain resilience, which promotes the resilience paradigm in terms of building adaptability and the ability to bounce back from disruptions and stressors. The horizontal supply chain requires additional controls and oversight to enable resiliency.

To successfully manage for resilience, it is important to define the appropriate metrics (Seager et al. 2007). Metrics are "measurable properties that quantify the degree to which objectives have been achieved" (Reichert et al. 2007). Guidance exists on what constitutes a "good" metric, including being linked directly to decision making goals, ease of measurement, degree to which they can be comprehended, and linkage to strategy and behaviors (Collier et al. 2016; McKay et al. 2012; Marr 2010; Eckerson 2009; Keeney and Gregory 2005; Neely et al. 1997).

Linkov et al. (2013) propose a novel method for generating resilience metrics for a number of systems. This aid, called the "Resilience Matrix", defines a 4x4 grid, where along one axis are the generic resilience capabilities (Plan/Prepare, Absorb, Recover, Adapt), and along the other axis are the system domains (Physical, Information, Cognitive, Social). Each cell in the Resilience Matrix prompts the analyst to consider the question: "How is the system's ability to [plan/prepare for, absorb, recover from, adapt to] a cyber disruption implemented in the [physical, information, cognitive, social] domain?" (Linkov et al. 2013). This approach has been implemented for cyber security (Linkov et al. 2013), energy security (Roege et al. 2014), and coastal flooding (Fox-Lent et al. 2015). A Resilience Matrix with sample metrics related to counterfeits in the electronics supply chain is presented in Table 2.

Once high-level metrics are identified, they can be further decomposed and quantified. For example, following a semi-quantitative approach, Roege et al. (2014) identify seven levels of the metric "Functional Redundancy":

1. Each system component has a distinct function with no known functional redundancy
2. Most system components are not complementary. Some could be substituted but with substantial delay or they have differing capacities
3. Most system components are not complementary. Some could be substituted with minimal delay and/or they have differing capacities
4. System components are mostly complementary and can be substituted with minimal delay, but have differing capacities
5. System components are mostly complementary, and can be substituted with minimal delay, and have similar capacity

Table 2 Resilience matrix with selected supply chain metrics

	Plan and prepare	Absorb	Recover	Adapt
Physical	Implement traceability infrastructure at fabs that includes chip scanning capability throughout supply chain	Install functionally redundant components in system	Rapidly identify and segregate non-conforming counterfeit parts from conforming parts	Update system configuration, phase out high-risk components
Information	Establish traceability throughout supply chain. Implement pilot (s) for certificate authority at microelectronics fabricators	Operate warning systems with location and quantity of suspect counterfeit parts through active engagement in government and industry reporting services (e.g., GIDEP, ERAI, ESCO anti-counterfeiting forum)	Maintain logs of component purchase history and alternate sources of supply in order to locate and replace	Document actions taken and lessons learned
Cognitive	Training and education in supply chain best practices, technology advances, industry trends	Establish decision making protocols for rapid response	Delegate decision making authority to proper team members and stakeholders	Integrate lessons learned into planning and training
Social	Only purchase components from trusted suppliers	Maintain flexibility in acquiring parts from multiple suppliers	Follow recovery plan and risk communications plan	Conduct after-action reviews with personnel on strengths and weaknesses of actions taken for lessons learned and continuous improvement

6. System components are mostly complementary, can be readily substituted, and have similar capacity
7. System components are fully complementary, can be readily substituted, and have similar capacity

Each level of Functional Redundancy can then be assigned a quantitative score, facilitating organizational action. For instance, the Resilience Matrix can be populated with metrics and scored at specific points in time to provide an organizational snapshot of resilience, identify gaps, and track progress. The Resilience Matrix can also serve as a platform for decision making through cost-benefit studies of resilience-enhancing initiatives. More advanced approaches involve the weighting and scoring the cells and performing trade-off analysis.

However, several caveats apply to the generation of resilience metrics. For instance, no single metric or set of metrics will be able to assess resilience across all systems and supply chains, therefore metrics will need to be defined relative to the specific system, mission, and stakeholders (Bodeau and Graubart 2016). Ford et al. (2012) note that reporting resilience as a single number can present a difficulty, obscuring relevant contributing factors.

4.2 Resilience Beyond the Supply Chain

The products, infrastructure, and services enabled by electronic hardware are part of a larger cyber-physical system (CPS). These systems are complex and can be thought of as analogous to an ecosystem (Gosine 2017) including electronics, humans, supply chains, flows of information, etc. The U.S. National Institute for Standards and Technology (2016) defines cyber-physical systems as "smart systems that include engineered interacting networks of physical and computational components".

Supply chain risk management is only one aspect of security and resilience within the broader cyber-physical system ecosystem. There are 10 areas of concern in these complex systems as defined by DiMase et al. (2015) in Table 3.

Any cyber-physical system may have a number of resilience profiles associated with each of the various areas of concern. For instance, one area may be very resilient against cyber attacks, but relatively weak along some other area of concern. Using the areas of concern can guide formal and structured thinking across a complex cyber-physical system, and identify the location of gaps in protection. More research is needed however on quantitative resilience models of each of the areas of concern.

4.3 Example of Medical Device Supply Chains

This section provides an example of a global supply chain for medical devices, specifically focusing on the production and distribution of automated implantable insulin pumps for diabetes treatment. This example demonstrates how the chapter's framework can be applied to propose measures to enhance resilience.

The supply chain for this hypothetical automated implantable insulin pump spans multiple countries and includes various actors: original component manufacturers (OCMs), contract manufacturers, original equipment manufacturers (OEMs), master distributors, and authorized distributors. This global supply chain for automated implantable insulin pumps faces several resilience shortcomings: single-source dependence, lack of traceability, limited operational redundancy, and poor

Table 3 Areas of concern and their relationship to resilience

Area of concern	Considerations	Link to resilience
1. Electronic and physical security	Insider threats, controls (physical, technical, administrative), preventing unauthorized access to physical and digital assets	Plan/prepare by preventative controls to limit unauthorized access to system
2. Information assurance (IA) and data security	Defense of information systems, assurance of data availability, integrity, authenticity, confidentiality, nonrepudiation, protection against unauthorized data modification, destruction, or disclosure	Plan/prepare against unauthorized data manipulation. Absorb potential attacks with minimal corruption of data
3. Asset management and access control	Management, inventory, and maintenance of critical assets, verification of credentials, software or firmware revisions	Plan/prepare through knowledge of system assets. Adaptation through updates to system vulnerabilities
4. Life cycle and diminishing manufacturing sources and material shortages (DMSMS)	Sustainment processes against loss of manufacturers or suppliers of system components, materials, or services	Absorb and recover by mitigating effect of disruption in supply, quickly establishing backup supply or using reserves
5. Anti-counterfeit and supply chain risk management (SCRM)	Protection against counterfeit parts, supply chain vulnerabilities, authentication of parts, pedigree of parts, anti-counterfeiting controls	Plan/prepare by using effective risk mitigation measures. Absorb by limiting likelihood and impacts of counterfeit parts on supply chain.
6. Software assurance and application security	Limit vulnerabilities introduced into software (unintentionally or intentionally), ensure software functions as intended, limit access to source code	Plan/prepare by limiting access to source code. Absorb and recover through automated detection and fail safe mechanisms
7. Forensics, prognostics, and recovery plans	Gathering information regarding incidents, establishing baseline operations, forensics post-incident, system prognostics, responding to root causes of cyber attacks	Adapt by tracking forensics to learn from past attacks and use of prognostics to anticipate future attacks. Absorb and recovery through established plans to quickly limit damage and reestablish critical functions
8. Track and trace	Determination of current and past locations of goods, security controls and logistics	Adapt by knowing where compromised parts originated and utilizing different sources
9. Anti-malicious and anti-tamper	Methods of preventing introduction of tampered hardware or malware through reverse engineering, integration of protective measures and features to mitigate attack impacts	Plan/prepare by limiting access to system vulnerabilities. Absorb through multiple layers of defense designed into the system

(continued)

Table 3 (continued)

Area of concern	Considerations	Link to resilience
10. Information sharing and reporting	Shared databases, reporting, and exchange of cyber attack events, communication plans and information sharing	Plan/prepare and adapt by remediating information and taking action regarding lessons learned from reports shared

Adapted from DiMase et al. (2015)

communication and coordination. To enhance resilience in this global supply chain for automated implantable insulin pumps, several measures can be implemented: diversify sourcing, implement strong traceability systems, develop redundancy and backup systems, and enhance communication and coordination.

The example illustrates how the different elements of the chapter contribute to understanding and enhancing supply chain resilience: the overview of the electronics supply chain and the threat of counterfeiting in Sect. 2 provides a crucial context for understanding the importance of traceability and authenticity verification for ensuring the safety and effectiveness of automated implantable insulin pumps; the resilience metrics outlined in Sect. 3 can be used to assess the vulnerabilities of the medical device supply chain and identify key areas for improvement; the discussion of cyber-physical systems security in Sect. 4 highlights the importance of protecting critical infrastructure and data systems from disruptions; and the chapter's discussion of resilience strategies, such as diversifying sourcing, enhancing communication, and developing redundancy, provides a framework for developing concrete action plans to mitigate the vulnerabilities and enhance the resilience of the global supply chain for automated implantable insulin pumps.

By applying the framework and the insights presented in this chapter, the global supply chain for automated implantable insulin pumps can be transformed from a highly vulnerable system to a more resilient and reliable system. This example demonstrates the value of integrating risk and resilience considerations at all levels of the supply chain to ensure the safe, effective, and timely delivery of essential medical devices to patients worldwide.

5 Conclusions and Future Work

The above has suggested how it is beneficial to look beyond a classic understanding of risk and develop standards and guidance for managing risk and ensuring resilience at all levels of the cyber hardware supply chain and the embedded hardware of critical infrastructure systems. This chapter has examined a variety of strategies for this risk assessment and mitigation for cyber hardware at the component level, in the supply chain, and within a system, and explored how to move from purely risk-driven approaches to strategies that embrace resilience as an organizing principle of systems engineering and risk management.

Future work includes the application of a number of risk and resilience analytics tools and methods, including network science modeling (Barker et al. 2017; Falasca et al. 2008; Lambert and Sarda 2005), decision analysis (Thorisson et al. 2017; Collier et al. 2017), statistical process control (Lambert et al. 2005), and risk-cost-benefit analysis (Lambert and Farrington 2006). Qualitative methods such as mental modeling (Wood et al. 2017) may also prove beneficial for eliciting technical and operational information from stakeholders and technical experts in order to discover shared understandings and relationships between key risk drivers in a supply chain or firm. Investigations of how global supply chains are vulnerable to the variety of stressors and to disruptions in the supply of materials are also important for understanding how to build a stronger and more resilient supply chain.

Finally, there must be recognition that within large enterprises, cyber security is intertwined with physical and human domains as physical products blended with software are connected, and is no longer solely the purview of the IT department, which tends to be organizationally separated from other management functions (Ernst and Young 2014). Managing cyber-physical risks and securing the supply chain will require clear understanding and mapping of internal business processes (Teng et al. 2012, 2013) and external supply chains (Gardner and Cooper 2003). However, not only large enterprises need to consider resilience. Organizations of all sizes, from startups to large firms, can benefit from adopting a cyber resilience mindset (D'Anna and Collier 2023).

The following are recommendations for practitioners:

- Integrate resilience thinking across the supply chain: Move beyond a risk-based approach and embrace a resilience perspective that considers all actors and their interconnectedness. This involves focusing on the ability to absorb, adapt, and recover from disruptions rather than simply mitigating risks.
- Develop a comprehensive set of resilience metrics: Utilize the Resilience Matrix and other metrics outlined in the chapter to quantify the resilience of the supply chain in various domains. This helps identify vulnerabilities and prioritize actions to enhance resilience.
- Develop quantitative supply chain models: Models and simulations can help supply chain managers to understand the relationships among stressors and supply chain performance metrics. Discrete event simulation and other similar approaches can be leveraged to model risk scenarios and assess the efficacy of countermeasures.
- Implement robust communication and coordination mechanisms: Establish digital platforms, common protocols, and collaborative relationships to enable swift and effective communication and coordination across the supply chain during disruptions.
- Leverage technology to enhance traceability and transparency: Implement technologies such as RFID, Certificate Authority, trust anchors, blockchain, and digital twins to ensure the authenticity and traceability of components throughout the supply chain. This minimizes the risk of counterfeiting and improves the ability to quickly identify and isolate faulty parts.

- Develop redundancy and backup systems: Invest in backup facilities, alternate sources of supply, and strategic stockpiling of critical components to ensure continued operation even during disruptions.
- Foster a culture of resilience: Create a culture of continuous learning and improvement, encouraging all stakeholders to actively identify and address potential vulnerabilities and proactively enhance resilience.
- Support and expedite the development of standards for cyber assurance and traceability that can be linked to policy.

By embracing these recommendations, practitioners of cyber resilience can play a vital role in strengthening the ability of supply chains to withstand disruptions and recover quickly, ensuring the secure and reliable operation of critical systems and the delivery of essential goods and services.

Acknowledgments The authors are grateful for the comments of Mr. Thomas L. Polmateer, Logistics Systems Data Analyst, University of Virginia, and Commonwealth Center for Advanced Logistics Systems (CCALS). The authors are grateful for support of Ms. Karen Jackson, Fellow of the CCALS. Some of the authors are engaged with the CCALS in a multi-year contract for systems modeling and risk management for the Port of Virginia, USA, which had the first automated freight container port in the Western Hemisphere.

References

Al-Mutairi A, AlKheder S, Alzwayid S, Talib D, Heji M, Lambert J (2022) Scenario-based preferences modeling to investigate port initiatives resilience. Technol Forecast Soc Chang 176:121498

Andrews D, Eddy T, Hollenback K, Sreekumar S, Loose D, Pennetti C, Polmateer T, Haug J, Oliver-Clark L, Williams J et al (2023) Enterprise risk management for automation in correctional facilities with pandemic and other stressors. Risk Anal 43(4):820–837

Baker R, Polmateer T, Marcellin M, Chen T, Riggs R, Iqbal T, Hendrickson D, Slutzky D, Lambert J (2023) Mixed-integer programming with Enterprise risk analysis for vehicle electrification at maritime container ports. In: 2023 IEEE symposium series on computational intelligence (SSCI), pp 1759–1766

Barker K et al (2017) Defining resilience analytics for interdependent cyber-physical-social networks. Sustain Resilient Infrastruct 2:59–67. https://doi.org/10.1080/23789689.2017.1294859

Bazemore B, Cha M, Goss Z, Haywood H, Dye B, Gunn M, Loose D, Polmateer T, Hendrickson D, Lambert J (2024) Electrification of utility tractors at maritime container ports. In: 2024 systems and information engineering design symposium (SIEDS), pp 221–226

Beteto A, Melo V, Lin J, Alsultan M, Dias E, Korte E, Johnson D, Moghadasi N, Polmateer T, Lambert J (2022) Anomaly and cyber fraud detection in pipelines and supply chains for liquid fuels. Environ Syst Decis 42(2):306–324

Bodeau D, Graubart R (2016) Cyber resilience metrics: key observations. The MITRE Corporation. https://www.mitre.org/sites/default/files/publications/pr-16-0779-cyber-resilience-metrics-key-observations.pdf

Bonato M, Sambo B, Sperotto A, Lambert J, Linkov I, Critto A, Torresan S, Marcomini A (2022) Prioritization of resilience initiatives for climate-related disasters in the Metropolitan City of Venice. Risk Anal 42(5):931–952

Carvalho H, Cruz-Machado V (2011) Integrating lean, agile, resilience and green paradigms in supply chain management (LARG_SCM). In: Li P (ed) Supply chain management. InTech, Rijeka, pp 27–48
Christopher M, Peck H (2004) Building the resilient supply chain. Int J Logist Manag 15(2):1–13
Clark D, Berson T, Lin H (2014) At the nexus of cybersecurity and public policy, some basic concepts and issues. National Research Council, The National Academies Press. http://www.nap.edu/catalog/18749/at-the-nexus-of-cybersecurity-and-public-policy-some-basic
Coalition for American Electronics Recycling (2016) Unregulated e-waste exports fuel counterfeit electronics that undermine us national security. http://americanerecycling.org/images/Counterfeiting_position_paper_3_1-16.pdf
Collier ZA, DiMase D, Walters S, Tehranipoor MM, Lambert JH, Linkov I (2014a) Cybersecurity standards: managing risk and creating resilience. Computer 47(9):70–76
Collier ZA, Walters S, DiMase D, Keisler JM, Linkov I (2014b) A semi-quantitative risk assessment standard for counterfeit electronics detection. SAE Int J Aerosp 7:171–181
Collier ZA, DiMase D, Heffner K, Linkov I (2015) Building a trusted and agile supply chain network for electronic hardware. In: Proceedings from the 20th international command and control research and technology symposium
Collier ZA, Panwar M, Ganin AA, Kott A, Linkov I (2016) Security metrics in industrial control systems. In: Colbert EJM, Kott A (eds) Cyber-security of SCADA and other industrial control systems. Springer, Switzerland, pp 167–185
Collier ZA, Connelly EB, Polmateer TL, Lambert JH (2017) Value chain for next-generation biofuels: resilience and sustainability of the product life cycle. Environ Syst Decis 37(1):22–33
Collier ZA, Guin U, Sarkis J, Lambert JH (2021) Decision model with quantification of buyer-supplier trust in advanced technology enterprises. BIJ 29(10):3033–3056
Collier ZA, Gaskins A, Lambert JH (2022) Business process modeling for semiconductor production risk analysis using IDEF0. IEEE Eng Manag Rev 51(1):183–188
Collier ZA, Briglia B, Finkelston T, Manasco M, Slutzky D, Lambert J (2023a) On metrics and prioritization of investments in hardware security. Syst Eng 26(4):425–437
Collier ZA, Loose D, Sellers E, Polmateer T, Behl M, Linkov I, Lambert JH (2023b) Stress testing for resilience of semiconductor supply chains. In: 2023 IEEE 14th annual ubiquitous computing, electronics & mobile communication conference (UEMCON), pp 0042–0049
D'Anna G, Collier ZA (2023) Cybersecurity for entrepreneurs. Warrendale, SAE International
Daly T, Singh S (2020) How global pandemic jeopardizes China's metals supply chain. Reuters. https://www.reuters.com/article/markets/currencies/how-global-pandemic-jeopardizes-chinas-metals-supply-chain-idUSKCN2240OO/
Dasaklis T, Pappis C (2013) Supply chain management in view of climate change: an overview of possible impacts and the road ahead. J Ind Eng Manag 6:1139–1161
Davis C, Sreekumar S, Altman R, Clarens A, Lambert J, Colosi L (2023) Geospatially explicit technoeconomic assessment of sustainable aviation fuel production: a case study in Virginia. Fuel Commun 19:100114
DiMase D, Zulueta P (2009) An industry united to fight counterfeiting. A counterfeit EEE parts solution. In: SMTA international conference, San Diego
DiMase D, Collier ZA, Heffner K, Linkov I (2015) Systems engineering framework for cyber physical security and resilience. Environ Syst Decis 35(2):291–300
DiMase D, Collier ZA, Carlson J, Gray RB, Linkov I (2016) Traceability and risk analysis strategies for addressing counterfeit electronics in supply chains for complex systems. Risk Anal 36(10): 1834–1843
DMEA (2002) Trusted foundry program. http://www.dmea.osd.mil/trustedic.html
DoD (2012) Protection of mission critical functions to achieve trusted systems and networks (TSN). Department of Defense Instruction Number 5200.44
DoD (2016) Defense federal acquisition regulation supplement: detection and avoidance of counterfeit electronic parts – further implementation (DFARS Case 2014–D005). Federal Register 81(148)

Eckerson WW (2009) Performance management strategies: how to create and deploy effective metrics. TDWI best practices report. The Data Warehousing Institute, Renton

Er M, Ghadge A, Bititci U (2020) Modelling the impact of climate change risk on supply chain performance. Int J Prod Res 59:1–19

Ernst & Young (2014) The DNA of the CIO: opening the door to the c-suite. http://www.ey.com/Publication/vwLUAssets/ey-the-dna-of-the-cio/$FILE/ey-the-dna-of-the-cio.pdf

Exec (2013) Order No. 13636. 78 Fed. Reg. 11739-11744

Exec (2017) Order No. 13800. 82 Fed. Reg. 22391-22397

Falasca M, Zobel CW, Cook D (2008) A decision support framework to assess supply chain resilience. In: Fiedrich F, Van de Walle B (eds) Proceedings of the 5th international ISCRAM conference. Washington, DC, pp 596–605

Fiksel J, Polyviou M, Croxton KL, Pettit TJ (2015) From risk to resilience: learning to deal with disruption. MIT Sloan Manag Rev 56(2):1–8

Ford R, Cavalho M, Mayron L, Bishop M (2012) Toward metrics for cyber resilience. In 21st EICAR (European Institute for Computer Anti-Virus Research) annual conference proceedings

Fox-Lent C, Bates ME, Linkov I (2015) A matrix approach to community resilience assessment: an illustrative case at rockaway peninsula. Environ Syst Decis 35(2):209–218

Gaffey C (2015) German Missiles hacked by foreign source. Newsweek. http://europe.newsweek.com/german-missiles-hacked-by-foreign-source-329980

Gardner JT, Cooper MC (2003) Strategic supply chain mapping approaches. J Bus Logist 24(2):37–64

Gosine A (2017) Building an ICS cybersecurity ecosystem. Control Engineering. http://www.controleng.com/industry-news/more-news/single-article/building-an-ics-cybersecurity-ecosystem/564abfc38c1592f752a5eaa8fbc2f664.html

Guin U, Huang K, DiMase D, Carulli J, Tehranipoor M, Makris Y (2014) Counterfeit integrated circuits: a rising threat in the global semiconductor supply chain. Proc IEEE 102(8):1207–1228

Hamilton MC, Lambert JH, Keisler JW, Linkov I, Holcomb FM (2013) Research and development priorities for energy islanding of military and industrial installations. ASCE J Infrastruct Syst 19(3):297–305

Hamilton MC, Lambert JH, Connelly EB, Barker K (2016) Resilience analytics with disruption of preferences and lifecycle cost analysis for energy microgrids. Reliab Eng Syst Saf 150:11–21

Horowitz BM, Lambert JH (2006) Assembling off-the-shelf components: learn as you go systems engineering. Trans Syst Man Cybern Part A 36(2):286–297

Johnson D, Wheeler R, Marcellin M, Moghadasi N, Altman R, Polmateer T, Lambert J (2023) Integration of risk sources and risk controls to SysML requirements diagrams with application to sustainable aviation fuels. In: 2023 IEEE international conference on industrial engineering and engineering management (IEEM), pp 0443–0449

Kanike U (2023) Factors disrupting supply chain management in manufacturing industries. J Supply Chain Manag Sci 4(1–2):1–24

Kaplan S, Garrick BJ (1981) On the quantitative definition of risk. Risk Anal 1(1):11–27

Karvetski CW, Lambert JH (2012) Evaluating deep uncertainties in strategic priority-setting with an application to facility energy investments. Syst Eng 15(4):483–493

Karvetski CW, Lambert JH, Linkov I (2011) Scenario and multiple criteria decision analysis for energy and environmental security of military and industrial installations. Integr Environ Assess Manag 7(2):228–236

Keeney RL, Gregory RS (2005) Selecting attributes to measure the achievement of objectives. Oper Res 53(1):1–11

Kelly E, Marchese K (2015) Supply chains and value webs. Deloitte University Press. https://dupress.deloitte.com/dup-us-en/focus/business-trends/2015/supply-chains-to-value-webs-business-trends.html

Kim Y, Kim H (2018) Disruption management in a supply chain: a natural disaster perspective. Int J Prod Econ 202:300–310

Lambert JH, Farrington MW (2006) Risk-based objectives for the allocation of chemical, biological, and radiological air emissions sensors. Risk Anal 26(6):1659–1674

Lambert JH, Sarda P (2005) Terrorism scenario identification by superposition of infrastructure networks. J Infrastruct Syst 11(4):211–220

Lambert JH, Schulte BL, Sarda P (2005) Tracking the complexity of interactions between risk incidents and engineering systems. Syst Eng 8(3):262–277

Lambert JH, Keisler JM, Wheeler WE, Collier ZA, Linkov I (2013) Multiscale approach to the security of hardware supply chains for energy systems. Environ Syst Decis 33(3):326–334

Linkov I, Eisenberg DA, Plourde K, Seager TP, Allen J, Kott A (2013) Resilience metrics for cyber systems. Environ Syst Decis 33(4):471–476

Linkov I, Bridges T, Creutzig F, Decker J, Fox-Lent C, Kröger W, Lambert JH, Levermann A, Montreuil B, Nathwani J, Nyer R, Renn O, Scharte B, Scheffler A, Schreurs M, Thiel-Clemen T (2014) Changing the resilience paradigm. Nat Clim Chang 4:407–409

Loose D, Eddy T, Polmateer T, Hendrickson D, Lambert J (2023a) Systems modeling and resilience of schedule at maritime container ports. In: International conference on systems engineering, pp 328–336

Loose D, Eddy T, Polmateer T, Hendrickson D, Moghadasi N, Lambert J (2023b) Reinforcement learning and automatic control for resilience of maritime container ports. In: 2023 9th international conference on control, decision and information technologies (CoDIT), pp 01–06

Marcellin M, Pavur G, Loose D, Cardenas J, Denehy D, Almashhadani M, Waheed S, Trump B, Polmateer T, Linkov I et al (2024) Systems analysis for energy assets of Iraq influenced by water scarcity. Environ Syst Decis 44:1–21

Marr B (2010) How to design key performance indicators. The Advanced Performance Institute, Milton Keynes

Martinez LJ, Lambert JH (2012) Risk-benefit-cost prioritization of independent protection layers for a liquefied natural gas terminal. Int J Crit Infrastruct 8(4):306–325

Martinez LJ, Thekdi SA, Lambert JH (2013) Modeling energy facility regulatory compliance with application to multi-scale liquefied natural gas facilities. Environment Systems & Decisions 33(3):404–412

Mason SJ, Cole MH, Ulrey BT, Yan L (2002) Improving electronics manufacturing supply chain agility through outsourcing. International Journal of Physical Distribution & Logistics Management 32(7):610–620

McKay SK, Linkov I, Fischenich JC, Miller SJ, Valverde LJ Jr (2012) Ecosystem restoration objectives and metrics, ERDC TN-EMRRP-EBA-12-16. U.S. Army Engineer Research and Development Center, Vicksburg

Moghadasi N, Lambert J (2023) On evaluating system resilience by the degree of order disruption. In: INCOSE international symposium, pp 739–751

Moghadasi N, Collier Z, Koch A, Slutzky D, Polmateer T, Manasco M, Lambert J (2022) Trust and security of electric vehicle-to-grid systems and hardware supply chains. Reliab Eng Syst Saf 225:108565

Moghadasi N, Piran M, Valdez R, Moghaddasi N, Loose D, Polmateer T, Lambert J (2024a) Systems modeling of trust in AI-enabled medical diagnosis. In: 2024 IEEE international systems conference (SysCon), pp 1–8

Moghadasi N, Valdez R, Piran M, Moghaddasi N, Linkov I, Polmateer T, Loose D, Lambert J (2024b) Risk analysis of artificial intelligence in medicine with a multilayer concept of system order. Systems 12(2):47

National Institute of Standards and Technology (2016) Framework for cyber-physical systems. Release 1.0. Cyber Physical Systems Public Working Group, NIST, Gaithersburg, MD

National Research Council (2012) Disaster resilience: a national imperative. National Academies Press, Washington

Neely A, Richards H, Mills J, Platts K, Bourne M (1997) Designing performance measures: a structured approach. Int J Oper Prod Manag 17(11):1131–1152

Park J, Seager TP, Rao PSC, Convertino M, Linkov I (2013) Integrating risk and resilience approaches to catastrophe management in engineering systems. Risk Anal 33:356–367
Pecht M, Tiku S (2006) Bogus: electronic manufacturing and consumers confront a rising tide of counterfeit electronics. IEEE Spectr 43(5):37–46
Reichert P, Borsuk M, Hostmann M, Schweizer S, Sporri C, Tockner K et al (2007) Concepts of decision support for river rehabilitation. Environ Model Softw 22:188–201
Roege PE, Collier ZA, Mancillas J, McDonagh JA, Linkov I (2014) Metrics for energy resilience. Energy Policy 72(1):249–256
Rojo FJR, Roy R, Shehab E (2010) Obsolescence management for long-life contracts: state of the art and future trends. Int J Adv Manuf Technol 49(9):1235–1250
SAE (2017) AS6171: test methods standard; general requirements, suspect/counterfeit, electrical, electronic, and electromechanical parts. http://standards.sae.org/as6171/
Sambo B, Bonato M, Sperotto A, Torresan S, Furlan E, Lambert J, Linkov I, Critto A (2023) Framework for multirisk climate scenarios across system receptors with application to the Metropolitan City of Venice. Risk Anal 43(11):2241–2261
Seager TP, Satterstrom FK, Linkov I, Tuler SP, Kay R (2007) Typological review of environmental performance metrics (with illustrative examples for oil spill response). Integr Environ Assess Manag 3(3):310–321
Sheffi Y (2005) Building a resilient supply chain. Harv Bus Rev 1(8):1–4
Sood B, Das D, Pecht M (2011) Screening for counterfeit electronic parts. J Mater Sci Mater Electron 22(10):1511–1522
Teng K, Thekdi SA, Lambert JH (2012) Identification and evaluation of priorities in the business process of a risk or safety organization. Reliab Eng Syst Saf 99:74–86
Teng K, Thekdi SA, Lambert JH (2013) Risk and safety program performance evaluation and business process modeling. IEEE Trans Syst Man Cybern Part A 42(6):1504–1513
Thorisson H, Lambert JH, Cardenas JJ, Linkov I (2017) Resilience analytics for power grid capacity expansion in a developing region. Risk Anal 37:1268–1286. https://doi.org/10.1111/risa.12711
US Department of Commerce (2010) Defense industrial base assessment: counterfeit electronics. Bureau of Industry and Security, Office of Technology Evaluation. https://www.bis.doc.gov/index.php/forms-documents/technology-evaluation/37-defense-industrial-base-assessment-of-counterfeit-electronics-2010/file
US Government Accountability Office (2010) Intellectual property: observations on efforts to quantify the economic effects of counterfeit and pirated goods. (GAO-10-423, April 2010)
Villasenor J (2013) Compromised by design? Securing the defense electronics supply chain. Center for Technology Innovation at Brookings Institute
Wheeler R, Pennetti C, Polmateer T, Jones G, Dhanpal Y, Lambert J (2022) Information management of a lifeline infrastructure for mobility of people and goods. In: 2022 IEEE international systems conference (SysCon), pp 1–4
Wood G (2016) Costly counterfeit electronic components in the supply chain can also be a safety concern. IHS Markit. http://blog.ihs.com/costly-counterfeit-electronic-components-in-the-supply-chain-can-also-be-a-safety-concern
Wood MD, Thorne S, Kovacs D, Butte G, Linkov I (2017) Mental modeling approach: risk management application case studies. Springer, New York

Active Defense Techniques for Enhancing Cyber Resilience

Nathaniel Evans and Aaron Werth

1 Introduction

Active defense includes a variety of automated and human-directed activities that attempt to address the deficiencies of static defensive measures. Many of the approaches attempt to thwart attacks by increasing the diversity, complexity, or variability of the systems and networks. This approach aims to limit the attacker's ability to gather intelligence or reduce its usable life-span. Other approaches focus on gathering intelligence on the attackers, either by attracting attackers to instrumented honeypots or by patrolling the systems and networks to hunt for attackers. The intelligence gathering approaches rely upon cybersecurity personnel using semiautomated techniques to respond and repel attackers.

Active defense has gathered attention of the research and cyber defense communities because of the weaknesses of static security controls. Static security controls are developed to thwart known attacks and often fail to block unknown attacks, such as "zero-day" vulnerabilities, that are actively exploited before their disclosure to software vendors and the security community. Social engineering attacks, such as phishing, often successfully evade static controls by enticing people to reveal their credentials or accidentally install malware inside their organization.

Active defense has grown in popularity over the past few years but remains a research and development focused activity. Various general-purpose products have started to emerge, but in order for widespread adoption, organizations need to tailor their application or develop their own customized active defense. To apply active defense effectively, organizations will need to understand their potential adversaries and the vulnerability of the assets they might target.

N. Evans · A. Werth (✉)
Oak Ridge National Laboratory, Oak Ridge, TN, USA
e-mail: werthaw@ornl.gov

© The Author(s), under exclusive license to Springer Nature Switzerland AG 2025

I. Linkov, A. Kott (eds.), *Cyber Resilience: Applied Perspectives*, Risk, Systems and Decisions, https://doi.org/10.1007/978-3-031-90109-6_13

Architecture should provide the fundamental security for a system. Design and implementation of security architectures can encompass elements from network segmentation to server hardening and patching as well as data access controls.

A successfully architected secure system or application should include passive defenses, which add protection without requiring human interaction, as well as active defenses. Active defense solutions should include some combination of the following attributes:

- Threat and/or vulnerability agnosticism
- Resilient failure modes that protect confidentiality, integrity, and availability
- Diversity or mutation of attack surface

1.1 Types of Active Defense

The literature generally divides active techniques into "proactive," "reactive," and "hybrid." Often, Moving Target Defense (MTD) is used analogously with active or proactive solutions or as a subcategory of active defense solutions, though there are reactive and hybrid MTD solutions. According to the Department of Homeland Security, MTD is the concept of controlling change across multiple system dimensions in order to increase uncertainty and apparent complexity for attackers, reduce the window of opportunity, and increase the cost of their probing and attack efforts (US Department of Homeland Security 2011). In general, when referring to MTD solutions, we will be considering them as a subset of active defense that focuses on diversity or mutation of the attack surface, though MTD solutions may contain some or all of our other properties. Additionally Distributed Block Chain Technologies have been bucketed in this type of defense as its popularity has grown to be an effective integrity focused defense.

1.2 The Active Defense Continuum

Although the security community agrees that active defense provides measures beyond the protections provided by architecture and passive defenses, they disagree on using the term to include offensive actions. The disagreement concerns the physical scope of the activities and the definition of the perimeter of defense.

When the term "active defense" was first applied to computer security, Holdaway (Eric 2001) defined it broadly as "any measures originated by the defender against the attacker" that "must at least thwart any attack in progress, and ideally make further attacks more difficult." Holdaway categorized active defense into three categories: active deception, counterattack, and preemptive attack. Many would disagree with the scope of this definition and consider counterattack—hacking back—and preemptive attack as offensive measures outside the perimeter of the organization. In fact, in a more recent publication (Elliott 2018), there is a concern that counterattack and preemptive attack could bring legal liability to the organization that would consider these approaches in response to an attack. With this context, our discussion will revolve around strategies within the perimeter, or narrower definition, of active defense. Dittrich and Himma define an "Active Defense Continuum" to assist with the ambiguities in defining active defense techniques. The strategies and techniques discussed below generally fall into the "involved" (i.e., antivirus software, firewall software and hardware, etc.), "interactive" (i.e., altering firewall settings, searching domain registry information, etc.), and "cooperative" (i.e., seeking the help of other organizations, authorities, etc.) categories (Himma and Dittrich 2005).

2 Strategies and Techniques

2.1 Hybrid Proactive/Reactive Techniques

2.1.1 Activated Protective Measures

Various static protective measures, such as real-time anti-malware, active blocking, and recovery during operations, are employing methods to adapt automatically to changing threats.

Many anti-malware tools are employing real-time methods to actively detect and destroy malware. For example, Windows-based anti-malware tools can use an operating system feature, Early Launch Anti-malware (ELAM) (Microsoft 2017), to detect malware before execution.

Blocking access to non-authorized entities remains a major protective measure. Blocks are typically enforced by access control lists at system and network entry points, such as firewalls. Typically, these access control lists are maintained manually through periodic reviews and updates. Because threats emerge quickly and frequently, the access control lists often fail to block active threats. To improve access control, organizations are sharing threat information and actively updating their controls more frequently, often automatically. For example, The Cybersecurity Risk Information Sharing Program (CRISP) developed a toolkit to rapidly share cyber threat intelligence and automatically update the access control lists of participating organizations.

Recognizing that compromises are very likely, organizations are developing recovery playbooks to recover effectively from cyber events while continuing operations. The playbooks specify critical resources along with their responsible personnel, technology, functional and security dependencies, and the processes, priorities, and sequences for recovering the critical resources (National Institute of Standards and Technology 2016).

2.1.2 Cyber Deception

Cyber deception forms the basis of many active defense strategies and technologies. Deception attempts to manipulate the perception of attackers rather than blocking their actions. By manipulating attacker's perceptions, the cyber defender aims to mislead or confuse them into taking actions that help the defender.

Deception manipulates information attempts to mislead, misdirect, or confuse opponents. Typically, the information is false. False information might be released publicly, such as fake user names and contact data or inaccurate system and network configuration. False information might be released internally when adversaries interact with the system and network architecture.

Cyber deception works best when organizations have intelligence about the goals and motivations of their adversaries. By understanding their adversaries, organizations can develop stories that can manipulate their adversaries and cause them to behave in ways that help the organization defend against them.

Cyber Deception and Denial

Heckman et al. (2015) recommend employing deception at all stages of a cyber deception and denial chain. The chain involves all cybersecurity components, including cybersecurity planners and operators as well as intelligence analysts.

After gathering intelligence about their likely adversaries and setting a goal to deceive them, the organization develops a cover story to provide to their adversaries. The cover story guides the decisions on what to reveal to and conceal from the adversaries. The cover story might reveal fake information about personnel to entice phishing attacks. If the adversary exploits the fake personnel information, the chain directs them to an isolated, controlled honeypot environment that contains a variety of fake, personal information, including e-mail and activity history. The cover story may also use true information about system architecture to guide adversaries to attack other parts of the organization where they can be monitored and mislead, perhaps by high-interaction honeypots.

Many deception methodologies employ fake systems and networks to deceive adversaries. To attract adversaries, organizations use lures, honey, and thus decoy systems are often named honey-something. One of the first decoys was known as honeypots. A discussion of honeypots and other honey systems is discussed next.

Honeypots

Honeypots can help organizations improve their resilience by using decoys to distract and slow down attackers while collecting intelligence on them and their exploits.

Honeypots are widely used to deceive attackers and "learn the tools, tactics and motives involved in computer and network attacks" (The Honeynet Project 2003). Honeypots operate by mimicking actual systems, which attackers discover and attempt to compromise by deploying their skills and tools. As the attackers attack, the honeypot learns about the attacker and his or her techniques.

To maximize the usability of the information they provide, honeypots serve no organizational role other than as a decoy to attract attackers. As described by Lance Spitzner (2003), a "honeypot is an information system resource whose value lies in unauthorized or illicit use of that resource." Because honeypots serve no mission function, all interaction with them can be considered suspicious, and the data that are collected from monitors and event logs can be analyzed for signs of malicious behavior. This frees the data analyst from worrying about filtering out false-positive data that may represent legitimate user activity.

Honeypots cannot directly detect attacks on actual production systems, but they can detect attackers that interact with honeypots and production systems. By correlating the honeypot activity with system-wide activity, incident responders can compare known suspicious activity on honeypots with potentially suspicious activity on production systems.

Initially developed to gather intelligence about attackers, honeypots now accurately emulate actual systems, which distract attackers from production (Verma 2003) systems and provide early warnings and more system-specific information, giving security administrators more time and information to develop a response.

Configurations of honeypots vary from full emulation of production systems to more focused emulations of specific system services to synthetic data. When deployed to emulate production systems, production honeypots are described as "full" and require highly interactive implementations to gain the trust of attackers. Implementations of full honeypots range from mock-ups of various essential aspects of a production to full replicas with artificial data and extensive instrumentation to detect and log attacker interactions.

Service-level honeypots focus on specific types of low-interaction services. For example, honeypots have been designed to emulate block devices, such as universal serial bus (USB), and e-mail servers. Because USB drives are often used to distribute malware, the Ghost USB Honeypot (National Institute of Standards and Technology 2016) project developed a service-level honeypot to emulate USB drives and attract attackers that use removable USB drives to distribute malware.

Honeypots are developed to emulate e-mail services in order to attempt to collect, analyze, and block potentially malicious e-mail. The Honeypot Project developed the Spam Honeypot with Intelligent Virtual Analyzer (SHIVA) to collect and analyze spam for the gathering of "information on phishing attacks, scamming campaigns, malware campaigns, spam botnets, [and spammers'] identity.

Recent research has also delved into idiosyncrasies of cyber-physicial systems and related IOT devices. These have limitations and vulnerabilities in hardware and software in comparison with general purpose computers and servers which may attract cyberattacks. Their increased usage in today's society has brought increased interest. A survey (Franco et al. 2021) discusses relevant works on honeypots for cyber-physical systems and IoT. Interestingly, there is also a concern that the attacker may discover it is in a honeypot by using new techniques in fingerprinting so that the attacker can try to avoid such a situation. Researchers (Maesschalck et al. 2024) have learned to take advantage of the attacker's tendency to avoid honeypots by make PLCs, devices used in cyber-physical systems, such as critical infrastructure, to appear to be honeypots.

Other recent research on honeypots includes advances in the area of Generative Adversarial Networks (GANs). A paper (Gabrys et al. 2024) discusses the possibilities of GANs generating configurations for the honeypots to seem realistic to attackers.

Honeynets

Honeynets are a collection of honeypots that are set up to attract as many attackers as possible in order to learn about their patterns, tactics, and behaviors (Han et al. 2016). Virtualization technology, increasingly used in modern production environments, makes setting up entire networks of honeypotted machines easy and cost-effective.

There are a number of research groups that operate various honeypots and honeynets. The aim of these groups is to share the collected data. Deployments that share data from honeypots and honeynets can generally be classified into four types of deployment (Watson and Riden 2008).

Honeytokens

The concept of honeypots is also applied to data by constructing fake digital entities that help detect improper behavior and improve intrusion detection and data loss protection programs. Known as honeytokens and first named in 2003 by Augusto Paes de Barros (DLP and Honeytokens 2003), these fake digital entities, ranging from a fake data file to a fake database entry, have a unique property that allows an organization to design filters to detect them. These unique properties can be used to aid in investigations and possibly in attribution.

Cyber attribution, traditionally, is incredibly difficult. IP masquerading, bouncing through intermediary hosts, and using bot networks are just some of the techniques that attackers have at their disposal to disguise their behavior. The use of honeytokens in honeypots and honeynets provides some unique opportunities for tracking adversaries through the variety of hidden networks and identities to which they typically have access.

Decoy MTD

Decoy-based strategies are a variation on honeypotting. Decoy strategies have several available nodes, of which only one is genuine (Clark et al. 2013). This family of techniques shares many characteristics with honeypotting and are often combined. However, decoy strategies may be used independently. Some researchers have successfully combined service-level honeypots with decoy-based MTDs to track attackers' responses to vulnerability disclosures (Araujo et al. 2014).

Other Cyber Deception Techniques

Various other types of cyber deception exist, including that in the work of Ishikawa and Sakurai (2017). The authors propose a "deception proxy" that creates additional decoy functionalities in a web application rather than modifying the original application itself. In this approach, the proxy is able to intercept and modify HTTP communication in order to add decoy parameters under tokenized names. The decoy parameter tokenization creates confusion with respect to parameter names and increases the difficulty of attacking a legitimate parameter.

Chiang et al. (2016) present an adaptive cyber deception system (ACyDS) that creates for each host a virtual view of its network. ACyDS not only modifies network-view properties but also periodically changes a host's entire network view. The dynamic network views prevent reconnaissance by invalidating any collected data that an attacker may obtain prior to a switch. In addition, this technique minimizes the effect of a compromised host and improves attack identification.

2.2 *Proactive Techniques*

2.2.1 Networking

Different dynamic security strategies can be employed to secure networks from unauthorized access. Researchers have suggested that priority be given to approaches that treat the network as an untrusted communication channel between known parties (Okhravi et al. 2014). In the majority of existing studies on proactive and moving target defenses, researchers recommend the use of dynamic network addressing to protect against attacks that target a specific address (Kewley et al. 2001; Groat et al. 2011; Al-Shaer 2011; Leyi et al. 2008). The earliest recommendation for dynamic addressing was made by Kewley et al. (2001), who proposed the use of dynamic network address translation (DYNAT) to defend against packet sniffing. DYNAT worked by obfuscating transmission control protocol/internet protocol (TCP/IP) packets; the obfuscation was achieved by applying an algorithm to alter the packet contents. At the receiving end, the recipient used a private, predetermined key to decrypt the packet and reveal the IP address, port numbers, and other identification information in the packet header. DYNAT achieves diversity and randomness through the dynamic assignment of IP addresses and through periodic modifications of the decryption key. The authors tested DYNAT to assess its ability to protect against three common threats: packet-sniffing, denial of service (DoS), and session hijacking. The conclusion from the test was that DYNAT provided acceptable network defense.

Although acceptable security for defending against the aforementioned attacks is possible with DYNAT, its effectiveness depends on the frequency at which the dynamic components are varied. Groat's 2011 and 2012 work (Groat et al. 2011, 2012) address this limitation by demonstrating how changing the addresses multiple times per session—using a large address space and a large number of hosts per subnet—increases the security of dynamic addressing. Increasing the address space and change frequency of an IP address reduces the likelihood of detecting the IP address of a particular host. Leyi et al. (2008) demonstrate another relevant network-hopping solution utilizing full-service hopping—involving a pseudorandom rotation of port numbers, IP addresses, cryptography algorithms, and service protocols—can achieve optimal dynamic defense.

Closely related are network defense solutions that depend on probabilistic metrics to detect and defend network resources against attacks (Al-Shaer et al. 2013; Shakarian et al. 2014; Wang et al. 2014; Zhu et al. 2014). Al-Shaer et al. (2013), for instance, introduced random host mutation to assign virtual IP addresses to hosts. An adversary may detect dynamic addressing techniques, which are also vulnerable to DoS attacks (Phatak 2005; Phatak et al. 2013). However, random IP mutation overcomes these limitations because it is designed not to disrupt services during mutations, to be transparent to the end-host, and to mutate quickly and periodically. By these mechanisms, it renders hosts invisible to adversaries and thus defends against reconnaissance attacks. It is noteworthy that the random host mutation program can be deployed easily over an existing network (Al-Shaer et al. 2013).

Stream splitting (SS)-MTD is a technique that splits data over multiple TCP streams or multiple mediums, making it difficult for the attacker to access the entire communication payload and gain meaningful information. TCP stream splitting (SS) splits a network stream into multiple streams (see Fig. 1), making an attack on the system difficult by eliminating the advantage of fixed-system configurations and network architecture.

SS-MTD is similar to multipath TCP (MPTCP), though it has several key differences (Paasch and Bonaventure 2014). SS-MTD does not aim for layer-4 backward compatibility but rather runs totally at the application layer or the set of software components in which behavior is described in logical time (Karsai et al. 2017). In addition, it has no flags in the TCP headers to signal that the stream is running over multiple paths.

The increasing use of Internet Protocol version 6 (IPv6) creates an immediate need to address the growing concern over its immature security measures. Authors have demonstrated several MTD solutions (Groat et al. 2011; Dunlop et al. 2011; Yackoski et al. 2013) where IPv6 addresses are dynamically obscured to prevent

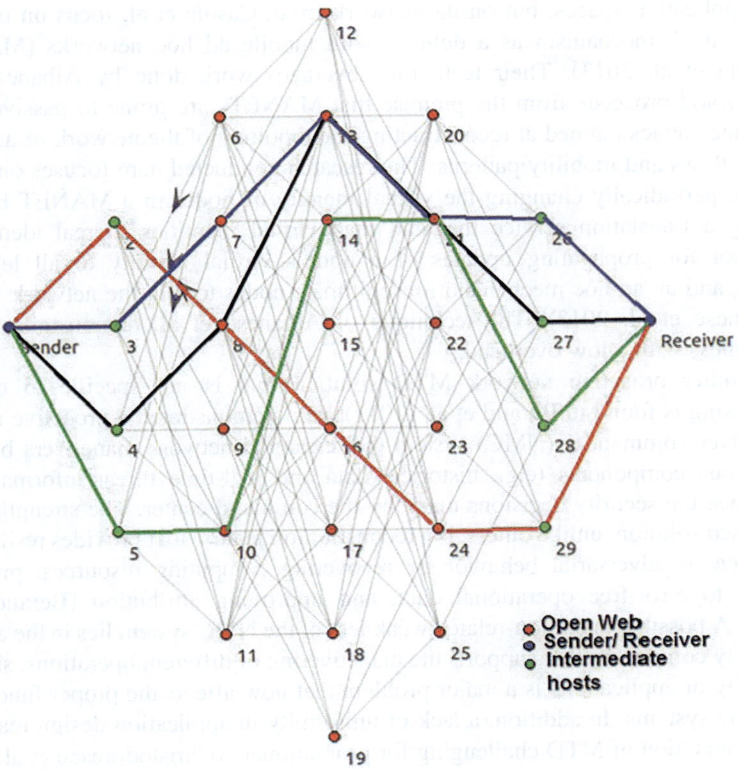

Fig. 1 Splitting TCP network streams

targeted network attacks. As an example, the work of Yackoski et al. (2013) demonstrated the feasibility of a dynamic network by implementing a dynamic network architecture together with a network hypervisor in an actual IPv6 networked environment. In the Moving Target IPv6 Defense (MT6D) approach, network source and destination addresses are rotated in mid-session to create a constantly shifting attack surface. Since its development, MT6D has been used as a basis for several other MTD approaches. Hardman et al. implemented MT6D in C (from the original Python) to make it more widely deployable (Hardman et al. 2013). Others have extended this set of techniques to include embedded devices, such as ICS and SCADA (Heydari and Yoo 2016). Similar work involving shifting network configuration for resource-constrained distributed devices, which includes many ICS devices as a general category, was done by Casola et al. (2013). Their approach utilizes fine-grained reconfiguration of different architectural layers within a network of distributed devices and shows that network-level MTD techniques could be valuable in this space. The authors note a lack of metrics to quantify the benefits accurately but emphasize that combining a number of MTD strategies can reinforce defensive posturing for these types of devices. Their efforts span network, server, and application spaces, but on the network front, Casola et al. focus on using the Sybil attack mechanism as a defense with mobile ad hoc networks (MANETs) (Casola et al. 2013). Their technique leverages work done by Albanese et al. (2013) and proceeds from the premise that MANETs are prone to passive reconnaissance attacks aimed at reconstructing the topology of the network or analyzing traffic flows and mobility patterns. The research conducted here focuses on several points: periodically changing the virtual identity of nodes in a MANET by introducing a translation service that can map virtual identities to real identities, a protocol for propagating updates of a node's virtual identity to all legitimate nodes, and an ad hoc mechanism for legitimate nodes to join the network securely (Albanese et al. 2013). The technique of Albanese et al. demonstrates a high robustness with a low overhead.

Another proactive network MTD solution that is not specific to dynamic addressing is found in Beraud et al. (2011), who demonstrated a proactive network maneuver commander (NMC) system that executed network maneuvers based on numerous components (e.g., historical data and real-time threat information) to influence the security decisions made by the command center. The strength of this proposed solution, unlike others, is a reconstitution feature that provides resilience in the face of adversarial behavior by recovering computing resources, providing access to error-free operational data, and improving attribution (Beraud et al. 2011). A possible operation-related weakness of the NMC system lies in the artificial diversity component that supports the maneuvering of different operations, since the diversity of applications is a major problem that now affects the proper functioning of MTD systems. In addition, a lack of uniformity in application design makes the implementation of MTD challenging for programmers (Christodorescu et al. 2011). A plausible recommendation, therefore, is to standardize applications and thus make MTD implementation feasible. One proposed solution involves using smart grid MTD against spoofing attacks like those that might not be easily detected by existing tools (such as specification-based intrusion detection). The proposed solution uses a deterministic and dynamic defense program, which mutates and randomizes the configuration parameters of smart grid meters to avoid adversarial detection (Al-Shaer et al. 2013). The Al-Shaer et al. evaluation found that this MTD approach provided better defense than static defense.

More recent surveys on MTD for network defenses discusses new developments within the field including artificial intelligence in decision making. Many also provides categorizations of some of the new research related to MTDs.

Many network approaches to MTD are now using software-defined networking (SDN) to rotate network configuration elements. SDN is an approach to computer networking that allows network administrators to manage network services through software abstractions of lower-level functionality. It also separates the various planes of the network, easing the rerouting of traffic between paths and providing a comprehensive view of the network to administrators (Curado et al. 2017). MacFarland and Shue (2015) explore an approach using SDN to modify IP and media access control (MAC) addresses in order to thwart attacker fingerprinting. They use an SDN controller combined with a DNS server and client-based configurations as part of a trusted computing base (TCB). The TCB is then utilized in an attempt to employ the principle of least privileges such that it can prevent adversaries from gaining access to the control plane of the MTD.

Graph-theoretic approaches to modeling and reconfiguring networks are also being applied to proactive defenses in a manner that is consistent with overall MTD methodology. The concept of honeypots, where an attacker is drawn into a fake host to study its behavior, has been extended to networks by Shakarian et al. (2014). In this technique, the authors analyze a graphical representation of a computer network's logical layout and an associated probabilistic model of an adversary's behavior with the goal of artificially modifying the network through addition of "distraction clusters"—collections of interconnected virtual machines—at key points to increase network complexity and delay intrusion (Beraud et al. 2011).

Recent works have investigated the use of the pro-active techniques with IoT devices. These devices present certain challenges unlike general purpose computers or servers on an IT network (Ge et al. 2021). IoT devices typically are limited in their computational power, they may even have limited battery life. Therefore, they must limit the amount of time that they run processes. Otherwise, they would drain the battery.

2.2.2 Servers

Like networks, servers are at an increased risk of security breaches and must be adequately secured against adversarial behavior. Existing server-specific MTD solutions include the proactive obfuscation technique of Roeder and Schneider (2010), which involves a predetermined periodic mutation of servers. The mutation is achieved through an obfuscation program that automatically generates new, executable code for each duplicate server to provide differentiation among server replicas (Roeder and Schneider 2010). The importance of this differentiation, as explained by the authors, is to reduce the number of common vulnerabilities shared among the server replicas, thereby hardening security. Another proposed server-MTD solution involves the use of diversified virtual servers to secure web services (Huang and Ghosh 2011). In this study, the random rotation of servers is based on either fixed or event-driven intervals. The security of the proposed technique lies in the random use of more than two servers simultaneously, which decreases predictability and increases the resilience of web services.

For a proactive, multisystem MTD, Cox et al. (2006) propose the use of the N-variant framework. This framework involves creating variants for a given system (such as a server or application) and replicating the input to all of the variants (Cox et al. 2006). Each variant processes the input and sends the output to a single monitoring system, which then compares the output for correctness, discrepancies, and early detection of abnormal activities.

Because variation can be achieved by randomly diversifying certain aspects of the system (e.g., memory locations, instruction sets, and address spaces), the use of the

N-variant framework could provide multiple MTD solutions, consequently reducing the likelihood of reconnaissance attacks. Subsequent work illustrates the effectiveness of the N-variant framework in defending against attacks that corrupt user identification data (Nguyen-Tuong et al. 2008).

Many moving target and proactive defense solutions are largely theoretical and have not been implemented in the industry. Although Jackson et al. demonstrated the feasibility of some of the solutions (Jackson et al. 2013), the lack of adoption of MTD by industry is very observable. It appears that only one firm now offers an implementable MTD solution; Coronado Group, Ltd. provides a self-cleaning, intrusion-tolerance (SCIT) MTD system that uses virtualized environments to provide dynamic defense against cyberattacks (Zank 2012). SCIT-MTD was designed to "rotate virtual servers and applications at configurable intervals. Typically, some servers are online while others are being wiped clean offline and eventually, every server is taken offline, cleaned and restored to its pristine state" (Zank 2012). Although SCIT-MTD seems to be a viable MTD solution, it lacks the critical element of diversity that would prevent zero-day attacks from propagating across all hosts in the configuration. Restoring the server does not remove the vulnerabilities that caused the attack in the first place. In addition, if all the dynamic virtual applications and servers are similar, their vulnerabilities will likely be similar and could still be exploited. Research by Thompson et al. (2014) addresses this problem using a technique known as Multiple OS Rotational Environment (MORE). MORE takes several different OS platforms and rotates them in a random configuration. In this way, only one platform is exposed at any given time to an attacker; however, the application that is running on the OS in the configuration remains available to a legitimate client at all times. Figure 2 shows an example of MORE's runtime configuration.

Similar in approach to SCIT and MORE, Atighetchi et al. (2004) propose a middleware solution that allows existing applications to use rotating elements at the network and server level.

Other server diversification efforts focus on cloud infrastructure. The resilient dynamic data-driven application systems (DDDAS), integrated into the resilient DDDAS-As-A Service (rDaaS) framework, is designed as a componentized architecture that uses MTD's resilience properties to insure application uptime during a crisis management situation. RDaaS uses a variation of an N-variant framework, as used by Cox et al., to handle multiple redundant inputs and ensure a trustworthy output in an execution environment that is replicated k times (Cox et al. 2006; Badr et al. 2015). Additionally, researchers have posited using genetic algorithms (GA) to create moving target environments (Crouse et al. 2012). GA implementations use some combination of selection, crossover, and mutation to search the solution space for desirable configurations that satisfy a particular application's mission requirements. This metaphor uses chromosomes to represent rotational elements in a system, as depicted in Fig. 3.

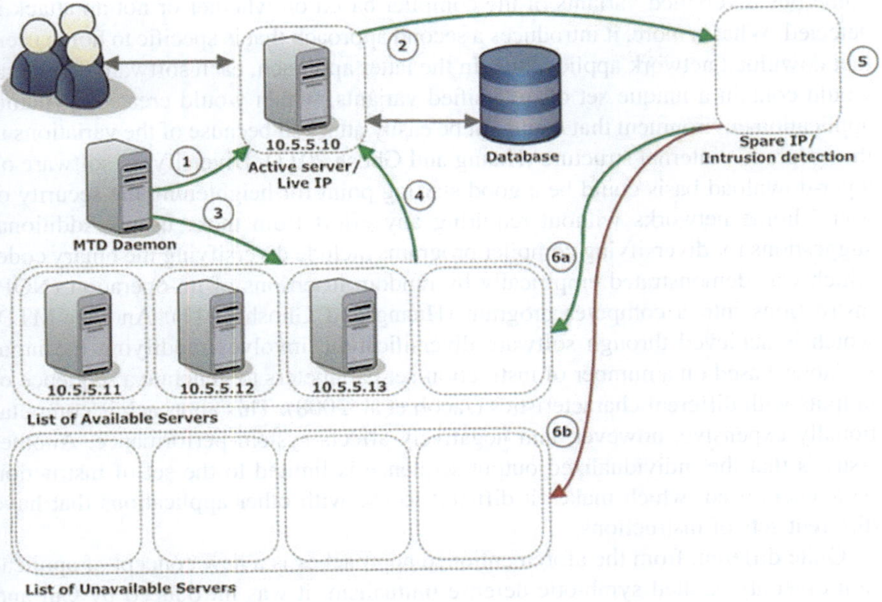

Fig. 2 The diagram shows an example of a MORE configuration

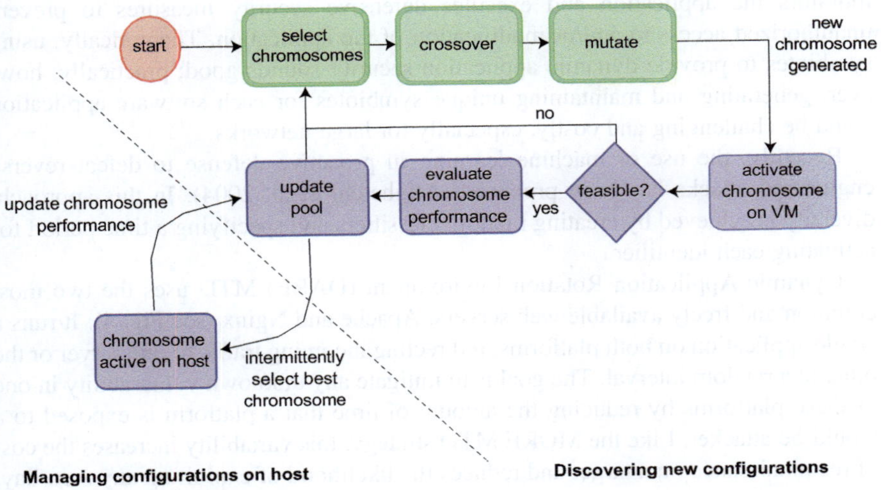

Fig. 3 Chromosomes represented as rotational elements

2.2.3 Applications

Jackson et al. (2013) introduce the use of diversified variants in compiler programs. This approach requires the use of a monitoring control center to execute the numerous diversified variants of the compiler based on whether or not an attack is detected. What is more, it introduces a second approach that is specific to home users that download network applications. In the latter approach, each software download would contain a unique set of diversified variants, which would create a dynamic application environment that could not be easily attacked because of the variations in the program's internal structure (Huang and Ghosh 2011). Diversifying software on a per-download basis could be a good starting point for heightening the security of users' home networks without requiring any effort from those users. Additional suggestions for diversifying compiler programs include diversifying the binary code, which was demonstrated empirically by random insertions of no-operation (NOP) instructions into a computer program (Huang and Ghosh 2011). Another MTD, which is achieved through software diversification, involves modifying the input variables based on a number of instruction-set parameters to generate a sequence of outputs with different characteristics (Jacob et al. 2008). This approach is computationally expensive, however, and negatively affects system performance. Another issue is that the individualized output sequence is limited to the set of instruction parameters used, which makes it difficult to use with other applications that have different sets of instructions.

Quite different from the aforementioned approaches is a new concept of application diversity. Called symbiotic defense mutualism, it was introduced by Cui and Stolfo (2011). The symbiotic relationship is between a diversifiable code (called a "symbiote" by the authors) and the application host into which it is injected. It monitors the application and executes defensive security measures to prevent unauthorized access to and/or modification of the application. Theoretically, using symbiotes to provide dynamic application security sounds good; practically, however, generating and maintaining unique symbiotes for each software application could be challenging and costly, especially for large networks.

Recently, the use of machine learning in proactive defense to defeat reverse engineered attacks has been proposed (Atighetchi et al. 2004). In this approach, diversity is achieved by creating random classifiers and specifying a time period for activating each identifier.

Dynamic Application Rotation Environment (DARE) MTD uses the two most common and freely available web servers, Apache and Nginx (see Fig. 4). It runs a single application on both platforms, redirecting incoming traffic to one server or the other at a random interval. The goal is to mitigate any unknown vulnerability in one of these platforms by reducing the amount of time that a platform is exposed to a would-be attacker. Like the MORE MTD strategy, this variability increases the cost of reconnaissance on a target and reduces the likelihood of exploiting any zero-day, or previously unknown, vulnerability (Thompson et al. 2016a).

DARE MTD takes incoming web traffic and redirects it to one of two ports: (1) nonstandard TCP port 81 for traffic served by Nginx or (2) nonstandard TCP port 82 for traffic served by Apache.

Carvalho et al. introduced the idea of using multiple, dynamic security mechanisms that incorporate both automated defense and human agent actions to make security decisions. When an attack is detected, human agents reconfigure the security

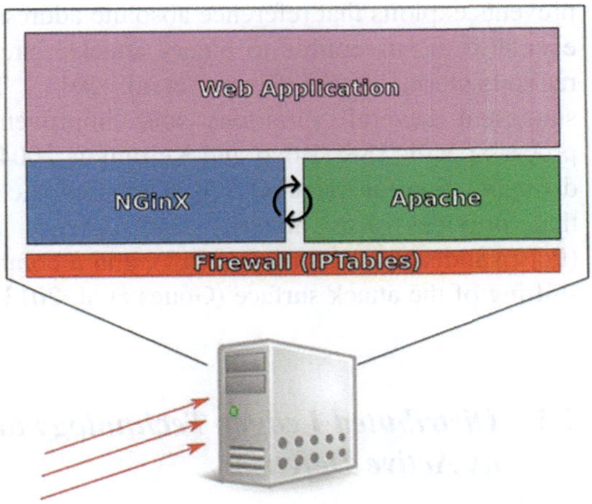

Fig. 4 DARE application stack

defense tools to prevent future attacks. Application diversity is achieved through a random redirection of file transfer protocol (FTP) requests (Carvalho et al. 2013). In a paper on application diversity, Azab and Eltoweissy propose using a dynamic platform that achieves diversity through periodic rotations of numerous functions, attributes, and code variants. The study shows that the platform could collect and analyze real-time data and consequently create network status reports that could be used to guide defensive strategy decisions (Azab et al. 2011).

Taguinod et al. present a web application that uses various rotating elements in its application space to improve defense at the application layer (Taguinod et al. 2015). In this methodology, they treat a web application as a series of layers and suggest applying moving target principles to each of these layers. A noted drawback of this method is that if any given layer is compromised, all layers above it become untrustworthy (Taguinod et al. 2015).

2.2.4 Processes and Processors

Another proactive application environment is processors. Numerous scholars (Kc et al. 2003; Boyd and Keromytis 2004; Portokalidis and Keromytis 2011) have investigated using code randomization techniques, such as instruction-set randomization (ISR), to defend against code-injection attacks. Relevant examples include applying active defense techniques by partitioning the address space of a system into multiple variants (Huang and Ghosh 2011). In this approach, when input is received from a user, it is copied to all the variants. When the variants receive unusual input, their behavior changes, allowing the built-in intrusion detection functionality to identify the anomaly and take mitigating actions. Each variant in the system is designed to have a unique memory space, which makes it difficult for an attacker to access any one of the available memory spaces stealthily and without causing detectable anomalies in the system's behavior. This solution successfully prevents exploits that reference absolute addresses (Huang and Ghosh 2011). However, ISR is susceptible to binary attacks, circumvention attacks, and incomplete randomization attacks (Evans et al. 2011). Therefore, subsequent works have suggested several variations and improvements. A few examples include processspecific ISR (Boyd and Keromytis 2004), structured query language (SQL) diversification for dynamic SQL databases (Kc et al. 2003), a global ISR technique that provides data integrity and defends against malicious binary attacks (Portokalidis and Keromytis 2011), and a combination of dynamic encryption and shifting of the attack surface (Goues et al. 2013).

2.3 Distributed Ledger Technology and Augmentations as Active Defense

2.3.1 Overview

Researchers, enthusiasts, and others have developed distributed ledger technology (DLT) including blockchain to secure transactions and records of information (Rauchs et al. 2018). The purpose of DLT is to ensure that records are immutable or at least that records are resistant to tampering. DLT may be considered a form of active defense because the nodes on the DLT network actively seek to come to a consensus on the true state of the distributed ledger despite an adversary that would attempt to hinder the communication or to send misleading information concerning the shared ledger which consists of the records or information. DLT is therefore an active defense for the integrity or immutability of the data in the ledger.

2.3.2 Basic Concept

These technologies have been used in such applications as cryptocurrency, Internet of Things (IoT), and many others (Farahani et al. 2021). DLT can be described as technology possessing certain characteristics:

> A ledger shared or copied among multiple nodes in a network. These nodes are generally individual computerized devices. The ledger contains a set of data, information, or transactions that are typically in a specific order, such as in chronological order.
>
> A consensus algorithm governs how new entries or blocks (a chunk of information) are added to the ledger. These algorithms include proof-of-work, proof-of-stake, and leader-based among others.

It should also be noted that DLT is a generic term that includes blockchain, but not all DLT is considered blockchain. Blockchain uses a specific data structure that involves links of blocks in a series, in which the blocks each store a certain amount of data that make up part of the ledger. The links typically involve hashes such that each successive block stores a hash of the previous block. If the previous block becomes corrupted, then the successive block would be able to compare its stored

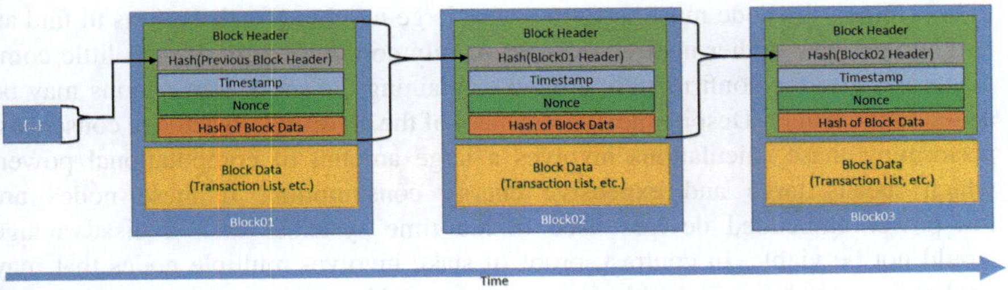

Fig. 5 Illustration of Blockchain with major components (NIST 2021)

hash with a rehash of the previous block. A hash in a one-way function that receives as input a message of varying size, and an output of finite size. A hash can be used to verify that the original message has not changed. A hash is analogous to a thumbprint used to identify a specific person but for a given set of data. If the original data is modified, creating a hash of the modified data will produce a different hash, which is an indication that the data changed. One standard hashing algorithm used in blockchain is SHA-256 (Alfrhan et al. 2021). The blocks themselves contain transactions, which in turn contain the information or data that is intended to be immutable. If the latest block becomes corrupted, it may be compared with the latest blocks of the other blocks of the network. Having multiple nodes in the network in which each node stores a copy of the ledger in the form of a data structure like a block chain ensures redundancy. This allows for resiliency in being able to secure and preserve the data in the face of failed, compromised, or malicious nodes. For certain consensus algorithms such as proof-of-work, a two thirds majority will establish what the true latest block and ledger should be. Figure 5 illustrates the structure of blockchain along with its various components and links. In the figure, for each block of the chain (Blue Box), the hash stored in the block header (grey box) is a hash of the previous block's header.

2.3.3 Consensus Algorithms

Since the introduction of blockchain, multiple consensus algorithms have been developed. The algorithms may have different advantages and disadvantages. It may be argued that the consensus algorithm itself constitutes a large part of the active defense for DLT. Proof-of-work consensus algorithms are studied in (Gervais et al. 2016), where the focus is on objectively comparing "the tradeoffs between their performance and security provisions." A recent survey of proof of stake algorithms can be found in (Nguyen et al. 2019). Both approaches are susceptible to attacks, and recent research has focused on thwarting attacks by implementing methods that require an adversary to achieve 100% network dominance (Akbar et al. 2021). Bitcoin (Nakamoto 2022), the first major cryptocurrency, uses a proof-of-work approach to consensus. Proof-of-work involves a given node, called a miner that arranges a set of transactions from the nodes of the blockchain network. For a node to be a miner, the node must perform a very large number of calculations to find an appropriate hash. Other nodes can subsequently confirm this hash with little computation. After the confirmation, a block containing the set of transactions may be added to the ledger. Despite the effectiveness of the algorithm to achieve consensus, performing these calculations involves a large amount of computational power, which means large and expensive energy consumption. If these nodes are low-power embedded devices used in real-time systems, such a disadvantage would not be viable. In contrast, proof of stake involves multiple nodes that may not have an equal vote in the blockchain but instead have a vote proportional to their "stake." For example, this stake may be the number of coins in some cryptocurrency. Other algorithms include those that are leader-based algorithms. One such algorithm is Raft, whose name is based on the phrase "Reliable, Replicated, Redundant, And Fault-Tolerant." Raft has the advantage of being lightweight computationally. But it could be more susceptible to a malicious cyberattack because its leader may become compromised. The leader is a node in the network whose purpose is to determine the transactions that will be included in the next block of the ledger (Ongaro and Ousterhout 2014).

2.3.4 Properties of DLT

Fault tolerance is a useful characteristic of DLT. This refers to the resilience of the DLT to maintain a shared ledger in the presence of one or more misbehaving or malfunctioning nodes in the network. There are several standards for fault tolerance, such as Byzantine fault tolerance (BFT) and crash fault tolerance (CFT). BFT is a high standard for DLT (Castro and Liskov 1999). The idea of BFT originates from the Byzantine generals problem and involves actively malicious nodes working to send deceptive communication or suppress communication. If the nodes of the network may be compromised, then BFT is recommended to use. Bitcoin uses a proof-of-work consensus algorithm that is Byzantine fault tolerant. Asynchronous BFT is an even higher standard than simply BFT. Generally, BFT assumes that messages from one node to another arrive within a finite period. Asynchronous BFT permits a larger or indefinite delay. CFT, which is not as stringent as DFT can tolerate one or more of the nodes are simply malfunctioning. In other words, one or more of the nodes have crashed, as the name CFT implies. Raft is an example of CFT.

An additional important property of DLT is whether the DLT is permissioned or permissionless. How the DLT employed, or the specific use case determines what is appropriate. Permissioned DLT is used when the set of nodes is known ahead of time. This is true with cyber-physical systems, such as power system or a group of drones with a fixed number of digital assets. New nodes must be explicitly added by a central authority for the DLT. Permissionless DLT is better to use when new nodes can be added without explicit permission from a central authority. An example of this is a cryptocurrency like Bitcoin or Ethereum.

2.3.5 Threats to DLT

There are certain threats that blockchain or DLT may face. These include routing attacks, sybil attacks, and 51% attacks. Routing attacks involve an attacker affecting the routing on the network, which ultimately affects the communication among nodes in the blockchain network. Sybil attacks involve malicious actors with false identities on the network that seek to perform Denial of Service (DoS) and Distributed Denial of Service (DDoS) attacks through volumetric flooding to block communication and render the blockchain network ineffective. A consensus majority attack, also known as 51% attack, involves a situation with blockchain that relies on the proof-of-work consensus algorithm, in which a number of miners together have more than 50% of the blockchain's networks computing power. In theory they can modify the content of the ledger. However, blockchain is largely resilient to bad actors who do intend to manipulate the ledger (Baninemeh et al. 2024).

2.3.6 Other Types of DLT Besides Blockchain

A relatively new DLT that is not like blockchain in its structure is hashgraph, which was invented by Leemon Baird. It involves multiple nodes connected by means of a network. Each of the nodes uses a gossip protocol to communicate with other nodes. Each node randomly selects another node to communicate with. The act of communication between the nodes is called a gossip event. In the communication, the two nodes share all the information they know from previous gossip events. From the information that each node updates through this communication, each node can build its own data structure of a hashgraph, which as the name implies has the structure of a graph. Of the various forms of DLT in this section, hashgraph has properties of interest including greater transactions per second, which may also be termed throughput.

Also, some surveys discuss these properties and others for various kinds of DLT and blockchain. For example, Fu discusses various blockchain technologies that use these algorithms and their advantages and disadvantages (Fu et al. 2020). Sankar surveys several consensus algorithms, such as Stellar Consensus Protocol, Corda, and Hyperledger Fabric (Sankar et al. 2017).

2.3.7 Augmentations to DLT for Cyber Defense

Besides the underlying nature of DLT and how it seeks to preserve its data despite faults or malicious entities, researchers have also developed technologies to augment DLT or have used DLT combined with other technology. These technologies serve as an active defense. The recent literature describes many of these: Liang et al. describe a cross-layer approach to prevent attacks that tamper with data (Liang et al. 2022). Xiaomeng Xu et al. uses smart contracts as part of a new security system for verification to prevent risky transactions. The system can use an off-chain environment to determine the execution of on-chain smart contracts to prevent risky transactions and protect assets for web3. Wang et al. propose an active defense for blockchain to thwart forking. In blockchain, forking from the main chain can occur when malicious nodes disagree with other nodes in the network.

3 Discussion of Proactive Techniques

3.1 Costs and Complexity

A nontrivial problem with active defense strategies, especially moving target defenses utilizing randomness and diversity, is quantifying and understanding the costs and benefits of these strategies. Cybersecurity expenses are difficult to justify to decision-makers in the face of competition, and even insurers have difficulties quantifying risks and benefits of individual approaches. Researchers have yet to coalesce around a single approach to quantification of either the complexity or the security improvements of MTD systems; however, many approaches are actively in use as of this writing.

3.1.1 System Complexity

"Complexity is the enemy of security" is a widely held belief in the security community (Geer 2008; Holstein 2009; Chan 2012; Krebs 2014; Hill 2007). Thus, it is not surprising that many security experts view a research area that aims to increase security by way of introducing complexity with skepticism. The complexity of active defense systems is problematic from a quantification standpoint (see Sect. 4) and from a deployment and maintenance standpoint (see Sects. 3.2 and 3.3). While introducing complexity increases costs for the defender, it holds the promise of reducing or eliminating the asymmetric advantages of the attacker. This makes complexity a necessary component of proactive and moving target defenses (Holstein 2009).

There have been many attempts to manage complexity. Hamlet and Lamb (2016) take the approach of first identifying dependencies at all points in a system and then performing cost analyses on those dependencies. By identifying what each portion of the system (users, tools, networks, etc.) needs in order to perform and to succeed, they propose a way to choose a subset of defenses and "minimally disrupt users while maximally impacting attackers." They formalize this notion by representing these dependencies in the form of a graph. Using a use case for one of the largely successful and adopted Moving Target Defenses (address space layout randomization), they identify various metrics not only for the users of the system but also for the potential attackers to the system. Using a scoring system and then comparing the changes in adversary-to-user cost of implementing a Moving Target Defense allows administrators to evaluate the effectiveness and the cost of choosing a certain MTD over another (Clark et al. 2013).

Taylor et al. (2016) propose two “task sets” that are necessary to be analyzed when evaluating the cost/benefit of installing a Moving Target Defense. The first of these sets is the “adversary activity” task space, which characterizes the defensive benefit of creating a Moving Target Defense instance, while the “mission activity” task space evaluates the impact that an MTD instance could potentially have on the overall system and on the cost of maintenance and development of that instance. In addition, they identify four areas of study for MTD: *integrity*, *confidentiality*, *success, and productivity*, within which both the attackers and defenders have objectives that affect which possible MTD will be chosen, if any.

3.1.2 Costs of Deployment

Enterprises tend to be prepared to invest more in upfront deployment costs if they have some quantifiable, demonstrable benefit to long-term safety and security. Since the majority of proactive and MTD solutions are currently either theoretical or in prototype stages, quantifying the cost of their deployment is crucial to the process of moving them from the research world into production environments.

Choudhury et al. (2015) attempt to create a way to measure the “defensive work factor,” that is, required for Moving Target Defense instances. Focusing mainly on the cost of deployment, with mention to studying impact on implementation, operation, and effectiveness, the authors identify areas that MTD deployments may influence. For example, in this paper, they primarily focus upon the system performance impacts of MTDs.

Their approach to analyzing system performance begins with baseline measurements, using application performance monitoring (APM) and network performance monitoring tools (NPM) on an instance of a system that is NOT running MTD. Finally, they perform the same performance test on an instance of the same system operating an MTD. To finish this paper, they perform the analyses described above on a virtualized system and report on the results (Van Leeuwen et al. 2015).

3.2 *Effectiveness of Proactive/MTD Methods*

Security has become a fundamental part of day-to-day operations for most public and private organizations as well as for most private individuals. The effectiveness of network security tools thus demands considerable attention from both the academic and the industrial sectors. Active defense strategies that involve diversity are often criticized for increasing the attack surface in a way that could expose more vulnerabilities. Proponents of these techniques cite the diverse attack surfaces’ ability to increase attacker uncertainty and reduce the asymmetric advantage as the counterbalance to any potential loss in security due to the increase in the attack surface. Clearly, this debate requires metrics and quantification to assist in establishing the mission value and appropriateness of these technologies.

Das et al. (2016) discuss the effectiveness of their framework called GuardOL, a real-time malware detection architecture. GuardOL utilizes the patterns and features of known legitimate and malicious software to create a model, and a "trained classifier" identifies each incoming sample as benign software or malware based on the model. Testing and evaluation of the system demonstrate a high classification success rate of 97%, rapid classification, energy efficiency (2.0762 mW consumption), and adaptability to ever-changing types of malware.

Proactive reset is an active defense technology by which systems are reset automatically and periodically in order to negate any malware that may or may not have infiltrated. Thompson et al. (2016b) assess the effectiveness of three particular proactive reset techniques on the integrity of autonomous systems: random reset, communication-based, and risk flow. Random reset, as the name suggests, resets each system at random intervals; the communication-based policy resets nodes based upon their communication flow, prioritizing the reset of nodes with relatively high communication; the risk flow policy assigns a "risk score" to each node and prioritizes the reset of nodes with relatively high scores. From an assessment of agent-based simulation, the results indicate that risk flow reset identifies and eliminates infection comparably to the performance of a perfect detector.

Several examples of MTD solutions display types of diversity that affect attackers without affecting users. MORE MTD describes a version of a multiple operating system rotational environment that uses virtual hosts rotating in a random fashion (Huang and Ghosh 2011). From an outside interface, the user sees a normal application as if it were being served on a traditional, static server. However, an attacker scanning the same system for vulnerabilities would see a constantly shifting platform. Although any vulnerabilities in the application itself would still exist, any zero-day vulnerabilities in the underlying systems would become difficult, if not impossible, to exploit due to the duration of their individual exposure windows.

The work of Evans et al. contains a thorough discussion of the effectiveness of MTD solutions against specific types of attacks (Evans et al. 2011). The authors determine that ISR and space-address-randomization defense mechanisms are vulnerable to circumvention attacks, deputy attacks, brute force, entropy reduction, and probing attacks (Evans et al. 2011). More specifically, the study concludes that dynamic defense is not effective in thwarting circumvention and deputy attacks, but it offers a modest to high success rate in defending against probing and incremental attacks (Evans et al. 2011). Note that the findings assume that there was a single diversity defense. Thus, using multiple diversity defenses might harden security by increasing the time, cost, and effort it would take to launch an attack.

Zhuang et al. (2012) developed an MTD prototype and evaluated its effectiveness in preventing malicious attacks. They assessed effectiveness by first simulating a real attack, then recording the probability of a successful attack. The results showed that increasing the frequency of reconfigurations greatly reduced the likelihood of successful attacks (Zhuang et al. 2012). Goues et al. (2013) recently confirmed this conclusion. Unlike earlier results found in Huang and Ghosh (2011), those found in Zhuang et al. (2012) further confirmed that dynamic defense offers better security against attacks than static defense does. In a similar vein, other scholars have demonstrated that proactive network defense is better at detecting malicious network activity than are standard detection approaches (Colbaugh and Glass 2013; Ali et al. 2013) and that improved database security is offered by SQL diversification (Christodorescu et al. 2011). In addition, the previously discussed N-variant technique shows effectiveness in thwarting memory corruption exploitations (Cox et al. 2006). As far as performance, Jackson et al. (Shakarian et al. 2014) made a significant discovery regarding the trade-off between improved security and system performance; specifically, they concluded that program diversity was subject to performance overhead.

There have been many attempts to create formal methods for studying and measuring the security effectiveness of Moving Target Defenses. Using current tools common for analyzing the vulnerability and the security of systems called attack representation models, the authors of (Hong and Kim 2014) abstract these tools for use in MTD environments. Using hierarchical attack representation models (HARM), the authors propose a better way to analyze large networks by segmenting areas of vulnerability and reachability. They can then combine separate attack graphs and trees to form a better representation of the entire network's security at a broad scale. Lastly, they propose using importance measures (IM) to guide the deployment of MTD techniques in "scalable and effective manners." They classify different types of MTD into *shuffle*, *diversity*, and *redundancy*. They perform HARM analyses on both the shuffle and diversity approaches to MTD and report on their findings as the bulk contribution of their paper. As a final contribution, they perform HARM analyses, once more taking into account the addition of IM. They implement MTD instances on these IMs and then perform HARM analyses.

Two analytical models developed in the work of Anderson et al. (2016) for comparing and measuring the effectiveness of MTD show promise. *Closed-form mathematical model* proposes an equation for calculating success of cyberattacks based on exploit availability, probability of implant detection, number of MTD configurations, attack time, and churn rate. Ultimately, this model attempts to combine many different factors that exist in an active attack situation and arrive at a conclusive data point that relates these factors to a success rate. The contribution in this model is the addition of the parameter of MTD configurations in relation to the existing factors of a system. They evaluate the current state of the MTD and the current position of the attacker in the attack chain. This model also takes into account the churn rate of the configurations as well as the ability of the defending host to detect the attacker.

Though multiple techniques for MTD and strategies for evaluation exist, a disconnect currently exists between quantification and implementation. Several efforts exist in both the public and private sector to experiment with and quantify the benefits and costs of MTD (Atighetchi et al. 2014; Soule et al. 2015; Eskridge et al. 2015; Cai et al. 2016). Most of these efforts would benefit from greater collaboration between researchers building MTD and those building frameworks for quantification and testing.

3.3 Resilience of Proactive/MTD Methods

Resilience within computer systems is generally taken to mean dependability and performance in the face of unanticipated changes (Meyer 2009). In a cyber-system or a cyber-physical system, our goal is to increase robustness while reducing brittleness. Figure 6 shows an idealized response curve of both a resilient system and a non-resilient system to demonstrate these properties of robustness and brittleness.

Active defense solutions, and particularly MTD solutions, often have high resilience as a necessary by-product of introducing diversity and redundancy into a platform and/or software system.

An example of resilience properties in an MTD in action, MORE MTD increases an application's resilience threshold, as described by Rieger (2010), by reducing the rotation window (Cyber Operations, Analysis, and Research 2017). Though this does not affect the brittleness of any individual OS, the application grows in overall resilience by residing on multiple platforms. The authors improve resilience through rotation of OSs with multiple different vulnerability footprints. They improve agility and robustness through increasing uncertainty and reacting to failure. For the sake of a security system, this means both attempting to prevent unanticipated execution of code and recovering sanely from any such unanticipated execution.

Moving target command and control (MTC2) presents an approach to resilience that relies on software agents to assure graceful, robust, and adaptive performance in the face of stressors and surprise (Hong and Kim 2014). The authors' notion of

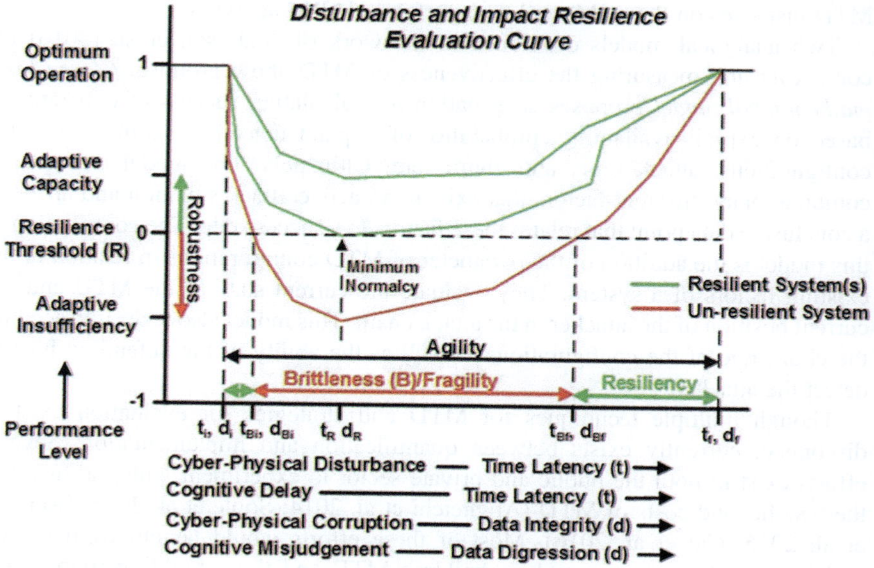

Fig. 6 Idaho National Laboratory, resilience of a cyber-physical system

organic resilience focuses on avoiding static and centralized single-point-of-failure solutions in system organization (Carvalho et al. 2010).

Stream splitting MTD also presents a potent example not only in introducing diverse channels for communications (via multiple communication mediums) but also by consistently testing redundant channels. In disaster situations, failure of backup systems after attempted failover from primary systems is typically traced to lack of testing of both the failover mechanism and the redundant system (Wong et al. 2010). Stream splitting MTD allows the operator to be continuously testing redundant channels and includes failover as part of the normal operating protocol (Cyber Operations, Analysis, and Research 2017).

While not directly related to failure and reaction to failure, work by Wang et al. studies the "MTD Timing Problem," which is closely related to the resiliency of MTD instances (Wang et al. 2016). In this study, they try to answer the question of "when is it the best time to make economical adaptations" by examining two "cost areas": attacked cost and adaptation cost. The main contribution of this paper is an algorithm that takes into account these two parameters and generates cost rate based upon various intervals of adaptations (changes to the system). For their study, they utilize various DDoS attack traces from various IP ranges across nearly a 6-month timespan.

Another algorithm, presented in the work of Choudhury et al. (2015), creates recommendations for an organization to achieve cyber resilience. The model uses a graph-theoretic approach that simulates the infrastructures and patterns of an organization as well as a wide range of cyberattacks with potentially damaging effects to its quality of service (QoS). For example, an algorithm that is generated from a modeled DoS attack recommends disabling the attacker from creating any more requests to the web application. Such action mitigates disruption of e-mail communication and of any other web-server traffic, thus maintaining QoS and, in turn, resilience.

In many cases, resilience has been a welcome by-product of the design of proactive defense systems, while in some cases authors and researchers have set out with resilience as an explicit design goal. In either case, it is clear that this area of research continues to be an important evolution in the cybersecurity toolkit and that more active s and MTD systems will lead to both more resilience challenges and opportunities for system designers, engineers, and operators.

4 Conclusion

Active defense consists of a diverse variety of techniques and strategies, all focused on protecting an organization's information or control infrastructure against adversaries. Some of the techniques are mature and ready to deploy, while others require much more research. Some operate automatically while others require varying amounts of human attention and involvement. Some rely almost entirely on skilled cybersecurity professionals or masters of the art of deception.

To benefit from active defense, an organization needs to assess their infrastructure, personnel, and threats then consider whether they can apply the various types of active defense methods. Most organizations can employ activated protective measures, such as active blocking and real-time anti-malware. Organizations with static information infrastructures might be able to exploit automated technologies such as moving target defense. Those with dynamic infrastructures might consider other methods such as dynamic networking. Organizations with experienced personnel might consider actively deceiving unauthorized intruders by deploying and monitoring honeypots or patrolling their infrastructure to hunt for threats.

Organizations that understand their public perception, adversaries, and threats might attempt to deceive their potential attackers. This requires manipulating publicly available information to lure attackers to behave in expected ways. This also requires monitoring the infrastructure to detect the expected, and perhaps unexpected, behavior of the deceived attackers.

Researchers continue to conceive and develop strategies and technologies to defend computers and networks. As the research matures, organizations may find new methods that fit into their information or control infrastructure to help them actively defend themselves against current and future adversaries.

References

Akbar NA, Muneer A, ElHakim N, Fati SM (2021) Distributed hybrid double-spending attack prevention mechanism for proof-of-work and proof-of-stake Blockchain consensuses. Future Internet 13(11):Art. no. 11. https://doi.org/10.3390/fi13110285

Albanese M, Benedictis AD, Jajodia S, Sun K (2013) A moving target defense mechanism for Manets based on identity virtualization. In: Proceedings of the first IEEE conference on communications and network security (CNS 2013). Washington, DC

Alfrhan A, Moulahi T, Alabdulatif A (2021) Comparative study on hash functions for lightweight blockchain in Internet of Things (IoT). Blockchain: Res Appl 2(4):100036

Ali MQ, Al-Shaer E, Duan Q (2013) Randomizing AMI configuration for proactive defense in smart grid. In: IEEE international conference on smart grid communications, Vancouver, BC

Al-Shaer E (2011) Toward network configuration randomization for moving target defense. In: Jajodia S et al (eds) Moving target defense, vol 54. Springer, New York, pp 153–159

Al-Shaer E, Duan Q, Jafarian J (2013) Random host mutation for moving target defense. In: Keromytis A, Pietro R (eds) Security and privacy in communication networks, vol 106. Springer, Berlin/Heidelberg, pp 310–327

Anderson N, Mitchell R, Chen IR (2016) Parameterizing moving target defenses. In: 2016 8th IFIP international conference on new technologies, mobility and security (NTMS). Larnaca, pp 1–6

Araujo F, Hamlen KW, Biedermann S, Katzenbeisser S (2014) From patches to honey- patches: lightweight attacker misdirection, deception, and disinformation. In: Proceedings of the 2014 ACM SIGSAC conference on computer and communications security (CCS '14). ACM, New York, pp 942–953

Atighetchi M, Pal P, Webber F, Schantz R, Jones C, Loyall J (2004) Adaptive cyberdefense for survival and intrusion tolerance. IEEE Internet Comput 8(6):25–33

Atighetchi M, Soule N, Watro R, Loyall J (2014) The concept of attack surface reasoning. In: The third international conference on intelligent systems and applications (Intelli 2014), pp 39–42

Azab M, Hassan R, Eltoweissy M (2011) ChameleonSoft: a moving target defense system. In: 2011 7th international conference on collaborative computing: networking, applications and Worksharing (CollaborateCom), pp 241–250

Badr Y, Hariri S, Al-Nashif Y, Blasch E (2015) Resilient and trustworthy dynamic data- driven application systems (DDDAS) services for crisis management environments. Procedia Comput Sci 51:2623–2637

Baninemeh E, Jansen S, Labunets K (2024) A security risk assessment method for Distributed Ledger Technology (DLT) based applications: three industry case studies. arXiv preprint arXiv:2401.12358

Beraud P, Cruz A, Hassell S, Meadows S (2011) Using cyber maneuver to improve network resiliency. In: 2011 military communications conference (MILCOM), pp 1121–1126

Boyd S, Keromytis A (2004) SQLrand: preventing SQL injection attacks. In: Jakobsson M et al (eds) Applied cryptography and network security, vol 3089. Springer, Berlin/Heidelberg, pp 292–302

Cai G, Wang B, Luo Y, Li S, Wang X (2016) Characterizing the running patterns of moving target defense mechanisms. In: *2016* 18th international conference on advanced communication technology (ICACT). IEEE, PyeongChang, pp 191–196

Carvalho M, Lamkin T, Perez C (2010) Organic resilience for tactical environments. In: 5th international ICST conference on bio-inspired models of network, information, and computing systems (Bionetics), Boston

Carvalho M, Eskridge TC, Bunch L, Dalton A, Hoffman R, Bradshaw JM, Shanklin T (2013) MTC2: a command and control framework for moving target defense and cyber resilience. In: 2013 6th international symposium on resilient control systems

Casola V, Benedictis AD, Albanese M (2013) A moving target defense approach for protecting resource-constrained distributed devices. In: Proceedings of the 14th international conference on information reuse and integration (IEEE IRI 2013), San Francisco, CA

Castro M, Liskov B (1999) Practical byzantine fault tolerance. In: Proceedings of the third USENIX symposium on Operating Systems Design and Implementation (OSDI), New Orleans, Louisiana, USA, February 22–25, 1999, pp 173–186. [Online]. Available: https://dl.acm.org/citation.cfm?id=296824

Chan CS (2012) Complexity the worst enemy of security. Available at https://www.schneier. com/news/archives/2012/12/complexity_the_worst.html

Chiang CJ et al (2016) ACyDS: an adaptive cyber deception system. In: Military Communications Conference (MILCOM) 2016 IEEE, pp 800–805

Choudhury S et al (2015) Action recommendation for cyber resilience. In: 2015 workshop on automated decision making for active cyber defense, Denver, pp 3–8

Christodorescu M, Fredrikson M, Jha S, Giffin J (2011) End-to-end software diversification of internet services. In: Jajodia S et al (eds) Moving target defense, vol 54. Springer, New York, pp 117–130

Clark A, Sun K, Poovendran R (2013) Effectiveness of IP address randomization in decoy-based moving target defense. In: 2013 I.E. 52nd annual conference on decision and control

Colbaugh R, Glass K (2013) Moving target defense for adaptive adversaries. In: 2013 I.E. International Conference on Intelligence and Security Informatics, Florence

Cox B, Evans D, Filipi A, Rowanhill J, Hu W, Davidson J, Knight J, Nguyen-Tuong A, Hiser J (2006) N-variant systems: a secretless framework for security through diversity. In: Defense technical information center. USENIX-SS'06 proceedings of the 15th conference on USENIX security symposium, vol 15. Vancouver, p 9

Crouse M, Fulp E, Canas D (2012) Improving the diversity defense of genetic algorithm-based moving target approaches. In: Proceedings of moving target research symposium

Cui A, Stolfo S (2011) Symbiotes and defensive mutualism: Moving target defense. In: Jajodia S et al (eds) *Moving target defense*, vol 54. Springer, New York, pp 99–108

Curado M, Madeira H, Rupino P, Cabral B, Abreu DP, Barata J, Roque L, Immich R (2017) Next generation cyber-physical systems: towards resilient software and internet services. In: Cyber resilience. Centre for Informatics and Systems Department of Informatics, Engineering University of Coimbra, p 100

Cyber Operations, Analysis, and Research (2017) Argonne national lab, moving target defense. Available at https://coar.risc.anl.gov/research/moving-target-defense

Das S et al (2016) Semantics-based online malware detection: towards efficient real-time protection against malware. IEEE Trans Inf Forensics Secur 11(2):289–302

DLP and Honeytokens (2003) Augusto Paes de Barros. Available at http://blog.securitybalance.com/2007_08_01_archive.html

Dunlop M, Groat S, Urbanski W, Marchany R, Tront J (2011) MT6D: a moving target IPv6 defense. In: 2011 military communications conference (MILCOM), Baltimore, pp 1321–1326

Elliott KA (2018) Active cyber defense and attribution in cyber attacks (Master's thesis, Utica College)

Eric J (2001) Holdaway: active computer network defense: an assessment. Maxwell Air Force Base, Alabama. Available at www.iwar.org.uk/iwar/resources/usaf/maxwell/students/ 2001/01-055.pdf

Eskridge TC, Carvalho MM, Stoner E, Toggweiler T, Granados A (2015) VINE: a cyber emulation environment for MTD experimentation. In: Proceedings of the second ACM workshop on moving target defense. ACM, pp 43–47

Evans D, Nguyen-Tuong A, Knight J (2011) Effectiveness of moving target defenses. In: Jajodia S et al (eds) Moving target defense, vol 54. Springer, New York, pp 29–48

Farahani B, Firouzi F, Luecking M (2021) The convergence of IoT and distributed ledger technologies (DLT): opportunities, challenges, and solutions. J Netw Comput Appl 177:102936. https://doi.org/10.1016/j.jnca.2020.102936

Franco J, Aris A, Canberk B, Uluagac AS (2021) A survey of honeypots and honeynets for internet of things, industrial internet of things, and cyber-physical systems. IEEE Commun Surv Tutor 23(4):2351–2383

Fu X, Wang H, Shi P (2020) A survey of Blockchain consensus algorithms: mechanism, design and applications. Sci China Inf Sci 64(2):121101. https://doi.org/10.1007/s11432-019-2790-1

Gabrys R, Silva D, Bilinski M (2024) HoneyGAN pots: a deep learning approach for generating honeypots. arXiv preprint arXiv:2407.07292

Ge M, Cho JH, Kim D, Dixit G, Chen IR (2021) Proactive defense for internet-of-things: moving target defense with cyberdeception. ACM Trans Internet Technol 22(1):1–31

Geer DE (2008) Complexity is the enemy. IEEE Secur Priv 6(6):88–88

Gervais A, Karame GO, Wüst K, Glykantzis V, Ritzdorf H, Capkun S (2016) On the security and performance of proof of work blockchains. In: Proceedings of the 2016 ACM SIGSAC conference on Computer and Communications Security, New York, pp 3–16. https://doi.org/10.1145/2976749.2978341

Goues C, Nguyen-Tuong A, Chen H, Davidson J, Forrest S, Hiser J, Knight J, Gundy M (2013) Moving target defenses in the helix self-regenerative architecture. In: Jajodia S et al (eds) Moving target defense II, vol 100. Springer, New York, pp 117–149

Groat S, Dunlop M, Marchany R, Tront J (2011) Using dynamic addressing for a moving target defense. In: Proceedings of the 6th international conference on information warfare and security. Academic Conferences Limited, Washington, DC, p 84

Groat S, Dunlop M, Urbanksi W, Marchany R, Tront J (2012) Using an IPv6 moving target defense to protect the smart grid. In: 2012 I.E. power & energy society general meeting, Innovative Smart Grid Technologies (ISGT), San Diego, pp 1–7

Hamlet JR, Lamb CC (2016) Dependency graph analysis and moving target defense selection. In: Proceedings of the 2016 ACM workshop on Moving Target Defense (MTD '16). New York, ACM, pp 105–116

Han W, Zhao Z, Doupé A, Ahn G-J (2016) HoneyMix: toward SDN-based intelligent honeynet. In: Proceedings of the 2016 ACM international workshop on security in Software Defined Networks & Network Function Virtualization (SDN-NFV security '16). ACM, New York, pp 1–6

Hardman O, Groat S, Marchany R et al (2013) Optimizing a network layer moving target defense for specific system architectures. In: Proceedings of the ninth ACM/IEEE symposium on architectures for networking and communications systems. IEEE Press, pp 117–118
Heckman KE, Stech FJ, Schmoker BS, Thomas RK (2015) Denial and deception in cyber defense. Computer 48:36–44. https://doi.org/10.1109/MC.2015.104
Heydari V, Yoo SM (2016) Securing critical infrastructure by moving target defense. In: 11th International Conference on Cyber Warfare and Security (ICCWS 2016), Boston
Hill B (2007) Complexity as the enemy of security. In: W3C Workshop on Next Steps for XML Signature and Encryption 25/25 September
Himma, K. E., & Dittrich, D. (2005). Active response to computer intrusions. Available at https://ssrn.com/abstract¼790585
Holstein DK (2009) A systems dynamics view of security assurance issues: the curse of complexity and avoiding chaos. In: 2009 42nd Hawaii international conference on system sciences. Big Island, pp 1–9
Hong JB, Kim DS (2014) Scalable security models for assessing effectiveness of moving target defenses. In: 2014 44th annual IEEE/IFIP international conference on dependable systems and networks, Atlanta, pp 515–526
Huang Y, Ghosh A (2011) Introducing diversity and uncertainty to create moving attack surfaces for web services. In: Jajodia S et al (eds) Moving target defense, vol 54. Springer, New York, pp 131–151
Ishikawa T, Sakurai K (2017) Parameter manipulation attack prevention and detection by using web application deception proxy. In: Eleventh international conference on ubiquitous IMCOM 2017, Beppu
Jackson T, Homescu A, Crane S, Larsen P, Brunthaler S, Franz M (2013) Diversifying the software stack using randomized NOP insertion. In: Jajodia S et al (eds) Moving target defense II, vol 100. Springer, New York, pp 151–173
Jacob M, Jakubowski MH, Naldurg P, Saw CWN, Venkatesan R (2008) The superdiversifier: peephole individualization for software protection. In: Advances in information and computer security. Springer, New York, pp 100–120
Karsai G, Koutsoukos X, Neema H, Volgyesi P, Sztipanovitz J (2017) Simulation-based analysis of cyber resilience in cyber-physical systems. In: Cyber Resilience, p 131
Kc GS, Keromytis AD, Prevelakis V (2003) Countering code-injection attacks with instruction-set randomization. In: Proceedings of the 10th ACM conference on computer and communications security (Washington, DC, October 27–30, 2003). ACM, New York, pp 272–280
Kewley D, Fink R, Lowry J, Dean M (2001) Dynamic approaches to thwart adversary intelligence gathering. In: 2001 DARPA information survivability conference & exposition II. DISCEX '01 proceedings, vol 1. Anaheim, pp 176–185
Krebs B (2014) Complexity as the enemy of security. Available at https://krebsonsecurity.com/2014/05/complexity-as-the-enemy-of-security/
Leyi S, Chunfu J, Shuwang L (2008) Full service hopping for proactive cyber- defense. In: 2008 I. E. international conference on networking, sensing and control (ICNSC), Sanya, pp 1337–1342
Liang H, Li Zhu F, Richard Y, Wang X (2022) A cross-layer defense method for blockchain empowered CBTC systems against data tampering attacks. IEEE Trans Intell Transp Syst 24(1): 501–515
MacFarland DC, Shue CA (2015) The SDN shuffle: creating a moving-target defense using host-based software-defined networking. In: Proc of MTD '15, pp 37–41
Maesschalck S, Fantom W, Giotsas V, Race N (2024) These aren't the PLCs you're looking for: obfuscating PLCs to mimic honeypots. IEEE Trans Netw Serv Manag 21:3623–3635
Meyer JF (2009) Defining and evaluating resilience: a performability perspective. In: Presentation at international workshop on performability modeling of computer and communication systems
Microsoft (2017) Early launch antimalware. Available at https://msdn.microsoft.com/ windows/compatibility/early-launch-antimalware

Nakamoto S (2022) Bitcoin: a peer-to-peer electronic cash system. Available online at https://bitcoin.org/bitcoin.pdf. Accessed 13 Dec 2022

National Institute of Standards and Technology (2016) NIST special publication 800–184. Guide for cybersecurity event recovery. Available at http://nvlpubs.nist.gov/ nistpubs/SpecialPublications/NIST.SP.800-184.pdf. Accessed 27 July 2017

Nguyen CT, Hoang DT, Nguyen DN, Niyato D, Nguyen HT, Dutkiewicz E (2019) Proof-of-stake consensus mechanisms for future blockchain networks: fundamentals, applications and opportunities. IEEE Access 7:85727–85745. https://doi.org/10.1109/ACCESS.2019.2925010

Nguyen-Tuong A, Evans D, Knight JC, Cox B, Davidson JW (2008) Security through redundant data diversity. In: IEEE international conference on dependable systems and networks with FTCS and DCC. Anchorage, pp 187–196

NIST (2021) Blockchain. *National Institute for Standards and Technology*. https://www.nist.gov/blockchain

Okhravi H, Hobson T, Bigelow D, Streilein W (2014) Finding focus in the blur of moving-target techniques. Secur Priv 12(2):16–26. https://doi.org/10.1109/MSP.2013.137

Ongaro D, Ousterhout J (2014) In search of an understandable consensus algorithm. In: 2014 USENIX annual technical conference (USENIX ATC 14), pp 305–319

Paasch C, Bonaventure O (2014) Multipath TCP. Queue 12(2):40. 12 pages

Phatak DS (2005) Spread-identity mechanisms for DOS resilience and security. In: First international conference on security and privacy for emerging areas in communications networks, Athens, pp 23–34

Phatak DS, Sherman AT, Joshi N, Sonawane B, Relan VG, Dawalbhakta A (2013) Spread identity: a new dynamic address remapping mechanism for anonymity and DDoS defense. J Comput Secur 21(2):233–281

Portokalidis G, Keromytis A (2011) Global ISR: toward a comprehensive defense against unauthorized code execution. In: Jajodia S et al (eds) Moving target defense, vol 54. Springer, New York, pp 49–76

Rauchs M et al (2018) Distributed ledger technology systems: a conceptual framework, Rochester. https://doi.org/10.2139/ssrn.3230013

Rieger CG (2010) Notional examples and benchmark aspects of a resilient control system. In: 3rd international symposium on resilient control systems, pp 64–71

Roeder T, Schneider FB (2010) Proactive obfuscation. ACM Trans Comput Syst 28(2):4–54

Sankar LS, Sindhu M, Sethumadhavan M (2017) Survey of consensus protocols on blockchain applications. In: 2017 4th international conference on Advanced Computing and Communication Systems (ICACCS), pp 1–5. https://doi.org/10.1109/ICACCS.2017.8014672

Shakarian P, Kulkarni N, Albanese M, Jajodia S (2014) Keeping intruders at bay: a graph-theoretic approach to reducing the probability of successful network intrusions. In: International conference on E-business and telecommunications, Vienna

Soule N, Simidchieva B, Yaman F, Watro R, Loyall J, Atighetchi M, Carvalho M, Last D, Myers D, Flatley B (2015) Quantifying & minimizing attack surfaces containing moving target defenses. In: Resilience week (RWS), 2015. IEEE, pp 1–6

Spitzner L (2003) Definitions and value of honeypots. Available at http://www.tracking-hackers.com/papers/honeypots.html

Taguinod M, Doupé A, Zhao Z, Ahn GJ (2015) Toward a moving target defense for web applications. In: Information Reuse and Integration (IRI)

Taylor J, Zaffarano K, Koller B, Bancroft C, Syversen J (2016) Automated effectiveness evaluation of moving target defenses: metrics for missions and attacks. In: *Proceedings of the 2016 ACM workshop on moving target defense (MTD '16)*. ACM, New York, pp 129–134

The Honeynet Project (2003) Trapping the hackers: IEEE security and privacy 1, 2 (March 2003), pp 15–23. Available at: https://doi.org/10.1109/MSECP.2003.1193207

Thompson M, Kisekka V, Evans N (2014) Multiple OS rotational environment: an implemented moving target defense. In: 2014 seventh ISRCS. 7th international symposium on resilient control systems (ISRCS) 2014, Denver, pp 1–6

Thompson M, Mendolla M, Muggler M, Ike M (2016a) Dynamic application rotation environment for moving target defense. In: 2016 resilience week (RWS), Chicago, pp 17–26

Thompson B, Morris-King J, Cam H (2016b) Effectiveness of proactive reset for mitigating impact of stealthy attacks on networks of autonomous systems. In: *2016 I.E. Conference on Communications and Network Security (CNS)*. Philadelphia, pp 437–441

US Department of Homeland Security (2011) Moving target defense. Available at https://www.dhs.gov/science-and-technology/csd-mtd

Van Leeuwen B, Stout WMS, Urias V (2015) Operational cost of deploying moving target defenses defensive work factors. In: MILCOM 2015–2015 I.E. military communications conference. Tampa, pp 966–971

Verma A (2003) Production honeypots: an organization's view. SANS Security Essentials

Wang L, Zhang M, Jajodia S, Singhal A, Albanese M (2014) Modeling network diversity for evaluating the robustness of networks against zero-day attacks. In: European symposium on research in computer security

Wang H, Li F, Chen S (2016) Towards cost-effective moving target defense against DDoS and covert channel attacks. In: Proceedings of the 2016 ACM workshop on moving target defense (MTD '16). ACM, New York, pp 15–25

Watson D, Riden J (2008) The honeynet project: data collection tools, infrastructure, archives and analysis. In: WOMBAT workshop on information security threats data collection and sharing, pp 24–30

Wong WE, Debroy V, Surampudi A, Kim H, Siok MF (2010) Recent catastrophic accidents: investigating how software was responsible. In: 2010 fourth international conference on secure software integration and reliability improvement, Singapore, pp 14–22

Yackoski J, Bullen H, Yu X, Li J (2013) Applying self-shielding dynamics to the network architecture. In: Jajodia S et al (eds) Moving target defense II, vol 100. Springer, New York, pp 97–115

Zank A (2012) Moving target defense. Coronado group. June 18. Available at http://www.coronadogroup.com/images/Moving-Target-Defense-Coronado.pdf

Zhu M, Hu Z, Liu P (2014) Reinforcement learning algorithms for adaptive cyber defense against Heartbleed. In: Proceedings of the first ACM workshop on moving target defense

Zhuang R, Zhang S, DeLoach SA, Ou X, Singhal A (2012) Simulation based approaches to studying effectiveness of moving-target network defense. In: National symposium on moving target research

Economic Resilience to Cyber Threats

Adam Rose, Alexander Mermelstein, and Noah Miller

The term *cyber* refers to the computers, information technology, virtual reality, artificial intelligence, etc., linked to electronic communication networks, primarily the internet (Merriam-Webster 2015; OED 2024). This domain has seen a phenomenal rise in its role in advanced economies, and, more recently, even in developing ones. Disruptions of the cyber system can emanate from a variety of sources, such as malicious actors, natural disasters, and technological accidents. They can impact businesses in terms of their reputation, production levels, and profits; critical facilities the quantity and quality of their services; civilian government agencies providing services and maintaining order; the military in carrying out their missions successfully; and households in terms of their quality of life.

Cyber disruptions are becoming increasingly numerous, but the vast majority are still small in terms of economic impacts. A major exception was the recent CrowdStrike (2024) incident, which affected millions of devices and caused significant business interruption. On July 19, 2024, CrowdStrike, a cybersecurity technology company, released an update containing a critical flaw to Falcon, a service that monitors hacking attempts, viruses, and other threats. The update affected an estimated 8.5 million Microsoft Windows devices in what one expert deemed "the largest IT outage in history" (Goswami 2024). CrowdStrike remediated the issue less

The authors are Research Professor, Price School of Public Policy, University of Southern California and Senior Faculty Fellow and Director Emeritus, Center for Risk and Economic Analysis of Threats and Emergencies (CREATE), USC; Graduate Research Assistant, Price School and CREATE, USC; and Research Associate, Price School and CREATE, USC.

A. Rose (✉) · A. Mermelstein · N. Miller
Price School of Public Policy, University of Southern California, Los Angeles, CA, USA

Center for Risk and Economic Analysis of Threats and Emergencies, University of Southern California, Los Angeles, CA, USA
e-mail: adamzros@usc.edu

© The Author(s), under exclusive license to Springer Nature Switzerland AG 2025

I. Linkov, A. Kott (eds.), *Cyber Resilience: Applied Perspectives*, Risk, Systems and Decisions, https://doi.org/10.1007/978-3-031-90109-6_14

than 2 h later, but the proposed fix was in some cases difficult for affected businesses to deploy at scale. The outage disrupted the operations of numerous sectors including airlines, banks, hotels, restaurants, colleges, as well as governments at all levels. Its direct effects then rippled throughout the economy, causing operational delays in seaports and disrupting credit card payments (Schechner et al. 2024). The complete economic impact of the CrowdStrike incident has not yet been estimated, but one cyber insurance risk platform suggested it could reach “tens of billions of dollars” (Hussain 2024). Delta Air Lines alone, the U.S. carrier most affected by the outages, stated that the incident cost it $500 million in lost revenue and passenger compensation (Sider 2024). Furthermore, economic experts suggested the CrowdStrike incident revealed the fragility of the economy to other outages and disruptions, such as cyber-attacks (Bhattarai and Siegel 2024).

Humans, both individually and with respect to the organizations they operate, do not respond passively to most disruptions in life, including major disasters. Rather, they have a number of pre-existing and improvised coping mechanisms, or tactics, that can reduce operational, and hence economic, losses. The term *resilience* embodies these reactions. In this paper, we define resilience in terms of actions taken once a disruption begins, as opposed to actions taken prior to an event. The latter comes under the heading of *mitigation* and involves a different set of actions. However, we emphasize that resilience is a *process*, and its capacity can be built up in advance.[1] This is not to denigrate mitigation, which can reduce the frequency and magnitude of disruptions in the first place. However, there is widespread realization that not all cyber threats can be prevented, and thus resilience represents the best “second line of defense.”[2]

Cyber disruptions are not only becoming increasingly more prevalent in recent years, but so are society’s ways of coping with them. This chapter provides a major update estimating the cost and effectiveness of various resilience tactics presented in our original paper (Rose et al. 2019). Given that this is such a fast-moving field, an update is definitely warranted.

To help guide the reader, we further demarcate the scope of this chapter. We take a broad view of cyber disruptions to include cyber-attacks, hacks, technical accidents or failures, and side-effects of broader events, such as natural disasters that can affect cyber systems directly, or through electricity or communication systems. Because cyber systems are almost entirely dependent on communication systems, we

[1] During the CrowdStrike outage, cyber providers and customers implemented some very sophisticated resilience tactics, as noted below. However, some businesses even resorted to low-tech workarounds such as handing out paper boarding passes, taking orders in-person rather than via apps or websites, and only accepting cash payments (Schechner et al. 2024; Bhattarai and Siegel 2024). See also examples in the following two sections.

[2] Those in the cyber-security sphere will sometimes distinguish “risk mitigation” and “impact mitigation”—with the former focused on ways of reducing the probability that a disruption occurs, while the latter focused on ways of reducing the negative effects when something does inevitably occur. In this sense, resilience can be thought of as wholly separate from “risk mitigation” but overlapping quite meaningfully with “impact mitigation.”

explicitly incorporate threats to the latter in our analysis. However, we otherwise limit our analysis to direct impacts, and refer the reader to other research on related resilience of electricity networks (Rose et al. 2007; Sue Wing and Rose 2020). Moreover, we do not consider the full supply-chain (general equilibrium or macroeconomic) effects of cyber threats or resilience to them (see, Rose and Chen 2019).

This chapter is divided into six sections. The following section provides resilience concepts and experience with measuring them. Section 3 presents a set of 11 resilience tactic categories, examples of each, and estimates of their cost and effectiveness. Section 4 discusses the application of the cost and effectiveness findings. Section 5 presents suggestions for future research, and Sect. 6 provides a brief summary of our analysis and its implications.

1 Resilience Concepts and Measurement

Researchers and decision-makers in the disaster field are generally divided on the definition of resilience. One group utilizes the concept to refer to any action taken to reduce disaster losses. This group, with a large representation by engineers, focuses primarily on mitigation with an eye to reducing the frequency and magnitude of disasters and strengthening the built environment and equipment to *prevent damage* (see, e.g., Bruneau et al. 2003). This broad definition has also been adopted and applied across the board by major panels assessing resilience research and practice, such as the National Research Council, which defines resilience as "The ability to prepare and plan for, absorb, recover from, or more successfully adapt to actual or potential adverse events" (NRC 2012, p. 16).

Another group, with a large representation by social scientists, focuses primarily on actions implemented after a disaster strikes (Rose 2007; Cutter 2017) to promote business continuity. This group takes the meaning of resilience more literally, referring to its Greek language root, whose definition is "to rebound." They also acknowledge that resilience is a process, whereby steps can be taken before the disaster to build resilience capacity, but resilient actions do not take place until afterward. Examples of this pre-positioning include emergency drills, purchase of back-up electricity generators, and lining up alternative suppliers of critical inputs. In each case, the action provides no benefit before a disaster but lowers the interruption of the production of goods and services in the event that a disaster occurred. Here the focus is not on property damage, which has already occurred, but rather the reduction in the loss of the *flow of goods and services* emanating from property, or *capital stock*. The former is often measured in terms of gross domestic product (GDP) and employment, and is typically referred to as business interruption, or BI (Tierney 1997). BI just begins at the point when the disaster strikes, but continues until the system has recovered or reached a "new normal," which is typically now considered a "sustainable" level of activity (meaning a healthy economy). Measuring BI is thus much more complicated, because it involves matters of the duration and time-path of recovery, both of which are strongly affected by the behavioral responses of public and private decision-makers (Rose 2015, 2022). This definition of resilience continues to gain traction, in part through the realization that BI losses have exceeded property damage in many major disasters, such as 9/11, Hurricane Katrina, and COVID-19 (Rose 2022), thereby placing a stronger emphasis on reliability and recovery (NRC 2012).

1.1 Defining Economic Resilience

Numerous definitions exist for resilience, but Rose (2009) and others have found more commonalities than differences. We begin by offering the following general definitions of resilience, which capture the essence of the concept, and then follow them up with definitions that capture the essence of economic considerations. Following Rose (2004, 2009) and Rose and Dormady (2018), we distinguish two major categories of resilience:

In general, Static Resilience refers to the ability of the system to maintain a high level of functioning when subjected to a shock (Holling 1973). *Static Economic Resilience* is the efficient use of remaining resources at a given point in time to minimize BI losses. It refers to the core economic concept of coping with resource scarcity, which is exacerbated under disaster conditions. This dire situation requires that resources be used as efficiently as possible.

In general, Dynamic Resilience refers to the speed and duration of the system to recover (Pimm 1984). *Dynamic Economic Resilience* is the efficient use of resources over time for investment in repair and reconstruction in order to minimize BI losses. Investment is a time-related phenomenon—the act of setting aside resources that could potentially be used for current consumption in order to re-establish productivity in the future. Static Economic Resilience does not completely restore damaged capacity and is therefore not likely to lead to complete recovery. Note that since the vast majority of cyber incidents do not incur property damage, our focus will be on static economic resilience.[3]

Another important delineation in economic resilience, and resilience in general, is the distinction between inherent and adaptive resilience (Rose and Liao 2005; Wein and Rose 2011; Cutter 2017). Inherent resilience refers to resilience capacity already built into the system or pre-positioned, such as the ability to utilize more than one fuel in an electricity-generating unit, emergency practice drills, backup battery storage. Adaptive resilience, on the other hand, is exemplified by undertaking conservation that was not previously thought possible, such as modifying technology or relocating business activity.

[3]Originally, dynamic resilience was thought to be promoted by simply reducing the duration of the disruption. However, Zobel (2011, 2014) Rose and Dormady (2018) and Xie et al. (2018) have pointed out that one needs to examine the entire time-path of recovery, and that "jump-starting" (accelerating the speed of) the recovery is typically more effective in reducing losses than reducing the duration.

2 An Operational Metric and Initial Measurement

The next stage is to translate these definitions into a measurable form. For static resilience, this can be done in terms of the amount of BI prevented by the implementation of a given resilience tactic or set of tactics comprising a resilience strategy. The reference point would be the maximum potential BI loss in the absence of the resilience tactic. Here, the resilience metric for an individual tactic (or a group of tactics comprising a strategy in an overall index) is usually presented as a percentage between 0% and 100% (Rose 2009; Rose and Krausmann 2013).

Several studies have measured resilience using this and related metrics. For example, Rose et al. (2009) found that potential BI losses were reduced by 72% due to the rapid relocation of businesses following the September 11, 2001 terrorist attacks on the World Trade Center. Wei et al. (2020) found that a reduction in potential BI from a three-month closure of a major US seaport could be as high as 66% from the implementation of several resilience tactics. Xie et al. (2018) estimated that BI losses could have been reduced by 30% and recovery time by 1 year with an increase in investment funds and acceleration of their timing in the aftermath of the Wenchuan earthquake in China. Specialized studies have developed methodologies for examining the potential of specific resilience strategies, such as use of inventories (Barker and Santos 2009).

Several efforts have measured economic resilience more broadly. Most successful are surveys of businesses that have experienced economic disruptions; however, studies that surveyed respondents about their preparedness have also been successful. Kajitani and Tatano (2009) surveyed the preparedness and recovery potential of Japanese businesses, distinguished by sector, to utility lifeline disruptions that could emanate from an earthquake. Dormady et al. (2019, 2022) surveyed business victims of 2011 SuperStorm Sandy and 2018 Hurricane Harvey, respectively. Their surveys segmented businesses according to industry and measured both the cost and effectiveness of 11 resilience tactics (see below). They found that some of these tactics were in fact cost-saving (the implementation of the tactic actually reduced the cost of overall business operations in the process of reducing BI), and most others had very high cost-effectiveness (essentially benefit-cost) ratios.

3 Resilience Tactics

In this section, we present each of the 11 resilience tactics in general, their applicability to cyber resilience, and specific examples, with an emphasis on the cost and effectiveness of more recent advances. The assessment is summarized in Table 1.

The tactics are derived from an expanded economic production function (Rose 2009). The most typical implicit production function in the discipline is simply written as output being a function of capital and labor, with natural resources sometimes included. In this case, output is essentially "value-added," the income

Table 1 Microeconomic resilience options for businesses

Category	Action/investment	Cost of resilience	Effect of resilience	Source
Conservation				
Reduce non-essential use	Re-organization of data and optimized distribution[a]	Low	Low to moderate	CyberSheath (2014)
Remove non-essential access	Remove non-essential administrator access[b]	Roughly pays for itself	Low to moderate	Wahanani (2017)
Input substitution				
Paper records, traditional couriers	Re-contract[b]	Low	Low	
Enhance flexibility of input combinations	Supply procurement flexibility[a]	Low	Low to moderate	Fayezi et al. (2017)
Wireless-to-wired, and wired-to wireless internet and phone access	Use text messaging or social media[c]	Low to moderate	Low to moderate	
	Cells on wheels (COWs)[c]	Moderate	Moderate to high	Parenti (2021)
	Satellite phones[c]	Low to moderate	Moderate to high	Verizon (2024a, b, c)
	Femtocells (small mobile cellular base station)[b]	Low	Moderate to high	AT&T (2024a, 2024b)
	Cellular signal boosters[b]	Low	Moderate to high	Zazo (2024)
	Voice over IP telephone lines[b]	Low to moderate	Moderate	Vonage (2024)
	Smart intrusion prevention system[a]	Low	Moderate	Cisco (2024)
Import substitution				
Re-routing of goods/services	Data-center failover[a]	Low	Moderate	Carías et al. (2020)
Supply-chain management	Multi-sourcing strategy[a]	Low to moderate	Moderate	Linthorst and Telgen (2006)
Rerouting internet traffic	Internet exchange points[a]	Low	Low	IP Infusion (2023)
Inventories (stockpiles)				
Stockpile products and other essentials	"Safety" stock[a]	Low to moderate	Low to moderate	
Stockpile product inputs	Build-to-order	Low	Low to moderate	Wong et al. (2020)
Batteries	Battery storage[a]	Moderate to high	Moderate to high	Schafiei (2024)

(continued)

Table 1 (continued)

Category	Action/investment	Cost of resilience	Effect of resilience	Source
Excess capacity				
Maintain in good order	Maintain in good order[a]	Low	Low to moderate	
System redundancy	Redundant array of independent disks (RAID)	Low	Low to moderate	Newegg (2017)
	Cloud-based backup servers[a]	Low	Moderate	Amazon (2024)
Maintain capacity	Multiple connection pathways or internet provider contracts[a]	Low	Moderate	PacketPassers (2022)
Maintain service	Uninterruptible internet service premiums[a]	Low	Low to moderate	AT&T Business (2024a, 2024b)
Input isolation				
Decrease dependence	Permanent or temporary shift to non-cyber means[c]	Low to high	Low to high	
System partitioning	Independent active runtime system Security manager	Moderate	Moderate	Loveless (2020)
Relocation				
Physical move	Arrange for facilities in advance[a]	Low to moderate	High	
Telecommuting	Remote desktop/ VPN connections[c]	Worker productivity change/business cost dependent on sector	Moderate	Cisco Systems (2024)
	E-mail and work-mirroring software[a]	Low	Moderate	
Technological change				
Change processes	Edge processing[a]	Low	Moderate to high	Dell (2024)
Alter product characteristics	Increase flexibility of delivery[a]	Low to moderate	Low to high	
Resource sharing				
Mutual aid agreement[a]	Cooperative agreement	Low	Low to moderate	Roa Henriquez (2022)
Pool resources[a]	Shared warehouses	Low	Low to moderate	
Management effectiveness				
Succession/ continuity	Train; increased versatility[a]	Low	Low to moderate	NIST (2024)

(continued)

Table 1 (continued)

Category	Action/investment	Cost of resilience	Effect of resilience	Source
Increased awareness/ info sharing	Homeland security information network[a]	Low	Moderate to high	DHS (2024)
Emergency lines of communication	National emergency communications plan[a]	Low	Moderate	CISA (2024)
Production recapture				
Overtime/extra shifts	Extra work after system is restored[b]	Low	High	Park et al. (2011)
Restarting procedures	Uninterruptible power supply (UPS)[a]	Low	Moderate	Van Geet and Sickinger (2024)

[a]Requires pre-event action to build up resilience capacity for post-event implementation
[b]While not strictly requiring pre-event action, would likely be more effective if built up beforehand
[c]Does not necessarily require pre-event action, but rather is only instituted post-event

counterpart of gross domestic product (GDP), referring to net output produced by the aforementioned "primary" factors of production. A major expansion of the production function includes "intermediate goods", goods used to produce other goods (e.g., energy, materials, services) with the dependent variable in this case referred to as "gross output" because of the inclusion of intermediate inputs (value-added and GDP essentially are considered "net output"). Our expanded production function adds several delineations and dimensions to the latter formulation. These include capacity, inventories, location, isolation, resource pooling, and time-phasing of production. Also identified are aspects of the production function parameters that contribute to resilience, including productivity improvements in terms of the use of intermediate inputs and labor and capital, and input substitution from both domestic and external sources.[4] We note that the Dormady et al. (2022) survey of Hurricane Harvey victims asked respondents which of the 11 tactics they implemented once the disaster struck, but also included an open-ended question allowing them to enter other tactics used. Interestingly, fewer than 3% of the respondents mentioned alternatives, and half of these fell within our 11 resilience categories.[5]

We utilize the following three-point Likert scales for the Cost (on an annualized basis) and Effectiveness (in terms of reduction in BI) estimates, based on a synthesis of the literature. An example of literature cited for each tactic is presented in the last column of Table 1. In some cases, such as the Conservation and Import Substitution

[4]These pertain to the productivity term and elasticity of substitution between inputs that are common to all production functions (see the following on their relationship to resilience: Rose and Liao, 2007; Dormady et al. 2019; Dormady 2022).

[5]We note that the resilience tactics just discussed pertain to the operation of ordinary businesses. For specialized facilities, other resilience tactics are available. A good example is ship-rerouting when seaport is disrupted, though this could be thought of as a subset of Relocation.

tactics, the cost and effectiveness are intuitively obvious or have not changed their character over the years and do not require any citation or any update to previously used values.

Cost scale:	Effectiveness scale:
Low: Less than $100,000	Low: Less than 25%
Moderate: $100,000 to $500,000	Moderate: 25% to 75%
High: Greater than $500,000	High: Greater than 75%

We note at the outset that some tactics serve to promote mitigation in addition to resilience. However, we do not include separate estimates of the cost and effectiveness aspects for them. The costs are likely to be the same in nearly all cases for these "joint product" examples, as they can readily perform both objectives without incurring any additional upfront, maintenance or operation costs. However, their overall effectiveness in reducing the risk of cyber threats would be higher if mitigation were to be included, and thus our effectiveness estimates for such cases should be considered lower bounds.

3.1 Conservation

Conservation in general refers to doing more, or at least as much, with less. Often it is for the express purpose of focusing resources towards the most essential tasks, in this case under stress. In the cyber domain, examples include reducing nonessential usage, restricting nonessential access, and recycling cyber equipment. For instance, removing nonessential access increases the ability and speed of recovery from a cyber-disruption by removing the attacker's access to the system and reducing the number of access points in the case of a persistent threat (CyberSheath 2014). Reducing nonessential usage can take many forms, such as cutting bandwidth or server-allocation to certain activities, or more carefully organizing and distributing data stores. It can also include reorganization of data and optimized distribution of networks. Note that Conservation is usually an especially attractive resilience tactic, since it often pays for, or even more than pays for, itself. However, in the case of cyber systems where, for example, internal bandwidth or server usage have a time cost only, rather than a monetary cost, conservation is likely to be slightly less cost-effective. There are a few specific technologies applicable in this case, but it is more a matter of implementing an internal business decision based on advanced assessments or ad hoc responses.

3.2 Input Substitution

In the context of cyber systems, Input Substitution offers a wide range of options, from going back to paper records to enhancing system, process, and input flexibility to having backup capabilities. System flexibility entails investment into business-relationships with multiple suppliers (often called multi-sourcing), which leads to combined efforts towards quickly overcoming supply-chain disruptions. There are many examples of back-up systems, such as satellite phones for electrical and telephone systems, which are often considered the more trustworthy method of communication in any case (Verizon 2024c). A more dramatic example is that of "Cells on Wheels," or mobile cell-towers, for when conventional cell towers go down (Parenti 2021).

Another example involves architecture designed to enhance the efficiency of Finite State Machine (FSM) traversal, which is critical for enabling real-time traffic analysis and security threat detection. This system performs real-time traffic analysis, packet logging, and deep packet inspection, which are critical for identifying and mitigating security threats in real-time. This enables the system to handle large volumes of network traffic more efficiently, reducing the time required to detect and respond to security threats. This capability enhances the overall resilience of the network by ensuring continuous protection even under a heavy load. This product is available through a subscription-based model. For example, a 1-year subscription license for the Snort Subscriber Rule Set is priced at $1055 per license. For a mid-sized company, multiple licenses might be required depending on the scale of deployment, potentially increasing the total cost to $10,000 to $20,000 annually for comprehensive coverage across the organization. The effectiveness of this technology is in the low to moderate range.

3.3 Import Substitution

Import Substitution refers to bringing in goods and services in short supply from outside the affected region. It pertains primarily to the manufacturing of cyber equipment and various supply-chain effects, and hence is more pertinent to dynamic economic resilience. Setting up alternatives in advance, or at the minimum, researching options, can ensure smoother substitution of imports in place of the usual inputs following a cyber disruption. Of course, it can be constrained by damage to transportation infrastructure, physical or digital. Non-equipment specific examples include internet exchange point (IXP) failover solutions (IP Infusion 2023).

3.4 Inventories (Storage)

Inventories refer to stockpiling critical inputs for the production of cyber equipment and new maintenance of cyber systems. It is important to note that the cost of inventories is not the actual value of the goods, but simply the carrying costs. The goods themselves simply replace the cost that would have been incurred had the regular delivery arrived. That said, it should be further noted that carrying costs of electronic goods in general, and even more so for cyber-specific equipment, are typically much higher than for other goods, as they depreciate quite quickly (and carrying cost includes interest, storage costs and the cost of inventory becoming obsolete).

The major recent innovation in this area, is improved battery storage of electricity. For example, the Tesla Powerwall provides energy storage solutions that ensure continuity of power to cyber users, including a solar hookup option. A single Powerwall costs around $8700, including installation (Tesla 2024). Depending on the organization's energy needs, a company might require multiple units, potentially totaling $400,000 to $500,000 for a medium-sized company with 250 employees. The effectiveness of this option could be as high as 90% by providing backup power (Shafiei et al. 2024).

"Storage" of back-up data from which to restore disrupted or corrupted systems can also be a relatively inexpensive and highly effective solution, but its implementation post-disaster is reliant on an otherwise functioning system to which the back-ups would be restored. The costs will vary depending on multiple factors, such as the frequency at which data are mirrored to the back-up system, but would roughly correlate with the importance of the data being backed-up, which in turn would be approximately correlated with the effectiveness of the tactic. From the standpoint of the business, reducing disruption to more system-critical data would more effectively allow the business to continue operations. For example, for a server that hosts the web-interface (which does not change much), a daily back-up from which to restore would likely be sufficient. For a server that processes real-time transactions (e.g., credit card transaction, or system-critical information), however, every single transaction should be mirrored to a safe back-up (or multiple ones). The former would require very little overhead infrastructure to maintain with a commensurately low cost, while the latter would require much more robust infrastructure and could add a significant cost.

3.5 Excess Capacity

Excess Capacity overlaps to a great extent with system redundancy, which is ordinarily considered a supply-side resilience tactic. Typically, it is viewed as a rather expensive option, as, for example, in the case of back-up transformers for electric power systems.[6] However, both in-house and cloud-based data storage and/or servers can provide easy redundancy (Khasymski et al. 2015; Amazon 2024), with the goal being active-active redundant systems that provide load-balancing during normal operations, and continuation of service during disruptions. Additionally, maintaining multiple redundant physical connection pathways, whether built into a single internet service provider (ISP) contract or through multiple different ISP contracts, can provide higher day-to-day speed and bandwidth, while also providing redundancy if an ISP or connection pathway experiences service loss (Peplink 2017; Spiceworks 2017). Another possibility related to this resilience tactic is the development of uninterruptible internet service contracts, which could give firms the option to pay a small fee for being priority customers in the event of shortages in internet access.[7]

3.6 *Input Isolation*

Input Isolation pertains to the ability to separate key aspects of the production process from dependence on lifeline utilities, such as cyber systems.[8] The Cybersecurity Framework, a federally developed set of guidelines for cyber standards and practices, is one of many useful resources for identifying which aspects of a system are considered essential and nonessential (NIST 2014). Input Isolation of cyber systems is increasingly becoming less of an option as the technological sophistication of our economy advances. While it is typically inherent in the system or production process, it can also be improvised in the aftermath of a cyber disruption by temporarily shifting any processes possible to use non-cyber related channels.

The Lynx-Software Technologies Independent Active Runtime System Security Manager, LynxOS, is a Real-Time Operating System (RTOS) designed for secure, embedded systems (Loveless and Reese 2020). It provides advanced security features, including partitioning and real-time performance, making it ideal for monitoring and securing runtime environments autonomously. LynxOS isolates critical functions, reduces attack surfaces, and ensures that systems can continue to operate securely, which is essential for maintaining business continuity. The base cost for LynxOS is $10,000 to $20,000 for the RTOS and one seat of development tools. For a mid-sized company, additional costs arise from multiple licenses, customization, ongoing maintenance, and scaling the system across the organization. These factors

[6]Some excess capacity is often planned in order to enhance normal business flexibility or to accommodate downtime for maintenance; these aspects should not be credited to disaster resilience.

[7]This option is not currently allowed under net-neutrality laws. However, given the recent proposed changes to those laws, and the success of these premiums in other domains, such as electricity service provision, it is worth considering.

[8]Similar to excess capacity, some instances of input isolation, where some production activities are separated from the need for one or more inputs, are inherent in the system and should likewise not be credited to resilience unless they are expressly done for that purpose.

can increase the total cost to $50,000 to $100,000. This estimate multiplies the base price by approximately 5 to account for the broader deployment, additional tools, and support necessary for comprehensive coverage. The effectiveness of this resilience option is in the moderate range.

3.7 Relocation

Relocation is a tactic that increases resilience by physically moving the business' operations to a location away from the affected area. This requires not only the arrangement for alternate facilities with sufficient capacity, but is also facilitated by the standardization of processes and operations to allow for movement. With regards to cyber systems and services, however, telecommuting can be a very attractive option if the nature of the business allows for it. As long as employees can connect to cloud-based computing and data storage, relocation of the business itself would likely be unnecessary. Telecommuting is especially important for instances where the disruption is not caused by a focused cyber threat, but rather from extenuating circumstances associated with other disasters. A major example is the boom in telework during the recent COVID pandemic (Walmsley et al. 2023). The ensuing success of this approach has resulted in some permanence in off-site work and will facilitate application of this resilience tactic in future emergencies.

3.8 Technological Change

Technological change is a tactic that can increase resilience capacity, especially if it imparts additional flexibility into production systems both before and after the disruption (Zolli and Healy 2012). It can also refer to important improvisations in the way goods and services are produced in the wake of a disaster. One of the biggest recent examples in the cyber domain is switching from on-site servers and computing power to off-site, cloud-based, storage and applications. For instance, because everything is stored in the cloud, a local disruption of cyber infrastructure would, at most, only affect workers' ability to connect to the cloud. Thus, if, for example, a business lost power, employees could return home and still access their files and work via the cloud. This tactic also applies to increased flexibility of the delivery of cyber services. This can take the form of both inherent and adaptive resilience. Both are possible for any single incident, as not all cyber issues can be anticipated, and hence ad hoc adjustments once the disruption has begun are likely to be warranted.

A good example is the Dell (2024) Edge Gateway 5200, which is designed for rugged environments and can process data at the edge, making it suitable for IoT applications. It enables continuous operation even during network disruptions, ensuring critical data processing occurs locally. Edge computing nodes reduce reliance on centralized data centers by processing data locally. This ensures that operations can continue even during a cyberattack or network failure, enhancing the overall resilience of the system. The estimated cost for a mid-sized company deploying edge computing nodes typically involves purchasing hardware, software, and possibly additional services. The Dell Edge Gateway 5200 starts at $2286.81 per unit. For a mid-sized company, assuming the need for 10 to 20 units, the total cost would range from $22,868 to $45,736. Additional costs for setup, maintenance, and software could bring the total to around $50,000 to $75,000. The effectiveness of this technology would be in the range of 65–75%. Edge computing nodes are effective in reducing business loss by enabling faster and more localized decision-making in response to cyber threats. This enhances network resilience by minimizing disruptions and maintaining business continuity. Wang et al. (2022) found that resilience mechanisms in Edge Computing, which involve finding substitute nodes at the network edge to handle tasks of failing nodes, can significantly reduce performance loss and ensure business continuity.

3.9 *Management Effectiveness*

Management effectiveness[9] refers to any improvements in decision-making and expertise that improve functionality, primarily by using existing scarce resources more efficiently. Much of it refers to improvisation, but some relates to established emergency-management plans and information services. The Cybersecurity Framework (NIST 2024) is one such service that provides a platform for businesses to share information on current known threats as well as the tools available to counter and rebound from those threats. Typically, this is a relatively inexpensive option, with costs limited solely to the implementation of the Framework and any additional management training. Other examples include raising awareness of cyber threats and responses in general, emergency drills in particular, and ensuring emergency lines of communication with relevant government agencies, other businesses, and consumers remain open.

An example of a cyber resilience simulation exercise capability is provided by KnowBe4 (2024) in the form of its Security Awareness Training & Simulations. It provides platforms for conducting regular simulation exercises to test and improve an organization's response to cyber incidents. These simulations prepare teams to handle real-world cyber threats effectively, reducing the potential impact of an actual attack and aiding in faster recovery. The estimated cost for a company with 250 employees, KnowBe4's training packages typically range from $2500 to $7500 annually, depending on the level of training and simulation complexity. Cyber resilience simulation exercises are crucial for organizations to effectively

[9]Management effectiveness can be thought of as a general rubric that would include improved productivity of the workforce through better management direction and assignment, as well as overall improvements in business organization in anticipation of and in response to cyber threats.

prepare for and reduce the impact of cyber incidents. Odo (2023) points out that companies with well-developed risk management practices, which include regular simulation exercises, experience 67% lower financial losses from cyberattacks compared to those without such practices. These exercises help identify potential vulnerabilities, improve response strategies, and ensure that all relevant personnel are equipped to respond to cyber threats.

3.10 Resource Pooling

This tactic refers to such actions as mutual aid agreements signed in advance or reached once the threat strikes. This can provide more general and specialized resources to companies that are directly hit. This could include, for example, temporarily using another company's cabling/wireless infrastructure. It can also overlap with other tactics, as in the sharing of stockpiles of cyber-related equipment. It has obvious cost advantages in terms of capitalizing on economies of scale (e.g., quantity discounts in purchasing equipment and software) and reducing total inventory requirements. Many studies have found that information sharing about cyber threats is especially useful.

3.11 Production Recapture

Production Recapture refers to the ability to make up lost production by working extra shifts or over time after communication services and other capabilities are restored. Depending on the cause of the disruption, it may involve replacement of damaged equipment (which likely would be required either way), but otherwise the cost is only that of overtime pay for workers (Park et al. 2011). With regards to cyber systems, it is further facilitated by hastening the restoration of services such as electricity and internet access. This in turn can be promoted by other resilience tactics, such as Uninterruptible Power Supplies (UPS), which provide an emergency power source while switching to back-up generators or until central power service is restored, removing any down-time due to electrical power loss. Studies have shown that the effectiveness of recapture to reduce BI losses can be well over 50% for some sectors. The effectiveness of this tactic is highest for manufacturing firms that do not operate 24/7, so some production rescheduling is possible. It is lower in service industries and agriculture, as well as in manufacturing that has no slack time in its operations.[10]

[10] Note that the first 10 resilience tactics focus on coping with input disruptions directly. This tactic is focused on the output side and thus about rescheduling and restoring lost production once the inputs become available.

4 Applications

To make prudent resource management decisions, one must consider the cost of a mitigation or resilience tactic as well as its effectiveness. The effectiveness represents the gains from implementing it, which in our case is both the property damage and the BI prevented. One tactic might be capable of reducing more than twice the combined property and BI losses of another, but if it costs 10 times as much to implement, the former is not the better option. Cost-effectiveness analysis is based on the ratio of this gain to a unit increase in cost. Benefit-cost analysis (BCA) is a generalization of this principle over a broad range of levels and is not restricted to a linear proportional relationship, as is cost-effectiveness. More subtly, though, the concept of "benefits" in BCA opens the door to consider benefits to society as a whole and not just a single entity, be it business, household, or government agency. This becomes important if there are significant spillover effects in either the implementation of or the gains from mitigation and resilience.

4.1 Cost-Effectiveness Analysis

As background to this approach we note that some resilience tactics are primarily *inherent*, i.e., they already exist or can be enhanced relatively inexpensively during the process of building resilience capacity. Their utilization is simply withheld until a disaster strikes. The most common inherent resilience tactics, as detailed in Table 1, are: inventories, excess capacity, and management effectiveness. Other resilience tactics are *adaptive,* meaning they involve improvisation after the disaster strikes. The most common adaptive resilience tactics are: conservation, relocation, technological change, resource pooling, and production recapture. Note also that some tactics have both inherent and adaptive counterparts as well, including input substitution, import substitution, and input isolation.

Cost-Effectiveness Analysis begins by dividing the effectiveness estimate of a given resilience tactic by its cost estimate. Although we have expressed effectiveness in percentage terms, it is preferable to translate it into terms of dollars of net revenue from business interruption loss prevention. The ratios should then be ranked from lowest cost to highest. This essentially traces out an increasing marginal cost function, with its limit being the maximum percentage or dollar amount of resilience possible by the combination of tactics. Note that, since much conservation more than pays for itself, the function begins in the negative cost range.

The cost of each resilience tactic is affected by the context in which it is implemented. First, for any given tactic, its cost is not likely to be constant over the range of application (effectiveness). Nearly all economic processes eventually exhibit diminishing returns, resulting in a marginal cost relationship that increases at an increasing rate. The context in which the cyber threat occurs and resilience is implemented also has an influence on the effectiveness side. Relevant factors include the threat type, magnitude, and recovery duration, as well as background conditions relating to the economy, such as its economic health at the time of the threat and its geographic location. For example, inventories are finite and more likely to run out if the duration of recovery is long. Production recapture also erodes over time, as customers begin to seek other suppliers. Excess capacity is dependent on the business cycle. For example, one reason that relocation was so effective after the World Trade Center attacks was because New York City was in the throes of a recession, which provided a great deal of vacant office and manufacturing space (see Rose et al. 2009).

4.2 *Profitability Analysis and Benefit-Cost Analysis*

Resilience can be couched in a Profitability Analysis Framework or more broadly a Benefit-Cost Analysis (BCA) framework by bringing its rewards formally into the picture. For purposes of simplification, one can think of the benefits as the net revenue of business interruption losses that are prevented. In more sophisticated analyses, benefits would include avoided losses by entities beyond the directly affected firm (both in terms of supply-chain effects or non-market effects, such as damage to the environment). The optimal level of resilience would be at the point at which the marginal cost and marginal benefit curve intersect. Even without a precise numerical example, one can draw some insights from the example. All cost-saving resilience options would be taken, because they yield guaranteed positive net benefits. Also, given the relatively low cost of many of the tactics, at least in some of their initial applications, it is likely that a fairly high level of resilience would be chosen (see Rose 2017; Rose et al. 2019; Dormady et al. 2019).

A further consideration needs to be taken into account on the cost side for redundant systems, as well as some demand-side tactics, such as inventories or back-up equipment. Rose (2009) makes the case that customer-side resilience tactics need not be implemented until the disaster strikes, which would appear to give them a cost advantage over mitigation and supplier-side tactics such as redundancy. However, most forms of inherent resilience, such as inventories and back-up equipment, are in place whether or not the disaster strikes. While they lack the flexibility that other customer-side tactics have, there is a positive ramification of this—they exist to protect against *many* threats over the course of their lifetime. Thus, their cost-effectiveness is much higher than if one considers only a single threat. The larger the number of customers served by a water utility with a redundant system, the greater the benefits of redundancy. Likewise, the more threats battery storage protects against, the greater its benefits. This would seem to give inherent resilience and mitigation options advantages over adaptive resilience ones.

However, one critical thing to note is that the broader societal benefits need to be adjusted for the probability of occurrence of a cyber disruption. For example, if we are concerned about the major earthquake having severely adverse effect on cyber-network that could result in $10 billion worth of business interruption, we would have to take into account the fact that such a major quake might only occur once in 20 years, or have a 5% chance of occurring in a given year. We would then have to adjust the $10 billion revenue or benefit figure downward to only $500 million to adjust for the uncertainty of its occurrence. This gives an edge to adaptive resilience strategies, because they are implemented after the disruption has started, the occurrence of the event is known with perfect certainty and no risk adjustment is necessary.

Thus, there are trade-offs between various resilience tactics, and between resilience as an overall risk reduction strategy and mitigation. The optimal mix is likely to be a combination of the various tactics.

5 Future Research Directions

This chapter has provided a framework for analysis and some exploratory assessments of the economics of cyber disruption resilience tactics. However, there is much work to be done to fill in the framework and improve the accuracy of the empirical estimates. We offer the following list of future priorities:

- Formalize the conceptual framework through the use of production, market, and macroeconomic theory. For example, at the micro level this would involve the formal derivation of conditions for the choice of the optimal mix of mitigation and resilience tactics and then the overall optimal mix of mitigation and resilience strategies.
- Identify additional options within the broad range of mitigation and resilience tactics. While this chapter offers many examples of such tactics, the list is by no means exhaustive. Additionally, the cyber sector is one of the fastest growing and is ever-evolving; as new threats, as well as corresponding mitigation and resilience tactics come about, they will need to be identified and assessed.
- Estimate the sectoral effectiveness and cost of mitigation and resilience tactics.
- Many of the tactics discussed here will vary in effectiveness by sector, as will the threats faced, thus sector level analyses will be necessary in order to accurately assess these tactics.
- Perform case studies in relation to highly-connected tech conglomerates (e.g., Google, Meta, Amazon). A cyber-attack on a vertically integrated firm is likely to cause far larger damages than attacks on a non-integrated firm, and thus the optimal set of mitigation and resilience tactics may differ among company types.
- Examine interdependencies along the upstream and downstream supply chains in relation to cyber vulnerabilities and their related mitigation and resilience tactics. This would require further empirical work, noted above, on estimating costs and benefits of resilience tactics.
- Examine the trade-offs and complementarities between (pre-disaster) mitigation and (post-disaster) resilience. For example, some mitigation and resilience tactics may both work well against certain threat vectors and not against others, and thus

the benefit of implementing both may face diminishing returns, while also leaving the business unprotected against a different threat.
- Identify government policies that can help promote the optimal level and configuration of these two major strategies. Specifically, as more research is conducted on this topic, existing government platforms such as the Homeland Security Information Network or National Institute of Standards and Technology (NIST) Cybersecurity Framework can better inform, or recommend, mitigation and resilience tactics to businesses in need.

6 Conclusion

Losses from cyber threats can be decreased not only through mitigation that prevents them in the first place but also through economic resilience that facilitates recovering efficiently and in an accelerated manner. Our assessment indicates that many highly cost-effective cyber resilience tactics are available to businesses of all sizes. As with most business strategies, depending on a single tactic is unlikely to be optimal, and cyber providers and users should consider a (least-cost) mix of them. The constituents of an overall strategy are likely to differ across economic sectors, but less so within a given sector, though here the relative weights across tactics are likely to differ.

The current challenge is that resilience and cyber mitigation strategies must be developed faster than the capabilities and advancements of cyber threats. The other challenge is a need to implement the array of economic resilience tactics as cost-effectively as possible. This can be enhanced through careful assessments, planning, simulations, and experience.

Acknowledgements The authors thank Anne Wein for sharing interview data and references and for helpful comments to Bailey Shen, Juan Machado, and Konstantinos Papaefthymiou for their help in examining some of the relevant literature on cyber resilience tactics. This paper is based upon research supported by the U.S. Department of Homeland Security under Grant Award Number 17STQAC00001-01-02. The views and conclusions contained in this document are those of the authors and should not be interpreted as necessarily representing the official policies, either expressed or implied, of the U.S. Department of Homeland Security.

References

Amazon Web Services (AWS) (2024) Amazon S3 pricing. https://aws.amazon.com/s3/pricing/. Accessed 4 Aug 2024

Barker K, Santos JR (2009) Measuring the efficacy of inventory with a dynamic input–output model. Int J Prod Econ 126(1):130–143

Bhattarai A, Siegel R (2024) Largest IT outage in history expected to barely register in the economy. The Washington Post, July 20. https://www.washingtonpost.com/business/2024/07/20/outage-microsoft-economy-business/. Accessed 12 Aug 2024

Bruneau M et al (2003) A Framework to Quantitatively Assess and Enhance Seismic Resilience of Communities. Earthquake Spectra, 19:733–75

Carías JF, Arrizabalaga S, Labaka L, Hernantes J (2020) Cyber resilience progression model. Appl Sci 10(21):7393

Cisco Systems (2024) Cisco secure client at-a-glance. https://www.cisco.com/c/en/us/products/collateral/security/anyconnect-secure-mobility-client/secure-client-aag.html. Accessed 1 Aug 2024

CrowdStrike (2024) Technical details: Falcon update for windows hosts. crowdstrike.com, July 20. https://www.crowdstrike.com/blog/falcon-update-for-windows-hosts-technical-details/. Accessed 12 Aug 2024

Cutter S (2017) The Landscape of Disaster Resilience Indicators in the USA, Natural Hazards 80:741–58

CyberSheath Services International (2014) The role of privileged accounts in high profile breaches. CyberSheath Services International, Great Falls. http://lp.cyberark.com/rs/cyberarksoftware/images/wp-cybersheath-role-of-privileged-accounts-6-2-14-en.pdf

Dell (2024) Dell Edge Gateway 5200: Dell USA. https://www.dell.com/en-us/shop/network-hardware-devices/dell-edge-gateway-5200/spd/dell-edge-gateway-5200/dn_edge5200_15513_vi. Accessed 8 Aug 2024

Dormady N, Rose A, Rosoff H, Roa-Henriquez A (2019) Estimating the cost-effectiveness of resilience to disasters: survey instrument design & refinement of primary data. In: Ruth M, Reisemann SG (eds) Handbook on resilience of socio-technical systems. Edward Elgar, Cheltenham

Dormady N, Rose A, Roa-Henriquez A, Morin CB (2022) The cost-effectiveness of economic resilience. Int J Prod Econ 244:108371. https://doi.org/10.1016/j.ijpe.2021.108371

Fayezi S et al (2017) Understanding and development of supply chain agility and flexibility: a structured literature review. Int J Manag Rev 19(4):379–407. https://doi.org/10.1111/ijmr.12096

Goswami R (2024) Ongoing tech disruption is set to be the 'largest IT outage in history. CNBC, July 19. https://www.cnbc.com/2024/07/19/latest-live-updates-on-a-major-it-outage-spreading-worldwide.html. Accessed 12 Aug 2024

Holling C (1973) Resilience and stability of ecological systems. Annu Rev Ecol Syst 4:1–23

Hussain NZ (2024) Insurers face business interruption claims after global tech outage. Reuters, July 19. https://www.reuters.com/technology/insurers-face-business-interruption-claims-after-global-tech-outage-2024-07-19/. Accessed 12 Aug 2024

IP Infusion (2023) OCNOS data center for meeting IXP network design challenges, August 30. https://www.ipinfusion.com/blogs/ip-infusion-ocnos-data-center-for-meeting-ixp-network-design-challenges/. Accessed 1 Aug 2024

Kajitani Y, Tatano H (2009) Estimation of lifeline resilience factors based on empirical surveys of Japanese industries. Earthquake Spectra 25(4):755–776

Khasymski A et al (2015) Realizing accelerated cost-effective distributed RAID. In: Handbook on data centers. Springer, New York, pp 729–752

KnowBe4 (2024) Pricing: security awareness training. https://www.knowbe4.com/pricing-security-awareness-training. Accessed 8 Aug 2024

Linthorst MM, Telgen J (2006) Public purchasing future: buying from multiple suppliers. In: Advancing public procurement: practices, innovation and knowledge-sharing. PrAcademics Press, Boca Raton, pp 471–482

Loveless T, Reese L (2020) What is the cost of a real-time operating system (RTOS)? Lynx Software Technologies, July 7. https://www.lynx.com/embedded-systems-learning-center/cost-of-a-real-time-operating-system-rtos. Accessed 1 Aug 2024

Merriam-Webster (2015) English dictionary. Springfield

National Institute of Standards and Technology (NIST) (2014) Framework for improving critical infrastructure cybersecurity. Gaithersburg. http://www.nist.gov/cyberframework/upload/cybersecurity-framework-021214.pdf

National Institute of Standards and Technology (NIST) (2024) The NIST Cybersecurity Framework (CSF) 2.0. Gaithersburg, February 26. https://doi.org/10.6028/NIST.CSWP.29
National Research Council (2012) Disaster Resilience: A National Imperative. Washington, DC: National Academies Press. https://doi.org/10.17226/13457
Newegg (2017) RAID. https://www.newegg.com/p/pl?d=RAID
Odo C (2023) Strengthening cybersecurity resilience: the importance of education, training, and risk management. SSRN. https://doi.org/10.2139/ssrn.4779289
Oxford English Dictionary (OED) (2024) OUP, Oxford
PacketPassers Blog (2022) Multiple ISP connectivity & redundancy options. https://packetpassers.com/multiple-isp-connectivity-redundancy/. Accessed 1 Aug 2024
Parenti D (2021) Everything you need to know about cell towers on wheels. MCCR, November 12. https://www.mccr.info/blog/everything-you-need-to-know-about-cell-towers-on-wheels
Park J, Cho J, Rose A (2011) Modeling a major source of economic resilience to disasters: recapturing lost production. Nat Hazards 58:163–182
Peplink (2017) Peplink router products. Peplink. https://www.peplink.com/. Accessed 28 Mar 2017
Pimm SL (1984) The complexity and stability of ecosystems. Nature 307:321–326
Rose A (2004) Defining and measuring economic resilience to disasters. Disaster Prev Manag 13(4):307–314
Rose A (2009) Economic resilience to disasters, Community and regional resilience institute report no. 8. Oak Ridge National Laboratory, Oak Ridge
Rose A (2015) Macroeconomic consequences of terrorist attacks: estimation for the analysis of policies and rules. In: Benefit–cost analyses for security policies. Edward Elgar Publishing, Cheltenham, pp 172–200
Rose A (2017) Benefit-cost analysis of economic resilience actions. In: Cutter S (ed) Oxford research encyclopedia of natural hazard science. Oxford University Press, New York/Oxford
Rose A (2022) Behavioral consequences of disasters: a basis for inclusion in benefit-cost analysis. Econ Disasters Clim Change 6:213–223
Rose A, Chen Z (2019) Resilience to a cyber-attack on the Detroit automobile industry: a computable general equilibrium approach. In: Nijkamp P, Glaeser E, Kourtit K (eds) Urban empires. Springer, Heidelberg, pp 377–394
Rose A, Dormady N (2018) Advances in analyzing and measuring dynamic economic resilience. In: Trump B, Florin M, Linkov I (eds) IRGC resource guide on resilience (vol. 2): domains of resilience for complex interconnected Systems. EPFL International Risk Governance Center, Lausanne
Rose A, Krausmann E (2013) An economic framework for the development of a resilience index for business recovery. International Journal of Disaster Risk Reduction 5:73–83
Rose A, Oladosu G, Liao SY (2007) Business interruption impacts of a terrorist attack on the electric power system of Los Angeles: customer resilience to a Total blackout. Risk Analysis 27(3):513–531. https://doi.org/10.1111/j.1539-6924.2007.00912.x
Rose A, Oladosu G, Lee B, Asay GB (2009) The economic impacts of the 2001 terrorist attacks on the world trade center: a computable general equilibrium analysis. Peace Econ Peace Sci Public Policy 15(2):4
Rose A, Miller N, Eyer J, Banks J (2019) Economic mitigation of and resilience to cyber threats. In: Kott A, Linkov I (eds) Cyber resilience of Systems and networks. Springer, Heidelberg
Schechner S, Vipers G, Lukpat A (2024) Major tech outage grounds flights, hits banks and businesses worldwide. Wall Street J, July 19, 2024. Accessed 12 Aug 2024. https://www.wsj.com/tech/microsoft-reports-major-service-outage-affecting-users-worldwide-328a2f40
Shafiei K, Ghassem Zadeh S, Tarafdar Hagh M (2024) Planning for a network system with renewable resources and battery energy storage, focused on enhancing resilience. J Energy Storage 87:111339
Sider A (2024) Delta CEO says CrowdStrike tech outage cost it $500 million. Wall Street J, July 31. https://www.wsj.com/business/airlines/delta-ceo-says-crowdstrike-tech-outage-costs-could-reach-500-million-3b7f5a13

Sue Wing I, Rose A (2020) Economic consequence analysis of electric power infrastructure disruptions: an analytical general equilibrium approach. Energy Econ 89:104756
Tesla (2024) Tesla support: how Powerwall works. https://www.tesla.com/support/energy/powerwall/learn/how-powerwall-works. Accessed 1 Aug 2024
Tierney KJ (1997) Business impacts of the Northridge earthquake, Journal of Contingencies and Crisis Management 5(2):87–97. https://doi.org/10.1111/1468-5973.00040
U.S. Cybersecurity & Infrastructure Security Agency (CISA) (2023) National emergency communications plan. Arlington, September 19, 2019. https://www.cisa.gov/sites/default/files/publications/19_0924_CISA_ECD-NECP-2019_1.pdf
U.S. DHS (2024). Homeland Security Information Network (HSIN). https://www.dhs.gov/homeland-security-information-network-hsin. Accessed 2024
Van Geet O, Sickinger D (2024) Best practice guide for energy-efficient data center design. Department of Energy, Washington, DC. https://www.energy.gov/sites/default/files/2024-07/best-practice-guide-data-center-design.pdf
Verizon (2024a) Inmarsat Isatphone Pro 2 rental: Verizon satellite. https://www.vzwsatellite.com/products/rental/inmarsat-isatphone-pro-2. Accessed 4 Aug 2024
Verizon (2024b) Iridium Extreme 9575 rental: Verizon Satellite https://www.vzwsatellite.com/products/rental/iridium-extreme-9575. Accessed 4 Aug 2024.
Verizon (2024c) Satellite FAQs. http://www.vzwsatellite.com/faqs. Accessed 1 Aug 2024
Vonage (2024) How much does VoIP cost? A VoIP pricing guide, June 8. https://www.vonage.com/resources/articles/business-voip-pricing-considerations/
Wahanani H (2017) Performance analysis of failover cluster for system recovery. In Proceedings of the international conference on information technology and business, pp 88–92
Walmsley T et al (2023) Macroeconomic consequences of the COVID-19 pandemic. Econ Model 120:106147. https://doi.org/10.1016/j.econmod.2022.106147
Wang Q, Mateo-Fornés J, Anagnostopoulos C, Kolomvatsos K (2022) Predictive model resilience in edge computing. IEEE 8th world forum on Internet of Things.
Wei D, Chen Z, Rose A (2020) Evaluating the role of resilience in reducing economic losses from disasters: a multi-regional analysis of a seaport disruption. Pap Reg Sci 99(6):1691–1723
Wein A, Rose A (2011) Economic resilience lessons from the shakeout earthquake scenario. Earthquake Spectra 27(2):559–573. https://doi.org/10.1193/1.3582849
Wong C, Lirn T-C, Yang C-C, Shang K-C (2020) Supply chain and external conditions under which supply chain resilience pays: an organizational information processing theorization. Int J Prod Econ 226:107610
Xie W, Rose A, Li S, He J, Li N, Ali T, Wei W (2018) Dynamic economic resilience and economic recovery from disasters: a quantitative assessment. Risk Anal 38(6):1306–1318
Zazo E (2024) The 7 best cell phone signal boosters in 2024 – cell phone signal boosters for home. Pop Mech, April 26
Zolli A, Healy AM (2012) Resilience: why things bounce back. Simon and Schuster, New York
Zobel CW (2011) Representing the multi-dimensional nature of disaster resilience, Decision Support Systems, 51(1):82–92. https://doi.org/10.1016/j.dss.2010.10.001
Zobel CW (2014) Quantitatively representing nonlinear disaster recovery, Decision Sciences, 45 (6):1053–1082. https://doi.org/10.1111/deci.12103

9783031901089